THE DESIGN AND IMPLEMENTATION OF AN EXPERT ALGORITHM TO RAPIDLY COMPUTE SIMPLE PHYSICAL RELATIONS

by

JIMMY ALLEN DAVIS

Spring
2006

Note for Librarians: A cataloguing record for this book is available from Library and Archives
Canada at www.collectionscanada.ca/amicus/index-e.html
ISBN 1-4120-9308-2

Offices in Canada, USA, Ireland and UK

Book sales for North America and international:
Trafford Publishing, 6E–2333 Government St.,
Victoria, BC V8T 4P4 CANADA
phone 250 383 6864 (toll-free 1 888 232 4444)
fax 250 383 6804; email to orders@trafford.com
Book sales in Europe:
Trafford Publishing (UK) Limited, 9 Park End Street, 2nd Floor
Oxford, UK OX1 1HH UNITED KINGDOM
phone +44 (0)1865 722 113 (local rate 0845 230 9601)
facsimile +44 (0)1865 722 868; info.uk@trafford.com
Order online at:
trafford.com/06-1062

10 9 8 7 6 5 4 3 2 1

ABSTRACT

Here, the design, creation and implementation of a physics expert system are discussed. The underlying goal is to discover as many physical relationships possible in the shortest amount of time. A secondary goal is to begin developing algorithms that take advantage of tireless processors to do work error prone humans would find tedious. Beginning with the basics of physics, a mathematical method of doing this is designed that is easily coded. This code was then written on a common notebook computer and allowed to run for four weeks. Relations involving three and four variable and/or constants were computed. The resulting data contained over 6000 simple physical relations. Some are easily recognized while others are not. The formulae presented here allow a glimpse into the possibilities of physics being researched more closely with computers and will hopefully help to explain the path to currently unexplained physical phenomenon.

As a reminder that being a scientist does not entail discovery for self gratification or profit alone, but that the responsibility to improve the quality of life for all of humanity comes first; this work is dedicated to all of those who suffer from illness (terminal or otherwise) and desperately await for the most elite of scholars to finally provide relief.

ACKNOWLEDGMENTS

I extend my deep gratitude to Dr. Eduardo Martin, who employed and advised me during the creation of this work. His lecture on publishing on the boldest of claims helped inspire its creation. It is my thought that the broad freedom he affords his graduate students has been realized as a benefit that manifested into this manuscript.

I would also like to thank Dr. Humberto Campins for his patience and unwavering support of his students and Dr. Daniel Britt for his steady guidance and assurance. I credit these two professors for deepening my understanding and love of astronomy.

Dr. Llewellyn has been an excellent advisor and counselor, throughout the best and worst times up to this point in my academic career. I am most pleased to include him along with Dr. Martins, Dr. Campins and Dr. Britt as members of this committee.

I am deeply grateful for the time Dr. Jeong-Hyun Kang spent solidifying with me the parts of the program I was most unsure of.

My sincerest thanks to Michael Robert Swain, whose contribution of much needed equipment cut the computing time of this endeavor by at least a factor of two.

TABLE OF CONTENTS

LIST OF TABLES

LIST OF ACRONYMS/ABBREVIATIONS

FSB   Front Side Bus

HP   Hewlett-Packard

NEV   Newton-Einstein Vectors

OS   Operating System

RAM   Random Access Memory

SODIMM   Small Outline Dual Inline Memory Module

SI   International System of Units

VB5   Visual Basic 5

(X)B   (M)egabyte, (G)igabyte, (T)errabyte…

(X(Hz)   (M)egahertz, (G)igahertz, (T)errahertz…

XP   eXPerience

CHAPTER ONE: INTRODUCTION

"Imagination is more important than knowledge."

Albert Einstein

Units are by far the most fundamental tool used in physics. Without them; physicists would only be working with numbers that carried no physical meaning and the field would be reduced to simple math. Undergraduate students are force fed unit analysis from the very beginning of their formal physics education. The usefulness of units has even filtered itself into the real world, as can be attested to by any number of people pulled over by a police officer and told, "Do you know how fast you were going?"

The question is fundamental, and (to a large degree) intuitively understood by nearly everybody. Physics students begin learning about new units, but soon realize that even the units being used are only a mixture of a fundamental set of units. This of course has been going on since time immemorial in physical studies. In the author's research, no person seems to be credited with the invention of units. There are however, lots of stories about the natural system of units, where various parts of the King's body were considered the standard. This varied from country to country as each King wanted his body to be the standard. Finally, in 1960 the international committee got their act straight and adopted the SI system of units (http://www.bipm.fr/en/si/).

The advent of computer technology has made many things in physics possible. Numerical analysis and computer modeling make modern applications in physics easier by significant factors. With everything moving into the future, nobody really takes the time to look back at what was already accomplished and attempt to improve on it. The most fundamental principles are notable places to begin exploring to find out if we missed anything.

Simple formulae are the stuff of legend in the scientific community and the bread and butter for founding principles. Each new discovery opens our eyes a little more to the universe. The men and women who help discover major principles of science, Newton, Einstein, Curie, Schrödinger, Heisenberg, Bohr, Pascal, Oppenheimer and scores of others form a pantheon of science from which other scientists draw inspiration.

By simple formulae it is meant that one should think of physics from the high school era. Einstein's household formula, $E=mc^2$ and Newton's force equation, $F=ma$. Such formulae form memories when physics was "easy," a notion cruelly shattered upon stepping into the shoes of an undergraduate student. Now, one had to learn the ways of the older masters and understand where it was they were taking you. No easy path exists, just problem after mind-numbing problem meant to enforce upon the student the importance of knowing what came before.

No few undergraduate and graduate students has theorized (ie: daydreamed) of looking into the universal solutions manual and seeing all of the answers that nobody else has thought of yet. The idea that physics should be so easy to somebody that the answers should come without work. Indeed, why even ask the question? Just write the answers down and ask later. After all, who can be bothered to ask questions when one is getting thoroughly crush by books containing near incomprehensible ideas.

So how would one do such a thing? If scientists were armed with all the answers, they could better understand what questions to ask; no doubt discoveries would accelerate dramatically. In science, even the basic questions are not so basic and getting the answers is even more difficult. These are the thoughts that occupied the authors mind for about a month before a possibility illuminated itself.

The imagination runs wild with the idea of a computer program hell bent on discovering every formula in the universe. A theory engine, so to speak. Sounds like science fictions, and yet so enticing that I wrote a prototype, and was quite please with the results.

In chapter two, discussions of several texts on the matter of units is presented. Reviews on how unit analysis can be used to derive simple relationships will quickly be done, and from there we shall proceed into the next chapter.

In chapter three, the mathematical and physical development of NEVs is made along with applications that they may be applied to. Chapter three will focus more on the logic than on the code, as the code is not particularly pertinent and could be written in any number of ways by a competent programmer. Similar treatment will be done with the OS of the computer used to perform this project, as it too is not fundamentally important to the discussion at hand.

The findings of this project will be summarized in chapter four with reasons to justify the validity of this program's ability to derive simple physical formulae. The conclusion of the author on this project is made. Finally, for those more interested in the results, the appendix contains a list of approximately 6000 physics and science formulae to browse through. They are in but represent a months worth of processing time.

CHAPTER TWO: LITERATURE REVIEW

"It is a great nuisance that knowledge can be acquired only by hard work."

Somerset Maugham

As mentioned in the previous chapter, when undergraduates begin learning physics, unit analysis is stressed ad nauseam for nearly two years. As they proceed through the field, the subject matter is less stressed and the proficiency of the students in this subject is usually taken for granted. Delving deeper into the potential power of units is useful in furthering this work and creates a foundation on the subject.

To begin; students are reintroduced to common basic units (mass, length, time and temperature) in most common introductory texts [1] [2]. Basic units describe every quantity used in the sciences and allows for an easy check on the potential validity of any formula or function. However there are more than just these basic units. Below the basic units are listed in Table 1 with their SI notation and basic symbols. These are easily found on the internet [3]. (Note: This work only uses SI unit.)

Table 1
Basic Units, SI Notations and Basic Notation

Length	Meter(m)	L
Mass	Kilogram(kg)	M
Time	Second(s)	T
Electric Current	Ampere(A)	C
Temperature	Kelvin(K)	Θ

Amount of a Substance	Mole(mol)	A
Luminous Intensity	Candela(cd)	I

All of the above units can be grouped in a manner that allows for the creation of derived quantities such as force, energy, work and so on. For example, force has SI units of kg m s^{-2} and basic units of MLT^{-2}.

The reduction of most formulae and the primary test for simple relationships is the ability to have these units equal out on both sides of an equation. In Newton's equation F=ma, the derived quantity of force (F) has units of kg m s^{-2}, mass (m) has the SI units of m and the acceleration (a) has SI units of m s^{-2}. Placed in the same formula, it is easy to see that kg m s^{-2} = kg m s^{-2}. Done in basic units the expected MLT^{-2} = MLT^{-2}.

Thanks to Buckingham's Theorem [4] [5], we are assured that this is true no matter what units are used. So in fact whether a unit standard is used or we prefer to remain in basic units, all equations must maintain homogenous forms with possible modifications to the constant coefficients that accompany them.

This leads to an important question. Can equations be derived using the above known facts? The answer is yes, and very easily. Menzel's text [6] delves into this subject matter and contains easy to understand examples on the subject. So to illustrate, let us derive Einstein's equation (E=mc^2) in a few lines. No assumption beyond the fact that the energy, mass and speed of light are related will be made. The problem shall be done in basic units.

The energy (E) has basic units of ML^{2}T^{-2}, while the mass (m) has basic units of M and the speed of light (c) has basic units of LT^{-1}. One can set the energy equal to the mass and spead of light each raised to a constant power.

5

$$ML^2T^{-2} = (M)^{\alpha}(LT^{-1})^{\beta}$$

$$ML^2T^{-2} = M^{\alpha}L^{\beta}T^{-\beta}$$

This leads to a system of equations, shown below.

$$\alpha = 1$$

$$\beta = 2$$

$$-\beta = -2$$

Thus, α is 1 and β is 2. So $E = kmc^2$, the famous rest energy equation of Einstein. The constant k is inserted to denote the fact that while this technique is good at determining the relationship between certain quantities, it will not derive constants. They must be found analytically from first principles or experimentally. Common knowledge in this case tells us that k=1 and Einstein's equation is obtained.

Solution difficulty goes up as more quantities are sought, but the basic idea is the same. Set one of the quantities (preferably the one with units in all the others) equal to the product of all other quantities raised to arbitrary powers. Multiply through by the arbitrary powers and create a system of equations. Solve this system and the powers are no longer arbitrary. Finally, it should be understood that all formulae derived in this manner contain a constant that is not obtainable without further experimental work.

CHAPTER THREE: METHODOLOGY

"All men by nature desire to know."

Aristotle

3.1 Newton-Einstein Vectors

It can be shown that a Newton-Einstein Vector (NEV) can be constructed to represent the exponents of the units in a particular quantity. An NEV would contain the basic units some particular (though not set) order. In this work, the following order is used.

$NEV = (L,M,T,C,\Theta,A,I)$

So long as the NEVs are all ordered the same in an operation as they are being manipulated, all should be well. This is a convenient notation that allows for exponents to be applied without much effort. Let us take the quantity of force (F).

$F = (1,1,-2,0,0,0,0)$

F^2 would be computed by simply multiplying the NEV through by the exponent.

$F^2 = (2,2,-4,0,0,0,0)$

$F^n = (n,n,-2n,0,0,0,0)$

$X^n = (nL,nM,nT,nC,n\Theta,nA,nI)$

The notation at this point is sloppy and should be cleared up. It is proposed that an underlined exponent defines taking a quantity into its NEV form multiplied by the exponent. So the above would look more like the following.

$F^{\underline{n}} = (n,n,-2n,0,0,0,0)$

$X^{\underline{n}} = (nL,nM,nT,nC,n\Theta,nA,nI)$

This notation will be used from this point onwards. The properties of NEV addition and subtraction should now be explored. If the exponents of the quantity denote an NEV, then when two NEV quantities are multiplied, the NEVs add to get the result. Take Einstein's equation again and convert to NEV form.

$E = mc^2$

$E^1 = m^1 c^2$

$(2,1,-2,0,0,0,0) = (0,1,0,0,0,0,0) + 2(1,0,-1,0,0,0,0)$

$(2,1,-2,0,0,0,0) = (0,1,0,0,0,0,0) + (2,0,-2,0,0,0,0)$

$(2,1,-2,0,0,0,0) = (2,1,-2,0,0,0,0)$

So in effect, the NEVs suffer a reduction in mathematical intensity so to speak. Exponents drop to products and products are now sums. NEV subtraction from here can easily be shown as division in the original relation. NEVs seem to have a lot of the same properties as logarithms. Now if Einstein's equation had all of its terms brought to one side.

$1^1 = m^1 c^2 E^{-1}$

$(0,0,0,0,0,0,0) = (0,1,0,0,0,0,0) + 2(1,0,-1,0,0,0,0) - (2,1,-2,0,0,0,0)$

$(0,0,0,0,0,0,0) = (0,0,0,0,0,0,0)$

Now this implies that on the same side of one equation, if the exponents are chosen correctly, the NEVs will add up to the zero vector. Thus, zero vectors are unitarily correct representations of formulae.

How NEVs came to be named is a matter of the equations that were used to test out this concept. The first was Newton's and the second was Einstein's. As the vector properties of the system became clear, the term vector was appended. Thus, Newton-Einstein Vectors.

3.2 Exponential Permutations

From the previous section, it should be fairly straightforward to see how equations may be derived without much work. Simply go through all possible NEV permutations of that particular set of quantities until they return a zero vector, and then stop. The following example uses Einstein's equation.

$k = E^{-10}m^{-10}v^{-10}$

$k = E^{-10}m^{-10}v^{-9}$

$k = E^{-10}m^{-10}v^{-8}$

.

.

.

$k = E^{-1}m^{1}v^{2}$

Once again, Einstein's equation is obtained. This work would be unendurable for a person to wade through; there are uncountable NEV problems to be worked out. This is a task ideally suited for a computer though. Now the following questions must be asked.

a.) What does this formula mean?

b.) What experiment can be done to prove this formula works?

c.) What is the constant?

With a computer this procedure takes a few seconds. The computer does need to be provided with a list of quantities (and their accompanying NEVs) to shuffle about. This work used the quantities and NEVs in table 2. And missed derived quantity can simply be added in

9

and the possible relationships recalculated. It would take a fraction of the time to obtain those new equations than the initial processing.

This work is restricted to formulae that have three or four derived quantities or constants. Factors of a hundred equations are known for three-quantity formulae while factors of thousands of equation are known for four quantity equations. Likely five quantity equations will number in the tens of thousands and thus will be done when better computer resources are available.

Table 2
Quantities and their corresponding NEVs

Permeability of Free Space	(1,1,-2,-2,0,0,0)
Permittivity of Free Space	(-3,-1,4,2,0,0,0)
Light speed (c)	(1,0,-1,0,0,0,0)
Boltzmann's Constant (k)	(2,1,-2,-1,0,0,0)
Planck's Constant (h)	(2,1,-1,0,0,0,0)
Gravitational Constant (Gc)	(3,-1,-2,0,0,0,0)
Luminous Intensity	(0,0,0,0,0,0,1)
Temperature	(0,0,0,0,1,0,0)
Amount of a Substance	(0,0,0,0,0,1,0)
Electric Current	(0,0,0,1,0,0,0)
Time	(0,0,1,0,0,0,0)
Length	(1,0,0,0,0,0,0)
Volume Velocity	(3,0,-1,0,0,0,0)
Force (F)	(1,1,-2,0,0,0,0)
Energy (E)	(2,1,-2,0,0,0,0)
Acceleration (a)	(1,0,-2,0,0,0,0)
Mass (M)	(0,1,0,0,0,0,0)
Velocity (v)	(1,0,-1,0,0,0,0)
Area (A)	(2,0,0,0,0,0,0)
Volume (V)	(3,0,0,0,0,0,0)
Wavenumber	(-1,0,0,0,0,0,0)
Specific Volume	(3,-1,0,0,0,0,0)
Current Density	(2,0,0,1,0,0,0)
Magnetic Field Strength	(-1,0,0,1,0,0,0)
Frequency	(0,0,-1,0,0,0,0)
Pressure,Stress	(-1,1,-2,0,0,0,0)
Energy,Work,Heat Quantity	(2,1,-2,0,0,0,0)
Power, Radiance Flux	(2,1,-3,0,0,0,0)
Electric Charge	(0,0,1,1,0,0,0)
Electric Potential Diff	(2,1,-3,-1,0,0,0)

10

Capacitance	(-2,-1,4,2,0,0,0)
Electric Resistance	(2,1,-3,-2,0,0,0)
Magnetic Flux Density	(0,1,-2,-1,0,0,0)
Inductance	(2,1,-2,-2,0,0,0)
Illuminance	(-2,0,0,0,0,0,1)
Absorbed Dose	(2,0,-2,0,0,0,0)
Catalytic Activity	(0,0,-1,0,0,1,0)
Magnetic Flux	(2,1,-2,-1,0,0,0)
Dynamic Viscosity	(-1,1,-1,0,0,0,0)
Kinematic Viscosity	(2,0,-1,0,0,0,0)
Mass Density	(-3,1,0,0,0,0,0)

To cut down on computational time, the exponents were restricted from -10 to 10. There is nothing to restrict formulae with numbers beyond those from existing. It should be noted that for every prime number exponent, reducibility becomes impossible in terms of quantity.

For example, Newton's equation F=ma is reducible by any exponent. $F^{10}=m^{10}a^{10}$ is after all, Newton's equation taken to the 10th power. If an equation appeared that took the form $x^4=y^8z^7$, there is nothing to reduce. With this in mind, and the list in table 2, a program is easily written.

3.3 Computer Architecture, OS and Programming Language

Here a brief summary of what was used to perform the processing will be outlined for completeness. It should be noted that this method may be performed on any number of platforms/OS/language combinations. The computer that was used was an HP® ze4800 laptop with a 2.13 GHz processor and 512MB SODIMM RAM (2x256 MB) with a 266 FSB. The OS is the common Microsoft© Windows XP® Home Edition. The language used was VB5, Professional Edition. The choice of VB five was the simplicity of the string manipulation commands which made formatting the output of this program easy.

3.4 The Program

The code used for the program was implemented in the following way. Initially, the quantities that are to be related must be chosed. In an automated program, this would follow from the top of the list in Table 2 in combinatorial steps. The quantities should be stored along with their NEVs.

1st Quantity: (x11,x12,x13,x14,x15,x16,x17)

2nd Quantity: (x21,x22,x23,x24,x25,x26,x27)

3rd Quantity: (x31,x32,x33,x34,x35,x36,x37)

Here, the first number denotes the quantity and the second number denotes the NEV position. Then a decision based on processing time available needs to be made. Because of the limited processing time available on the computer used, the maximum number of the exponents to be scanned through was set to 10. So every possible combination with exponents of -10 to 10 was analyzed.

Reducing the time for computation can be done by writing an algorithm that reduces all exponenets to their lowest common denominators once the NEVs add to zero. For the Einstein's equation, the program should stop at $k=E^{-5}m^5c^{10}$ which is then reduced to $k=E^{-1}m^1c^2$. Cycling through the exponents can be done by implementing any number of loop structures. Each exponent is cycled independently, and the nested loops ensure that all combinations are attempted until either a formula is found and reduced or no relation is discovered. The following is pseudo-code:

For L1 = -10 to 10

For L2 = -10 to 10

   …

   $x11 = x11*L1$

   $x12 = x12*L1$

   .

   .

   .

   $x21 = x21*L2$

   $x22 = x22*L2$

   .

   .

   .

   …

   Add all NEVs

  End

End

In the inner loop, all the NEVs should be added to see if they add up to zero. It should also be noted that if one NEV has a value where all other NEVs have a zero, then one may skip the analysis as there are no linear combinations of that set that may cancel out to give the 0 NEV. This "orthogonal" property of NEVs seems to allow a good reduction in processing. If a match is made or not, record what was found, move on to the next set of quantities and start the process over again.

Due to the manner in which this program processes information, it is recommended if any progress tracker is used that Appendix B be referenced for a derivation of the one used in this program.

CHAPTER FOUR: FINDINGS AND CONCLUSIONS

"We are at the very beginning of time for the human race. It is not unreasonable that we grapple with problems. But there are tens of thousands of years in the future. Our responsibility is to do what we can, learn what we can, improve the solutions, and pass them on."

Richard Feynman

Running this program produced thousands of simple relationships. Some of these are quickly recognized ($E=mc^2$, $F=ma$). Others are known only to those who are somewhat in the field like Maxwell's derivation of the speed of light with electromagnetic constants. Still others are being written about as this is being written ($F=c^4G^{-1}$) and others may either be relegated to the categories of "true but useless" and "untrue."

An example of the "true but useless" category would apply to any equation that gave values consistently below anything measurable. Consider a correction to Newton's force equation which consistently returns a value of 10^{-850} N. This correction may be true in some particular case, but is it important? Probably not, but it is left to those that investigate these formulae to determine this.

"Untrue" formulae are those relations that are conveniently unitarily correct but are meaningless. No speculations as to which of the equations presented in the appendix is untrue will be made here. It would be an arrogant researcher indeed to assume that any equation is absolutely untrue. Let the experimentalists determine their validity.

It is cited that simply by the derivation of well known simple relations such as Newton's and Einstein's equation, that credence is lent to this work. The number of equations found using this method stirs the imagination as to what discoveries may be made in the future. It is interesting to note that the time between discovering Newton's equation and Einstein's equation was approximately 4 days and 6 hours. The time between Newton's birth and Einstein's birth was 236 years. The "discovery" of these equations by a laptop was accomplished ~20,000 faster than mankind did it. Assuming learning is linear (a bad assumption I'm sure, but entertaining nonetheless,) if left running for 236 years, this method will discover by 2242AD what will take man alone until 22372AD.

It is believed by the author that a sort of "Ladder Rule" applies to these formulae. It appears that if all equations with three quantities are found, that they will lead to all four-quantity equations, which in turn will lead to all five-quantity equations and so on. It would therefore be more important to begin work to uncover all possible three-quantity equations and work out from there.

It is also recommended that faster computers are applied to this project and the work be continued into the domain of five-quantity relations and beyond. Future compilations of this work should be published online in a more "accessible" format than presented here. Hopefully, experimentalist will now have a nearly endless supply of formulae to prove or disprove.

APPENDIX A: THREE AND FOUR QUANTITY EQUATIONS

Luminous Intensity = Length^2 · Illuminance

Luminous Intensity = Area · Illuminance

Luminous Intensity^3 = Volume^2 · Illuminance^3

Illuminance = Luminous Intensity · Wavenumber^2

Amount of a Substance = Time · Catalytic Activity

Catalytic Activity = Amount of a Substance · Frequency

Electric Charge = Electric Current · Time

Current Density = Electric Current · Length^2

Electric Current = Length · Magnetic Field Strength

Energy = Electric Current^2 · Inductance

Energy = Electric Current · Magnetic Flux

Current Density = Electric Current · Area

Electric Current^2 = Area · Magnetic Field Strength^2

Current Density^3 = Electric Current^3 · Volume^2

Electric Current^3 = Volume · Magnetic Field Strength^3

Electric = Wavenumber^2 · Current Density

Magnetic Field Strength = Electric Current · Wavenumber

Electric Current^3 = Current Density · Magnetic Field Strength^2

Electric Current = Electric Charge · Frequency

Energy,Work,Heat Quantity = Electric Current^2 · Inductance

Energy,Work,Heat Quantity = Electric Current · Magnetic Flux

Power, Radiance Flux = Electric Current · Electric Potential Difference

Power, Radiance Flux = Electric Current^2 · Electric Resistance

Electric Potential Difference = Electric Current · Electric Resistance

Magnetic Flux = Electric Current · Inductance

$Length^3$ = Time · Volumetric Velocity

Length = $Time^2$ · Acceleration

Length = Time · Velocity

Length = $Time^2$ · Absorbed Dose

$Length^2$ = Time · Kinematic Viscosity

Volumetric Velocity = $Time^5$ · $Acceleration^3$

Volumetric Velocity = $Time^2$ · $Velocity^3$

$Area^3$ = $Time^2$ · Volumetric $Velocity^2$

Volume = Time · Volumetric Velocity

A Constant = Time · Volumetric Velocity · $Wavenumber^3$

Volumetric $Velocity^2$ = $Time^4$ · Absorbed $Dose^3$

Volumetric $Velocity^2$ = Time · Kinematic $Viscosity^3$

Energy = Power · Time

Velocity = Time · Acceleration

Area = $Time^4$ · $Acceleration^2$

Volume = $Time^6$ · $Acceleration^3$

A Constant = $Time^2$ · Acceleration · Wavenumber

Absorbed Dose = $Time^2$ · $Acceleration^2$

Kinematic Viscosity = $Time^3$ · $Acceleration^2$

Area = $Time^2$ · $Velocity^2$

Volume = Time^3 · Velocity^3

A Constant = Time · Velocity · Wavenumber

Kinematic Viscosity = Time · Velocity^2

Area = Time^2 · Absorbed Dose

Area = Time · Kinematic Viscosity

Volume^2 = Time^6 · Absorbed Dose^3

Volume^2 = Time^3 · Kinematic Viscosity^3

A Constant = Time^2 · Wavenumber^2 · Absorbed Dose

A Constant = Time · Wavenumber^2 · Kinematic Viscosity

Dynamic Viscosity = Time · Pressure,Stress

Energy,Work,Heat Quantity = Time · Power,Radiance Flux

Magnetic Flux = Time · Electric Potential Difference

Time = Capacitance · Electric Resistance

Time^2 = Capacitance · Inductance

Inductance = Time · Electric Resistance

Kinematic Viscosity = Time · Absorbed Dose

Volumetric Velocity^2 = Length^5 · Acceleration

Volumetric Velocity = Length^2 · Velocity

Volumetric Velocity = Length^3 · Frequency

Volumetric Velocity^2 = Length^4 · Absorbed Dose

Volumetric Velocity = Length · Kinematic Viscosity

Energy = Force · Length

Force = Length^2 · Pressure,Stress

Energy = Length^3 · Pressure,Stress

Velocity^2 = Length · Acceleration

Acceleration = Length · Frequency^2

Absorbed Dose = Length · Acceleration

Kinematic Viscosity^2 = Length^3 · Acceleration

Length^3 = Mass · Specific Volume

Mass = Length^3 · Mass Density

Kinematic Viscosity = Length · Velocity

Current Density = Length^3 · Magnetic Field Strength

Absorbed Dose = Length^2 · Frequency^2

Kinematic Viscosity = Length^2 · Frequency

Energy,Work,Heat Quantity = Length^3 · Pressure,Stress

Magnetic Flux = Length^2 · Magnetic Flux Density

Kinematic Viscosity = Length^2 · Absorbed Dose

Energy^2 = Force^2 · Area

Energy^3 = Force^3 · Volume

Force = Energy · Wavenumber

Force^3 = Energy^2 · Pressure,Stress

Force = Mass · Acceleration

Power,Radiance Flux = Force · Velocity

Force = Area · Pressure,Stress

Energy,Work,Heat Quantity = Force^2 · Area

Force^3 = Volume^2 · Pressure,Stress^3

Energy^3 = Force^3 · Volume

Pressure,Stress = Force · Wavenumber^2

Force = Wavenumber · Energy,Work,Heat Quantity

Force = Specific Volume · Dynamic Viscosity^2

Kinematic Viscosity^2 = Force · Specific Volume

Force = Magnetic Field Strength · Magnetic Flux

Force^3 = Pressure,Stress · Energy,Work,Heat Quantity^2

Power, Radiance Flux^2 = Force^2 · Absorbed Dose

Force = Dynamic Viscosity · Kinematic Viscosity

Force = Kinematic Viscosity^2 · Mass Density

Dynamic Viscosity^2 = Force · Mass Density

Energy = Mass · Velocity^2

Energy = Mass · Absorbed Dose

Energy^2 = Area^3 · Pressure,Stress^2

Energy = Volume · Pressure,Stress

Pressure,Stress = Energy · Wavenumber^3

Energy = Current Density · Magnetic Flux Density

Power,Radiance Flux = Energy · Frequency

Energy = Electric Charge · Electric Potential Difference

Electric Charge^2 = Energy · Capacitance

Energy = Electric Potential Difference^2 · Capacitance

Magnetic Flux^2 = Energy · Inductance

$$\text{Velocity}^4 = \text{Acceleration}^2 \cdot \text{Area}$$

$$\text{Velocity}^6 = \text{Acceleration}^3 \cdot \text{Volume}$$

$$\text{Acceleration} = \text{Velocity}^2 \cdot \text{Wavenumber}$$

$$\text{Acceleration} = \text{Velocity} \cdot \text{Frequency}$$

$$\text{Velocity}^3 = \text{Acceleration} \cdot \text{Kinematic Viscosity}$$

$$\text{Acceleration}^2 = \text{Area} \cdot \text{Frequency}^4$$

$$\text{Absorbed Dose}^2 = \text{Acceleration}^2 \cdot \text{Area}$$

$$\text{Kinematic Viscosity}^4 = \text{Acceleratio}^2 \cdot \text{Area}^3$$

$$\text{Acceleration}^3 = \text{Volume} \cdot \text{Frequency}^6$$

$$\text{Absorbed Dose}^3 = \text{Acceleration}^3 \cdot \text{Volume}$$

$$\text{Kinematic Viscosity}^2 = \text{Acceleration} \cdot \text{Volume}$$

$$\text{Frequence}^2 = \text{Acceleration} \cdot \text{Wavenumber}$$

$$\text{Acceleration} = \text{Wavenumber} \cdot \text{Absorbed Dose}$$

$$\text{Acceleration} = \text{Wavenumber}^3 \cdot \text{Absorbed Dose}^2$$

$$\text{Acceleration}^2 = \text{Frequency}^2 \cdot \text{Absorbed Dose}$$

$$\text{Acceleration}^2 = \text{Frequency}^3 \cdot \text{Kinematic Viscosity}$$

$$\text{Absorbed Dose}^3 = \text{Acceleration}^2 \cdot \text{Kinematic Viscosity}^2$$

$$\text{Energy,Work,Heat Quantity} = \text{Mass} \cdot \text{Velocity}^2$$

$$\text{Area}^3 = \text{Mass}^2 \cdot \text{Specific Volume}^2$$

$$\text{Mass}^2 = \text{Area}^3 \cdot \text{Mass Density}^2$$

$$\text{Volume} = \text{Mass} \cdot \text{Specific Volume}$$

$$\text{Mass} = \text{Volume} \cdot \text{Mass Density}$$

$$\text{A Constant} = \text{Mass} \cdot \text{Wavenumber}^3 \cdot \text{Specific Volume}$$

Mass Density = Mass $\cdot$ Wavenumber^3

Energy,Work,Heat Quantity = Mass $\cdot$ Absorbed Dose

Velocity^2 = Area $\cdot$ Frequency^2

Kinematic Viscosity^2 = Velocity^2 $\cdot$ Area

Velocity^3 = Volume $\cdot$ Frequency^3

Kinematic Viscosity^3 = Velocity^3 $\cdot$ Volume

Frequency = Velocity $\cdot$ Wavenumber

Velocity = Wavenumber $\cdot$ Kinematic Viscosity

Velocity^2 = Specific Volume $\cdot$ Pressure,Stress

Velocity^2 = Frequency $\cdot$ Kinematic Viscosity

Pressure,Stress = Mass Density $\cdot$ Velocity^2

Current Density^2 = Area^3 $\cdot$ Magnetic Field Strength^2

Absorbed Dose = Area $\cdot$ Frequency^2

Kinematic Viscosity = Area $\cdot$ Frequency

Energy,Work,Heat Quantity^2 = Area^3 $\cdot$ Pressure,Stress^2

Magnetic Flux = Area $\cdot$ Magnetic Flux Density

Kinematic Viscosity^2 = Absorbed Dose $\cdot$ Area

Current Density = Volume $\cdot$ Magnetic Field Strength

Absorbed Dose^3 = Volume^2 $\cdot$ Frequency^6

Kinematic Viscosity^3 = Volume^2 $\cdot$ Frequency^3

Energy,Work,Heat Quantity = Volume $\cdot$ Pressure,Stress

Magnetic Flux^3 = Volume^2 $\cdot$ Magnetic Flux Density^3

Kinematic Viscosity^6 = Volume^2 · Absorbed Dose^3

Magnetic Field Strength = Wavenumber^3 · Current Density

Frequency^2 = Wavenumber^2 · Absorbed Dose

Pressure,Stress = Wavenumber^3 · Energy,Work,Heat Quantity

Magnetic Flux Density = Wavenumber · Magnetic Flux^2

Absorbed Dose = Wavenumber^2 · Kinematic Viscosity^2

Absorbed Dose = Specific Volume · Pressure,Stress

Kinematic Viscosity = Specific Volume · Dynamic Viscosity

Energy,Work,Heat Quantity = Current Density · Magnetic Flux Density

Current Density = Electric Charge · Kinematic Viscosity

Pressure,Stress = Magnetic Field Strength · Magnetic Flux Density

Pressure,Stress = Frequency · Dynamic Viscosity

Power,Radiance Flux = Frequency · Energy,Work,Heat Quantity

Electric Potential Difference = Frequency · Magnetic Flux

A Constant = Frequency · Capacitance · Electric Resistance

A Constant = Frequency^2 · Capacitance · Inductance

Electric Resistance = Frequency · Inductance

Absorbed Dose = Frequency · Kinematic Viscosity

Pressure,Stress = Absorbed Dose · Mass Density

Energy,Work,Heat Quantity = Electric Charge · Electric Potential Difference

Electric Charge^2 = Energy,Work,Heat Quantity · Capacitance

Energy,Work,Heat Quantity = Electric Potential Difference^2 · Capacitance

Magnetic Flux^2 = Energy,Work,Heat Quantity · Inductance

Electric Potential Difference^2 = Power,Radiance Flux · Electric Resistance

Electric Charge = Electric Potential Difference · Capacitance

Magnetic Flux = Electric Charge · Electric Resistance

Electric Potential Difference = Magnetic Flux Density · Kinematic Viscosity

Inductance = Capacitance · Electric Resistance^2

Dynamic Viscosity = Kinematic Viscosity · Mass Density

Velocity^4 = G_c · Force

Power,Radiance Flux = G_c · Force^4

Absorbed Dose^2 = G_c · Force

Acceleration^2 = G_c · Pressure,Stress

Velocity^5 = G_c ·Power,Radiance Flux

G_c = Specific Volume · Frequency^2

Frequency^2 = G_c · Mass Density

Absorbed Dose^5 = G_c^2 · Power,Radiance Flux^2

Energy = Volumetric Velocity · Dynamic Viscosity

Velocity^5 = Volumetric Velocity · Acceleration

Volumetric Velocity^4 = Acceleration^2 · Area^5

Volumetric Velocity^6 = Acceleration^3 · Volume^5

Acceleration^2 = Volumetric Velocity^4 · Wavenumber^{10}

Acceleration^6 = Volumetric Velocity^2 · Frequency^{10}

Absorbed Dose^{10} = Volumetric Velocity^4 · Acceleration^8

Kinematic Viscosity^{10} = Volumetric Velocity^6 · Acceleration^2

Volumetric Velocity = Velocity $\cdot$ Area

Volumetric Velocity^3 = Velocity^3 $\cdot$ Volume^2

Velocity^2 = Volumetric Velocity^2 $\cdot$ Wavenumber^4

Velocity^3 = Volumetric Velocity $\cdot$ Frequency^2

Kinematic Viscosity^2 = Volumetric Velocity $\cdot$ Velocity

Volumetric Velocity^2 = Area^3 $\cdot$ Frequency^2

Volumetric Velocity^2= Area^2 $\cdot$ Absorbed Dose

Volumetric Velocity^2 = Area $\cdot$ Kinematic Viscosity^2

Volumetric Velocity = Volume $\cdot$ Frequency

Volumetric Velocity^6 = Volume^4 $\cdot$ Absorbed Dose^3

Volumetric Velocity^3 = Kinematic Viscosity^3 $\cdot$ Volume

Frequency = Volumetric Velocity $\cdot$ Wavenumber^3

Absorbed Dose = Volumetric Velocity^2 $\cdot$ Wavenumber^4

Kinematic Viscosity = Volumetric Viscosity $\cdot$ Wavenumber

Absorbed Dose^3 = Volumetric Velocity^2 $\cdot$ Frequency^4

Kinematic Viscosity^3 = Volumetric Velocity^2 $\cdot$ Frequency

Power, Radiance Flux = Volumetric Velocity $\cdot$ Pressure,Stress

Energy,Work,Heat Quantity = Volumetric Velocity $\cdot$ Dynamic Viscosity

Kinematic Viscosity^4 = Volumetric Velocity^2 $\cdot$ Absorbed Dose

h = Energy $\cdot$ Time

h = Time^2 $\cdot$ Power,Radiance Flux

h = Length^3 $\cdot$ Dynamic Viscosity

Energy = h $\cdot$ Frequency

$\text{Energy}^2 = h \cdot \text{Power, Radiance Flux}$

$h = \text{Mass} \cdot \text{Kinematic Viscosity}$

$h^2 = \text{Area}^3 \cdot \text{Dynamic Viscosity}$

$h = \text{Volume} \cdot \text{Dynamic Viscosity}$

$\text{Dynamic Viscosity} = h \cdot \text{Wavenumber}^3$

$\text{Energy,Work,Heat Quantity} = h \cdot \text{Frequency}$

$\text{Power,Radiance Flux} = h \cdot \text{Frequency}$

$\text{Energy,Work,Heat Quantity}^2 = h \cdot \text{Power,Radiance Flux}$

$h = \text{Electric Charge}^2 \cdot \text{Electric Resistance}$

$h = \text{Electric Charge} \cdot \text{Magnetic Flux}$

$\text{Magnetic Flux}^2 = h \cdot \text{Electric Resistance}$

$h = k \cdot \text{Electric Charge}$

$k^2 = h \cdot \text{Electric Resistance}$

$\text{Energy} = k \cdot \text{Electric Current}$

$k = \text{Electric Current} \cdot \text{Inductance}$

$k = \text{Electric Potential Difference} \cdot \text{Time}$

$c^4 = Gc \cdot \text{Force}$

$c^5 = Gc \cdot \text{Power, Radiance Flux}$

$\text{A Constant} = \varepsilon_0 \cdot \mu_0 \cdot c^2$

$\text{A Constant} = \varepsilon_0 \cdot \mu_0 \cdot \text{Velocity}^2$

$\mu_0 = \varepsilon_0 \cdot \text{Electric Resistivity}^2$

$\text{A Constant} = \varepsilon_0 \cdot \mu_0 \cdot \text{Absorbed Dose}$

Electric Resistance $= \mu_0 \cdot c$

Force $= \mu_0 \cdot$ Electric Current^2

Inductance $= \mu_0 \cdot$ Length

A Constant $= \mu_0 \cdot$ Acceleration $\cdot$ Capacitance

Electric Resistance $= \mu_0 \cdot$ Velocity

Inductance$^2 = \mu_0{}^2 \cdot$ Area

Inductance$^3 = \mu_0{}^3 \cdot$ Volume

$\mu_0 =$ Wavenumber $\cdot$ Inductance

Pressure,Stress $= \mu_0 \cdot$ Magnetic Field Stength^2

Magnetic Flux Density $= \mu_0 \cdot$ Magnetic Field Strength

Magnetic Flux Density$^2 = \mu_0 \cdot$ Pressure,Stress

Electric Resistance$^2 = \mu_0{}^2 \cdot$ Absorbed Dose

A Constant $= \varepsilon_0 \cdot c \cdot$ Electric Resistance

Capacitance $= \varepsilon_0 \cdot$ Length

Force $= \varepsilon_0 \cdot$ Electric Potential Difference^2

A Constant $= \varepsilon_0 \cdot$ Acceleration $\cdot$ Inductance

A Constant $= \varepsilon_0 \cdot$ Velocity $\cdot$ Electric Resistance

Capacitance$^2 = \varepsilon_0{}^2 \cdot$ Area

Capacitance$^3 = \varepsilon_0{}^3 \cdot$ Volume

$\varepsilon_0 =$ Wavenumber $\cdot$ Capacitance

A Constant $= \varepsilon_0 \cdot$ Specific Volume $\cdot$ Magnetic Flux Density^2

A Constant $= \varepsilon_0{}^2 \cdot$ Electric Resistance$^2 \cdot$ Absorbed Dose

Mass Density $= \varepsilon_0 \cdot$ Magnetic Flux Density^2

A Constant $= \mu_0{}^2 \cdot \varepsilon_0{}^2 \cdot c^2 \cdot$ Absorbed Dose

$h^2 \cdot \mu_0 = k^4 \cdot \varepsilon_0$

Electric Charge$^2 \cdot \mu_0 = \varepsilon_0 \cdot k^2$

Electric Charge$^4 \cdot \mu_0 = \varepsilon_0 \cdot h^2$

$h^2 \cdot \mu_0 = \varepsilon_0 \cdot$ Magnetic Flux^4

A Constant $= \mu_0{}^3 \cdot \varepsilon_0{}^2 \cdot G_c \cdot$ Electric Current$^2 \cdot$

A Constant $= \mu_0{}^2 \cdot \varepsilon_0{}^2 \cdot G_c \cdot$ Force

A Constant $= \mu_0{}^5 \cdot \varepsilon_0{}^5 \cdot G_c{}^2 \cdot$ Power,Radiance Flux^2

A Constant $= \mu_0{}^2 \cdot \varepsilon_0{}^3 \cdot G_c{}^2 \cdot$ Electric Potential Difference^2

$\mu_0 \cdot$ Electric Current$^4 = \varepsilon_0 \cdot$ Power,Radiance Flux^2

$\mu_0 \cdot$ Electric Current$^2 = \varepsilon_0 \cdot$ Electric Potential Difference^2

Time$^2 = \mu_0 \cdot \varepsilon_0 \cdot$ Length^2

Time$^4 = \mu_0{}^3 \cdot \varepsilon_0{}^3 \cdot$ Volumetric Velocity^2

A Constant $= \mu_0 \cdot \varepsilon_0 \cdot$ Time$^2 \cdot$ Acceleration^2

Time$^4 = \mu_0{}^2 \cdot \varepsilon_0{}^2 \cdot$ Area^2

Time$^6 = \mu_0{}^3 \cdot \varepsilon_0{}^3 \cdot$ Volume^2

Time$^4 \cdot$ Wavenumber$^4 = \mu_0 \cdot \varepsilon_0$

Time$^2 \cdot \varepsilon_0 = \mu_0 \cdot$ Capacitance^2

Time$^2 \cdot \mu_0 =$ Inductance$^2 \cdot \varepsilon_0$

Time $= \mu_0 \cdot \varepsilon_0 \cdot$ Kinematic Viscosity

Length$^4 = \mu_0 \cdot \varepsilon_0 \cdot$ Volumetric Velocity^2

A Constant $= \mu_0 \cdot \varepsilon_0 \cdot$ Length $\cdot$ Acceleration

A Constant $= \mu_0 \cdot \varepsilon_0 \cdot$ Length$^2 \cdot$ Frequency^2

$\text{Length}^2 = \mu_0 \cdot \varepsilon_0 \cdot \text{Kinematic Viscosity}^2$

$\text{A Constant} = \mu_0{}^5 \cdot \varepsilon_0{}^5 \cdot \text{Volumetric Velocity}^2 \cdot \text{Acceleration}^4$

$\text{Area}^2 = \mu_0 \cdot \varepsilon_0 \cdot \text{Volumetric Velocity}^2$

$\text{Volume}^4 = \mu_0{}^3 \cdot \varepsilon_0{}^3 \cdot \text{Volumetric Velocity}^6$

$\text{A Constant} = \mu_0 \cdot \varepsilon_0 \cdot \text{Volumetric Velocity}^2 \cdot \text{Wavenumber}^4$

$\text{A Constant} = \mu_0{}^3 \cdot \varepsilon_0{}^3 \cdot \text{Volumetric Velocity}^2 \cdot \text{Frequency}^4$

$\text{Capacitance}^4 = \mu_0 \cdot \varepsilon_0{}^5 \cdot \text{Volumetric Velocity}^2$

$\text{Inductance}^4 = \mu_0{}^5 \cdot \varepsilon_0 \cdot \text{Volumetric Velocity}^2$

$\text{Volumetric Velocity}^2 = \mu_0 \cdot \varepsilon_0 \cdot \text{Kinematic Viscosity}^4$

$\text{Force}^2 = \mu_0 \cdot \varepsilon_0 \cdot \text{Power,Radiance Flux}^2$

$\text{Mass} = \mu_0 \cdot \varepsilon_0 \cdot \text{Energy}$

$\text{A Constant} = \mu_0{}^2 \cdot \varepsilon_0{}^2 \cdot \text{Acceleration}^2 \cdot \text{Area}$

$\text{A Constant} = \mu_0{}^3 \cdot \varepsilon_0{}^3 \cdot \text{Acceleration}^3 \cdot \text{Volume}$

$\text{Wavenumber} = \mu_0 \cdot \varepsilon_0 \cdot \text{Acceleration}$

$\text{Frequency}^2 = \mu_0 \cdot \varepsilon_0 \cdot \text{Acceleration}^2$

$\text{A Constant} = \mu_0{}^3 \cdot \varepsilon_0{}^3 \cdot \text{Acceleration}^2 \cdot \text{Kinematic Viscosity}^2$

$\text{Mass} = \mu_0 \cdot \varepsilon_0 \cdot \text{Energy,Work,Heat Quantity}$

$\text{A Constant} = \text{Absorbed Dose} \cdot \text{Velocity}^2 \cdot \mu_0{}^2 \cdot \varepsilon_0{}^2 \quad \text{(Strange)}$

$\text{A Constant} = \mu_0 \cdot \varepsilon_0 \cdot \text{Area} \cdot \text{Frequency}^2$

$\text{Area} = \mu_0 \cdot \varepsilon_0 \cdot \text{Kinematic Viscosity}^2$

$\text{A Constant} = \text{Frequency}^6 \cdot \text{Volume}^2 \cdot \mu_0{}^3 \cdot \varepsilon_0{}^3$

$\text{Volume}^2 = \mu_0{}^3 \cdot \varepsilon_0{}^3 \cdot \text{Kinematic Viscosity}^6$

31

$\text{Wavenumber}^2 = \mu_0 \cdot \varepsilon_0 \cdot \text{Frequency}^2$

$\text{A Constant} = \text{Wavenumber}^2 \cdot \text{Kinematic Viscosity}^2 \cdot \mu_0 \cdot \varepsilon_0$

$\text{A Constant} = \text{Magnetic Field Strength}^2 \cdot \text{Specific Volume} \cdot \mu_0{}^2 \cdot \varepsilon_0$

$\text{A Constant} = \text{Pressure,Stress} \cdot \text{Specific Volume} \cdot \mu_0 \cdot \varepsilon_0$

$\text{Mass Density} = \text{Magnetic Field Strength}^2 \cdot \mu_0{}^2 \cdot \varepsilon_0$

$\varepsilon_0 = \mu_0 \cdot \text{Frequency}^2 \cdot \text{Capacitance}^2$

$\mu_0 = \varepsilon_0 \cdot \text{Frequency}^2 \cdot \text{Inductance}^2$

$\text{A Constant} = \text{Frequency} \cdot \text{Kinematic Viscosity} \cdot \mu_0 \cdot \varepsilon_0$

$\text{Mass Density} = \mu_0 \cdot \varepsilon_0 \cdot \text{Pressure,Stress}$

$\varepsilon_0 \cdot \text{Electric Potential Difference}^4 = \mu_0 \cdot \text{Power,Radiance Flux}^2$

$\varepsilon_0 \cdot \text{Magnetic Flux}^2 = \mu_0 \cdot \text{Electric Charge}^2$

$\mu_0 \cdot \text{Capacitance} = \varepsilon_0 \cdot \text{Inductance}$

$\text{Capacitance}^2 = \mu_0 \cdot \varepsilon_0{}^3 \cdot \text{Kinematic Viscosity}^2$

$\text{Inductance}^2 = \mu_0{}^3 \cdot \varepsilon_0 \cdot \text{Kinematic Viscosity}^2$

$k^2 = \mu_0 \cdot c \cdot h$

$k = \mu_0 \cdot c \cdot \text{Electric Charge}$

$h = \mu_0 \cdot c \cdot \text{Electric Charge}^2$

$\text{Magnetic Flux}^2 = \mu_0 \cdot c \cdot h$

$c^4 = \mu_0 \cdot G_c \cdot \text{Electric Current}^2$

$G_c \cdot \text{Electric Potential Difference}^2 = \mu_0 \cdot c^6$

$\text{Power,Radiance Flux} = \mu_0 \cdot c \cdot \text{Electric Current}^2$

$\text{Electric Potential Difference} = \mu_0 \cdot c \cdot \text{Electric Current}$

$\text{Time} = \text{Capacitance} \cdot \mu_0 \cdot c$

Inductance = Time $\cdot$ μ_0 $\cdot$ c

Length = μ_0 $\cdot$ c^2 $\cdot$ Capacitance

Volumetric Velocity = $\mu_0{}^2$ $\cdot$ c^5 $\cdot$ Capacitance^2

Inductance^2 $\cdot$ c = Volumetric Velocity $\cdot$ $\mu_0{}^2$

Electric Potential Difference^2 = μ_0 $\cdot$ c^2 $\cdot$ Force

Acceleration $\cdot$ Inductance = μ_0 $\cdot$ c^2

Area = Capacitance^2 $\cdot$ $\mu_0{}^2$ $\cdot$ c^4

Volumetric Velocity = Capacitance^3 $\cdot$ $\mu_0{}^3$ $\cdot$ c^6

A Constant = μ_0 $\cdot$ c^2 $\cdot$ Capacitance $\cdot$ Wavenumber

c^2 = μ_0 $\cdot$ Specific Volume $\cdot$ Magnetic Field Strength^2

Specific Volume $\cdot$ Magnetic Flux Density^2 = μ_0 $\cdot$ c^2

Mass Density $\cdot$ c^2 = μ_0 $\cdot$ Magnetic Field Strength^2

A Constant = μ_0 $\cdot$ c $\cdot$ Frequency $\cdot$ Capacitance

Frequency $\cdot$ Inductance = μ_0 $\cdot$ c

Electric Potential Difference^2 = μ_0 $\cdot$ c $\cdot$ Power,Radiance Flux

Magnetic Flux = Electric Charge $\cdot$ μ_0 $\cdot$ c

Inductance = Capacitance $\cdot$ $\mu_0{}^2$ $\cdot$ c^2

Kinematic Viscosity = μ_0 $\cdot$ c^3 $\cdot$ Capacitance

Kinematic Viscosity $\cdot$ μ_0 = Inductance $\cdot$ c

k^2 = μ_0 $\cdot$ h $\cdot$ Velocity

k^4 = $\mu_0{}^2$ $\cdot$ h^2 $\cdot$ Absorbed Dose

k^2 = μ_0 $\cdot$ G_c $\cdot$ Mass^2

$k = \mu_0 \cdot$ Electric Current $\cdot$ Length

$k^2 = \mu_0{}^2 \cdot$ Electric Current$^2 \cdot$ Area

$k^3 = \mu_0{}^3 \cdot$ Electric Current$^3 \cdot$ Volume

$k \cdot$ Wavenumber $= \mu_0 \cdot$ Electric Current

$k^2 = \mu_0{}^2 \cdot$ Electric Current $\cdot$ Current Density

$k \cdot$ Magnetic Field Strength $= \mu_0 \cdot$ Electric Current^2

$k^2 \cdot$ Pressure,Stress $= \mu_0{}^3 \cdot$ Electric Current^4

$k \cdot$ Magnetic Flux Density $= \mu_0{}^2 \cdot$ Electric Current^2

$k^2 = \mu_0 \cdot$ Length$^2 \cdot$ Force

$k^2 = \mu_0 \cdot$ Length $\cdot$ Energy

$k \cdot$ Length $= \mu_0 \cdot$ Current Density

$k = \mu_0 \cdot$ Length$^2 \cdot$ Magnetic Field Strength

$k^2 = \mu_0 \cdot$ Length$^4 \cdot$ Pressure,Stress

$k^2 = \mu_0 \cdot$ Length $\cdot$ Energy,Work,Heat Quantity

$k^2 \cdot$ Specific Volume $= \mu_0 \cdot$ Volumetric Velocity^2

$k^2 = \mu_0 \cdot$ Volumetric Velocity$^2 \cdot$ Mass Density

$k^2 \cdot$ Force $= \mu_0 \cdot$ Energy^2

$k^2 = \mu_0 \cdot$ Force $\cdot$ Area

$k^6 = \mu_0{}^3 \cdot$ Force$^3 \cdot$ Volume^2

$k^2 \cdot$ Wavenumber$^2 = \mu_0 \cdot$ Force

$k^4 = \mu_0{}^3 \cdot$ Force $\cdot$ Current Density^2

$k^2 \cdot$ Pressure,Stress $= \mu_0 \cdot$ Force^2

$k^2 \cdot$ Force $=$ Energy,Work,Heat Quantity$^2 \cdot \mu_0$

$k \cdot \text{Magnetic Flux Density} = \mu_0 \cdot \text{Force}$

$\text{Force} \cdot \text{Inductance}^2 = \mu_0 \cdot k^2$

$k^4 = \mu_0{}^2 \cdot \text{Energy}^2 \cdot \text{Area}$

$k^6 = \mu_0{}^3 \cdot \text{Energy}^3 \cdot \text{Volume}$

$k^2 \cdot \text{Wavenumber} = \mu_0 \cdot \text{Energy}$

$k^3 = \mu_0{}^2 \cdot \text{Energy} \cdot \text{Current Density}$

$k^3 \cdot \text{Magnetic Field Strength} = \mu_0 \cdot \text{Energy}^2$

$k^6 \cdot \text{Pressure,Stress} = \mu_0{}^3 \cdot \text{Energy}^4$

$k^3 \cdot \text{Magnetic Flux Density} = \mu_0{}^2 \cdot \text{Energy}^2$

$k = \mu_0 \cdot \text{Electric Charge} \cdot \text{Velocity}$

$k^2 \cdot \text{Area} = \mu_0{}^2 \cdot \text{Current Density}^2$

$k = \mu_0 \cdot \text{Area} \cdot \text{Magnetic Field Strength}$

$k^2 = \mu_0 \cdot \text{Area}^2 \cdot \text{Pressure,Stress}$

$k^4 = \mu_0{}^2 \cdot \text{Area} \cdot \text{Energy,Work,Heat Quantity}^2$

$k^3 \cdot \text{Volume} = \mu_0{}^3 \cdot \text{Current Density}^3$

$k^3 = \mu_0{}^3 \cdot \text{Volume}^2 \cdot \text{Magnetic Field Strength}^3$

$k^6 = \mu_0{}^3 \cdot \text{Volume}^4 \cdot \text{Pressure,Stress}^3$

$k^6 = \mu_0{}^3 \cdot \text{Volume} \cdot \text{Energy,Work,Heat Quantity}^3$

$k = \mu_0 \cdot \text{Wavenumber} \cdot \text{Current Density}$

$k \cdot \text{Wavenumber}^2 = \mu_0 \cdot \text{Magnetic Field Strength}$

$k^2 \cdot \text{Wavenumber}^4 = \mu_0 \cdot \text{Pressure,Stress}$

$k^2 \cdot \text{Wavenumber} = \mu_0 \cdot \text{Energy,Work,Heat Quantity}$

$k^3 = {\mu_0}^3 \cdot \text{Current Density}^2 \cdot \text{Magnetic Field Strength}$

$k^6 = {\mu_0}^5 \cdot \text{Current Density}^4 \cdot \text{Pressure,Stress}$

$k^3 = {\mu_0}^2 \cdot \text{Current Density} \cdot \text{Energy,Work,Heat Quantity}$

$k^3 = {\mu_0}^2 \cdot \text{Current Density}^2 \cdot \text{Magnetic Flux Density}$

$k \cdot \text{Inductance} = {\mu_0}^2 \cdot \text{Current Density}$

$k^3 \cdot \text{Magnetic Field Strength} = \mu_0 \cdot \text{Energy,Work,Heat Quantity}$

$\mu_0 \cdot k = \text{Magnetic Field Strength} \cdot \text{Inductance}^2$

$k^6 \cdot \text{Pressure,Stress} = {\mu_0}^3 \cdot \text{Energy,Work,Heat Quantity}^4$

${\mu_0}^3 \cdot k^2 = \text{Pressure,Stress} \cdot \text{Inductance}^4$

${\mu_0}^2 \cdot \text{Energy,Work,Heat Quantity}^2 = k^3 \cdot \text{Magnetic Flux Density}$

$k^2 = {\mu_0}^2 \cdot \text{Electric Charge}^2 \cdot \text{Absorbed Dose}$

$\text{Magnetic Flux Density} \cdot \text{Inductance}^2 = k \cdot {\mu_0}^2$

$h \cdot \text{Volumetric Velocity} = \mu_0 \cdot \text{Current Density}^2$

$h = \mu_0 \cdot \text{Velocity} \cdot \text{Electric Charge}^2$

$\text{Magnetic Flux}^2 = \mu_0 \cdot h \cdot \text{Velocity}$

$h^2 = {\mu_0}^2 \cdot \text{Electric Charge}^4 \cdot \text{Absorbed Dose}$

$\text{Inductance}^3 \cdot \text{Dynamic Viscosity} = {\mu_0}^3 \cdot h$

$\text{Magnetic Flux}^4 = {\mu_0}^2 \cdot h^2 \cdot \text{Absorbed Dose}$

$\text{Velocity}^4 = \mu_0 \cdot G_c \cdot \text{Electric Current}^2$

$\text{Power,Radiance Flux}^4 = {\mu_0}^5 \cdot G_c \cdot \text{Electric Current}^0$

$\text{Electric Potential Difference}^4 = {\mu_0}^5 \cdot G_c \cdot \text{Electric Current}^6$

$\text{Electric Resistance}^4 = {\mu_0}^5 \cdot G_c \cdot \text{Electric Current}^2$

$\text{Absorbed Dose}^2 = \mu_0 \cdot G_c \cdot \text{Electric Current}^2$

$$\text{Electric Potential Difference}^4 = \mu_0{}^2 \cdot G_c \cdot \text{Force}^3$$

$$\text{Electric Resistance}^4 = \mu_0{}^4 \cdot G_c \cdot \text{Force}$$

$$\text{Acceleration}^2 = \mu_0 \cdot G_c \cdot \text{Magnetic Field Strength}^2$$

$$G_c \cdot \text{Magnetic Flux Density}^2 = \mu_0 \cdot \text{Acceleration}^2$$

$$\text{Magnetic Flux}^2 = \mu_0 \cdot G_c \cdot \text{Mass}^2$$

$$\text{Electric Potential Difference}^2 \cdot Gc = \mu_0 \cdot \text{Velocity}^6$$

$$\text{Current Density}^2 = \mu_0 \cdot Gc \cdot \text{Electric Charge}^4$$

$$\text{Kinematic Viscosity}^4 = \mu_0 \cdot Gc \cdot \text{Current Density}^2$$

$$\text{A Constant} = \mu_0{}^3 \cdot G_c \cdot \text{Magnetic Field Strength}^2 \cdot \text{Capacitance}^2$$

$$\text{A Constant} = \mu_0{}^2 \cdot G_c \cdot \text{Pressure,Stress} \cdot \text{Capacitance}^2$$

$$\text{Electric Potential Difference}^0 = \mu_0{}^5 \cdot G_c \cdot \text{Power,Radiance Flux}^6$$

$$\text{Electric Resistance}^5 = \mu_0{}^5 \cdot G_c \cdot \text{Power,Radiance Flux}$$

$$\text{Kinematic Viscosity}^2 = \text{Electric Charge}^2 \cdot \mu_0 \cdot G_c$$

$$\text{Electric Resistance}^6 = \mu_0{}^5 \cdot G_c \cdot \text{Electric Potential Difference}^2$$

$$G_c \cdot \text{Electric Potential Difference}^2 = \mu_0 \cdot \text{Absorbed Dose}^3$$

$$\text{A Constant} = \mu_0 \cdot G_c \cdot \text{Capacitance}^2 \cdot \text{Magnetic Flux Density}^2$$

$$\text{Inductance}^2 \cdot \text{Illuminance} = \mu_0{}^2 \cdot \text{Luminous Intensity}$$

$$\text{Energy} = \mu_0 \cdot \text{Electric Current}^2 \cdot \text{Length}$$

$$\text{Pressure,Stress} \cdot \text{Length}^2 = \mu_0 \cdot \text{Electric Current}^2$$

$$\text{Energy,Work,Heat Quantity} = \mu_0 \cdot \text{Electric Current}^2 \cdot \text{Length}$$

$$\text{Magnetic Flux Density} \cdot \text{Length} = \mu_0 \cdot \text{Electric Current}$$

$$\text{Magnetic Flux} = \mu_0 \cdot \text{Electric Current} \cdot \text{Length}$$

$$\text{Energy}^2 = \text{Area} \cdot \mu_0{}^2 \cdot \text{Electric Current}^4$$

$$\text{Energy}^3 = \text{Volume} \cdot \mu_0{}^3 \cdot \text{Electric Current}^6$$

$$\text{Wavenumber} \cdot \text{Energy} = \mu_0 \cdot \text{Electric Current}^2$$

$$\text{Energy}^2 = \text{Current Density} \cdot \mu_0{}^2 \cdot \text{Electric Current}^3$$

$$\text{Energy} \cdot \text{Magnetic Field Strength} = \mu_0 \cdot \text{Electric Current}^3$$

$$\text{Energy}^2 \cdot \text{Pressure,Stress} = \mu_0{}^3 \cdot \text{Electric Current}^6$$

$$\text{Energy} \cdot \text{Magnetic Flux Density} = \mu_0{}^2 \cdot \text{Electric Current}^3$$

$$\text{Acceleration} \cdot \text{Mass} = \mu_0 \cdot \text{Electric Current}^2$$

$$\text{Mass} = \text{Capacitance} \cdot \mu_0{}^2 \cdot \text{Electric Current}^2$$

$$\text{Power,Radiance Flux} = \mu_0 \cdot \text{Electric Current}^2 \cdot \text{Velocity}$$

$$\text{Electric Potential Difference} = \mu_0 \cdot \text{Electric Current} \cdot \text{Velocity}$$

$$\text{Area} \cdot \text{Pressure,Stress} = \mu_0 \cdot \text{Electric Current}^2$$

$$\text{Energy,Work,Heat Quantity}^2 = \text{Area} \cdot \mu_0{}^2 \cdot \text{Electric Current}^4$$

$$\text{Magnetic Flux Density}^2 \cdot \text{Area} = \mu_0{}^2 \cdot \text{Electric Current}^2$$

$$\text{Magnetic Flux}^2 = \mu_0{}^2 \cdot \text{Electric Current}^2 \cdot \text{Area}$$

$$\text{Pressure,Stress}^3 \cdot \text{Volume}^2 = \mu_0{}^3 \cdot \text{Electric Current}^6$$

$$\text{Enery,Work,Heat Quantity}^3 = \mu_0{}^3 \cdot \text{Electric Current}^6 \cdot \text{Volume}$$

$$\text{Magnetic Flux Density}^3 \cdot \text{Volume} = \mu_0{}^3 \cdot \text{Electric Current}^3$$

$$\text{Pressure,Stress} = \mu_0 \cdot \text{Electric Current}^2 \cdot \text{Wavenumber}^2$$

$$\text{Energy,Work,Heat Quantity} \cdot \text{Wavenumber} = \mu_0 \cdot \text{Electric Current}^2$$

$$\text{Magnetic Flux Density} = \mu_0 \cdot \text{Electric Current} \cdot \text{Wavenumber}$$

$$\text{Magnetic Flux} \cdot \text{Wavenumber} = \mu_0 \cdot \text{Electric Current}$$

$$\text{Dynamic Viscosity}^2 \cdot \text{Specific Volume} = \mu_0 \cdot \text{Electric Current}^2$$

Kinematic Viscosity^2 = μ_0 · Electric Current^2 · Specific Volume

Pressure,Stress · Current Density = μ_0 · Electric Current^3

Energy,Work,Heat Quantity^2 = $\mu_0{}^2$ · Electric Current^3 · Current Density

Magnetic Flux Density^2 · Current Density = $\mu_0{}^2$ · Electric Current^3

Inductance^2 · Electric Current = $\mu_0{}^2$ · Current Density

Magnetic Flux^2 = $\mu_0{}^2$ · Electric Current · Current Density

Energy,Work,Heat Quantity · Magnetic Field Strength = μ_0 · Electric Current^3

Inductance · Magnetic Field Strength = μ_0 · Electric Current

Magnetic Flux · Magnetic Field Strength = μ_0 · Electric Current^2

Energy,Work,Heat Quantity^2 · Pressure,Stress = $\mu_0{}^3$ · Electric Current^6

Inductance^2 · Pressure,Stress = $\mu_0{}^3$ · Electric Current^2

Magnetic Flux^2 · Pressure,Stress = $\mu_0{}^3$ · Electric Current^4

Magnetic Flux Density · Energy,Work,Heat Quantity = $\mu_0{}^2$ · Electric Current^3

Power,Radiance Flux^2 = $\mu_0{}^2$ · Electric Current^4 · Absorbed Dose

Electric Potential Difference^2 = $\mu_0{}^2$ · Electric Current^2 · Absorbed Dose

Inductance · Magnetic Flux Density = $\mu_0{}^2$ · Electric Current

Magnetic Flux · Magnetic Flux Density = $\mu_0{}^2$ · Electric Current^2

Kinematic Viscosity · Dynamic Viscosity = μ_0 · Electric Current^2

Dynamic Viscosity^2 = Mass Density · μ_0 · Electric Current^2

Mass Density · Kinematic Viscosity^2 = μ_0 · Electric Current^2

Time^2 = μ_0 · Length · Capacitance

Time · Electric Resistance = μ_0 · Length

$\text{Time}^5 = \mu_0{}^3 \cdot \text{Volumetric Velocity} \cdot \text{Capacitance}^3$

$\text{Time}^2 \cdot \text{Electric Resistance}^3 = \mu_0{}^3 \cdot \text{Volumetric Velocity}$

$\text{Inductance}^3 = \mu_0{}^3 \cdot \text{Time} \cdot \text{Volumetric Velocity}$

$\text{Time}^2 \cdot \text{Force} = \mu_0 \cdot \text{Electric Charge}^2$

$\text{Electric Resistance} = \mu_0 \cdot \text{Time} \cdot \text{Acceleration}$

$\text{Inductance} = \mu_0 \cdot \text{Time}^2 \cdot \text{Acceleration}$

$\text{Time} = \mu_0 \cdot \text{Velocity} \cdot \text{Capacitance}$

$\text{Inductance} = \mu_0 \cdot \text{Time} \cdot \text{Velocity}$

$\text{Time}^4 = \mu_0{}^2 \cdot \text{Area} \cdot \text{Capacitance}^2$

$\text{Time}^2 \cdot \text{Electric Resistance}^2 = \mu_0{}^2 \cdot \text{Area}$

$\text{Time}^6 = \mu_0{}^3 \cdot \text{Volume} \cdot \text{Capacitance}^3$

$\text{Time}^3 \cdot \text{Electric Resistance}^3 = \mu_0{}^3 \cdot \text{Volume}$

$\text{Time}^2 \cdot \text{Wavenumber} = \mu_0 \cdot \text{Capacitance}$

$\mu_0 = \text{Time} \cdot \text{Wavenumber} \cdot \text{Electric Resistance}$

$\text{Dynamic Viscosity} = \mu_0 \cdot \text{Time} \cdot \text{Magnetic Field Strength}^2$

$\text{Time}^2 = \mu_0{}^2 \cdot \text{Capacitance}^2 \cdot \text{Absorbed Dose}$

$\text{Time}^3 = \mu_0{}^2 \cdot \text{Capacitance}^2 \cdot \text{Kinematic Viscosity}$

$\text{Time} \cdot \text{Electric Resistance}^2 = \mu_0{}^2 \cdot \text{Kinematic Viscosity}$

$\mu_0 \cdot \text{Dynamic Viscosity} = \text{Time} \cdot \text{Magnetic Flux Density}^2$

$\text{Inductance}^2 = \mu_0{}^2 \cdot \text{Time}^2 \cdot \text{Absorbed Dose}$

$\text{Inductance}^2 = \mu_0{}^2 \cdot \text{Time} \cdot \text{Kinematic Viscosity}$

$\text{Length}^5 = \mu_0 \cdot \text{Volumetric Velocity}^2 \cdot \text{Capacitance}$

$\text{Length}^2 \cdot \text{Electric Resistance} = \mu_0 \cdot \text{Volumetric Velocity}$

Length^4 · Force = μ_0 · Current Density^2

Force = μ_0 · Length^2 · Magnetic Field Strength^2

Length^2 · Magnetic Flux Density^2 = μ_0 · Force

Magnetic Flux^2 = μ_0 · Length^2 · Force

Length^3 · Energy = μ_0 · Current Density^2

Energy = μ_0 · Length^3 · Magnetic Field Strength^2

Length^3 · Magnetic Flux Density^2 = μ_0 · Energy

Magnetic Flux^2 = μ_0 · Length · Energy

Electric Resistance^2 = μ_0^2 · Length · Acceleration

Length · Mass = μ_0 · Electric Charge^2

Length = μ_0 · Velocity^2 · Capacitance

Length^4 = μ_0 · Specific Volume · Electric Charge^2

Length^6 · Pressure,Stress = μ_0 · Current Density^2

Length^3 · Energy,Work,Heat Quantity = μ_0 · Current Density^2

Length^3 · Magnetic Flux Density = μ_0 · Current Density

Length · Magnetic Flux = μ_0 · Current Density

Energy,Work,Heat Quantity = μ_0 · Length^3 · Magnetic Field Strength^2

Magnetic Flux = μ_0 · Length^2 · Magnetic Field Strength

A Constant = μ_0 · Length · Frequency^2 · Capacitance

Electric Resistance = μ_0 · Length · Frequency

Magnetic Flux^2 = μ_0 · Length^4 · Pressure,Stress

Length^3 · Magnetic Flux Density^2 = μ_0 · Energy,Work,Heat Quantity

Magnetic Flux^2 = μ_0 · Length · Energy,Work,Heat Quantity

Length^4 · Mass Density = μ_0 · Electric Charge^2

Capacitance · Electric Resistance^2 = μ_0 · Length

Length = μ_0 · Capacitance · Absorbed Dose

Length^3 = μ_0 · Capacitance · Kinematic Viscosity^2

Length · Electric Resistance = μ_0 · Kinematic Viscosity

Electric Resistance^5 = μ_0^5 · Volumetric Velocity · Acceleration^2

μ_0^5 · Volumetric Velocity^2 = Acceleration · Inductance^5

μ_0^2 · Velocity^5 · Capacitance^2 = Volumetric Velocity

μ_0^2 · Volumetric Velocity = Velocity · Inductance^2

μ_0^2 · Volumetric Velocity^4 · Capacitance^2 = Area^5

μ_0 · Volumetric Velocity = Area · Electric Resistance

μ_0^3 · Volumetric Velocity^6 · Capacitance^3 = Volume^5

μ_0^3 · Volumetric Velocity^3 = Volume^2 · Electric Resistance^3

μ_0 · Volumetric Velocity^2 · Wavenumber^5 · Capacitance = A Constant

μ_0 · Volumetric Velocity · Wavenumber^2 = Electric Resistance

μ_0 · Volumetric Velocity^2 = Specific Volume · Magnetic Flux^2

μ_0 · Volumetric Velocity · Magnetic Field Strength^2 = Power, Radiance Flux

μ_0^3 · Volumetric Velocity · Frequency^5 · Capacitance^3 = A Constant

μ_0^3 · Volumetric Velocity · Frequency^2 = Electric Resistance^3

μ_0^3 · Volumetric Velocity = Frequency · Inductance^3

μ_0 · Power, Radiance Flux = Volumetric Velocity · Magnetic Flux Density^2

μ_0^3 · Volumetric Velocity = Capacitance^2 · Electric Resistance^5

$\mu_0{}^6 \cdot \text{Volumetric Velocity}^2 \cdot \text{Capacitance} = \text{Inductance}^5$

$\mu_0{}^4 \cdot \text{Capacitance}^4 \cdot \text{Absorbed Dose}^5 = \text{Volumetric Velocity}^2$

$\mu_0 \cdot \text{Capacitance} \cdot \text{Kinematic Viscosity}^5 = \text{Volumetric Velocity}^3$

$\mu_0{}^3 \cdot \text{Volumetric Velocity} = \text{Electric Resistance} \cdot \text{Inductance}^2$

$\mu_0 \cdot \text{Kinematic Viscosity}^2 = \text{Volumetric Velocity} \cdot \text{Electric Resistance}$

$\mu_0{}^4 \cdot \text{Volumetric Velocity}^2 = \text{Inductance}^4 \cdot \text{Absorbed Dose}$

$\mu_0 \cdot \text{Volumetric Velocity} = \text{Inductance} \cdot \text{Kinematic Viscosity}$

$\mu_0 \cdot \text{Volumetric Velocity}^2 \cdot \text{Mass Density} = \text{Magnetic Flux}^2$

$\mu_0 \cdot \text{Force}^3 \cdot \text{Current Density}^2 = \text{Energy}^4$

$\mu_0 \cdot \text{Energy}^2 \cdot \text{Magnetic Field Strength}^2 = \text{Force}^3$

$\mu_0 \cdot \text{Force}^3 = \text{Energy}^2 \cdot \text{Magnetic Flux Density}^2$

$\mu_0 \cdot \text{Energy} = \text{Force} \cdot \text{Inductance}$

$\mu_0 \cdot \text{Energy}^2 = \text{Force} \cdot \text{Magnetic Flux}^2$

$\mu_0 \cdot \text{Force} \cdot \text{Capacitance} = \text{Mass}$

$\mu_0 \cdot \text{Force} \cdot \text{Velocity}^2 = \text{Electric Potential Difference}^2$

$\mu_0 \cdot \text{Current Density}^2 = \text{Force} \cdot \text{Area}^2$

$\mu_0 \cdot \text{Area} \cdot \text{Magnetic Field Strength}^2 = \text{Force}$

$\mu_0 \cdot \text{Force} = \text{Area} \cdot \text{Magnetic Flux Density}^2$

$\mu_0 \cdot \text{Force} \cdot \text{Area} = \text{Magnetic Flux}^2$

$\mu_0{}^3 \cdot \text{Current Density}^6 = \text{Force}^3 \cdot \text{Volume}^4$

$\mu_0{}^3 \cdot \text{Volume}^2 \cdot \text{Magnetic Field Strength}^6 = \text{Force}^3$

$\mu_0{}^3 \cdot \text{Force}^3 = \text{Volume}^2 \cdot \text{Magnetic Flux Density}^6$

$$\mu_0{}^3 \cdot \text{Force}^3 \cdot \text{Volume}^2 = \text{Magnetic Flux}^6$$

$$\mu_0 \cdot \text{Wavenumber}^4 \cdot \text{Current Density}^2 = \text{Force}$$

$$\mu_0 \cdot \text{Magnetic Field Strength}^2 = \text{Force} \cdot \text{Wavenumber}^2$$

$$\mu_0 \cdot \text{Force} \cdot \text{Wavenumber}^2 = \text{Magnetic Flux Density}^2$$

$$\mu_0 \cdot \text{Force} = \text{Wavenumber}^2 \cdot \text{Magnetic Flux}^2$$

$$\mu_0{}^3 \cdot \text{Current Density}^2 \cdot \text{Magnetic Field Strength}^4 = \text{Force}^3$$

$$\mu_0 \cdot \text{Current Density}^2 \cdot \text{Pressure,Stress}^2 = \text{Force}^3$$

$$\mu_0 \cdot \text{Force}^3 \cdot \text{Current Density}^2 = \text{Energy,Work,Heat Quantity}^4$$

$$\mu_0 \cdot \text{Force}^3 = \text{Current Density}^2 \cdot \text{Magnetic Flux Density}^4$$

$$\mu_0{}^5 \cdot \text{Current Density}^2 = \text{Force} \cdot \text{Inductance}^4$$

$$\mu_0{}^3 \cdot \text{Force} \cdot \text{Current Density}^2 = \text{Magnetic Flux}^4$$

$$\mu_0 \cdot \text{Magnetic Field Strength}^2 \cdot \text{Energy,Work,Heat Quantity}^2 = \text{Force}^3$$

$$\mu_0 \cdot \text{Force} = \text{Magnetic Field Strength}^2 \cdot \text{Inductance}^2$$

$$\mu_0 \cdot \text{Frequency}^2 \cdot \text{Electric Charge}^2 = \text{Force}$$

$$\mu_0{}^2 \cdot \text{Force} = \text{Pressure,Stress} \cdot \text{Inductance}^2$$

$$\mu_0 \cdot \text{Force}^2 = \text{Pressure,Stress} \cdot \text{Magnetic Flux}^2$$

$$\mu_0 \cdot \text{Force}^3 = \text{Energy,Work,Heat Quantity}^2 \cdot \text{Magnetic Flux Density}^2$$

$$\mu_0 \cdot \text{Energy,Work,Heat Quantity} = \text{Force} \cdot \text{Inductance}$$

$$\mu_0 \cdot \text{Energy,Work,Heat Quantity}^2 = \text{Force} \cdot \text{Magnetic Flux}^2$$

$$\mu_0 \cdot \text{Power, Radiance Flux}^2 = \text{Force} \cdot \text{Electric Potential Difference}^2$$

$$\mu_0 \cdot \text{Power, Radiance Flux} = \text{Force} \cdot \text{Electric Resistance}$$

$$\mu_0 \cdot \text{Electric Potential Difference}^2 = \text{Force} \cdot \text{Electric Resistance}^2$$

$$\mu_0 \cdot \text{Force} \cdot \text{Absorbed Dose} = \text{Electric Potential Difference}^2$$

$\mu_0{}^3 \cdot$ Force = Magnetic Flux Density$^2 \cdot$ Inductance^2

$\mu_0 \cdot$ Force = Magnetic Flux Density $\cdot$ Magnetic Flux

$\mu_0 \cdot$ Magnetic Flux^2 = Force $\cdot$ Inductance^2

$\mu_0 \cdot$ Acceleration $\cdot$ Electric Charge^2 = Energy

$\mu_0 \cdot$ Energy $\cdot$ Acceleration = Electric Potential Difference^2

$\mu_0{}^2 \cdot$ Energy = Mass $\cdot$ Electric Resistance^2

$\mu_0{}^2 \cdot$ Current Density^4 = Energy$^2 \cdot$ Area^3

$\mu_0{}^2 \cdot$ Area$^3 \cdot$ Magnetic Field Strength^4 = Energy^2

$\mu_0{}^2 \cdot$ Energy^2 = Area$^3 \cdot$ Magnetic Flux Density^4

$\mu_0{}^2 \cdot$ Energy$^2 \cdot$ Area = Magnetic Flux^4

$\mu_0 \cdot$ Current Density^2 = Energy $\cdot$ Volume

$\mu_0 \cdot$ Volume $\cdot$ Magnetic Field Strength^2 = Energy

$\mu_0 \cdot$ Energy = Volume $\cdot$ Magnetic Flux Density^2

$\mu_0{}^3 \cdot$ Energy$^3 \cdot$ Volume = Magnetic Flux^6

$\mu_0 \cdot$ Wavenumber$^3 \cdot$ Current Density^2 = Energy

$\mu_0 \cdot$ Magnetic Field Strength^2 = Energy $\cdot$ Wavenumber^3

$\mu_0 \cdot$ Energy $\cdot$ Wavenumber^3 = Magnetic Flux Density^2

$\mu_0 \cdot$ Energy = Wavenumber $\cdot$ Magnetic Flux^2

$\mu_0 \cdot$ Current Density $\cdot$ Magnetic Field Strength = Energy

$\mu_0 \cdot$ Current Density$^2 \cdot$ Pressure,Stress = Energy^2

$\mu_0{}^4 \cdot$ Current Density^2 = Energy $\cdot$ Inductance^3

$\mu_0{}^2 \cdot$ Energy $\cdot$ Current Density = Magnetic Flux^3

$\mu_0^2 \cdot$ Energy = Magnetic Field Strength$^2 \cdot$ Inductance^3

$\mu_0 \cdot$ Energy^2 = Magnetic Field Strength $\cdot$ Magnetic Flux^3

$\mu_0^3 \cdot$ Energy = Pressure,Stress $\cdot$ Inductance^3

$\mu_0^3 \cdot$ Energy^4 = Pressure,Stress $\cdot$ Magnetic Flux^6

$\mu_0^4 \cdot$ Energy = Magnetic Flux Density$^2 \cdot$ Inductance^3

$\mu_0^2 \cdot$ Energy^2 = Magnetic Flux Density $\cdot$ Magnetic Flux^3

$\mu_0 \cdot$ Velocity^2 = Acceleration $\cdot$ Inductance

$\mu_0^4 \cdot$ Acceleration$^2 \cdot$ Area = Electric Resistance^4

$\mu_0^6 \cdot$ Acceleration$^3 \cdot$ Volume = Electric Resistance^6

$\mu_0^2 \cdot$ Acceleration = Wavenumber $\cdot$ Electric Resistance^2

$\mu_0 \cdot$ Acceleration = Frequency $\cdot$ Electric Resistance

$\mu_0 \cdot$ Acceleration = Frequency$^2 \cdot$ Inductance

$\mu_0 \cdot$ Acceleration $\cdot$ Electric Charge^2 = Energy,Work,Heat Quantity

$\mu_0 \cdot$ Acceleration $\cdot$ Energy,Work,Heat Quantity = Electric Potential Difference^2

$\mu_0 \cdot$ Acceleration $\cdot$ Electric Charge = Electric Potential Difference

$\mu_0 \cdot$ Acceleration $\cdot$ Inductance = Electric Resistance^2

$\mu_0^3 \cdot$ Acceleration $\cdot$ Kinematic Viscosity = Electric Resistance^3

$\mu_0 \cdot$ Absorbed Dose = Acceleration $\cdot$ Inductance

$\mu_0^3 \cdot$ Kinematic Viscosity^2 = Acceleration $\cdot$ Inductance^3

$\mu_0^2 \cdot$ Electric Charge^4 = Mass$^2 \cdot$ Area

$\mu_0^3 \cdot$ Electric Charge^6 = Mass$^3 \cdot$ Volume

$\mu_0 \cdot$ Wavenumber $\cdot$ Electric Charge^2 = Mass

$\mu_0^3 \cdot$ Electric Charge^6 = Mass$^4 \cdot$ Specific Volume

$\mu_0^3 \cdot$ Mass $\cdot$ Specific Volume = Inductance^3

$\mu_0^2 \cdot$ Energy,Work,Heat Quantity = Mass $\cdot$ Electric Resistance^2

$\mu_0^2 \cdot$ Electric Charge^2 = Mass $\cdot$ Inductance

$\mu_0^3 \cdot$ Electric Charge$^6 \cdot$ Mass Density = Mass^4

$\mu_0^3 \cdot$ Mass = Inductance$^3 \cdot$ Mass Density

$\mu_0^2 \cdot$ Velocity$^4 \cdot$ Capacitance^2 = Area

$\mu_0^3 \cdot$ Velocity$^6 \cdot$ Capacitance^3 = Volume

$\mu_0 \cdot$ Velocity$^2 \cdot$ Wavenumber $\cdot$ Capacitance = A Constant

$\mu_0 \cdot$ Specific Volume $\cdot$ Magnetic Field Strength^2 = Velocity^2

$\mu_0 \cdot$ Velocity^2 = Specific Volume $\cdot$ Magnetic Flux Density^2

$\mu_0 \cdot$ Magnetic Field Strength^2 = Velocity$^2 \cdot$ Mass Density

$\mu_0 \cdot$ Velocity $\cdot$ Frequency $\cdot$ Capacitance = A Constant

$\mu_0 \cdot$ Velocity = Frequency $\cdot$ Inductance

$\mu_0 \cdot$ Velocity $\cdot$ Power, Radiance Flux = Electric Potential Difference^2

$\mu_0 \cdot$ Velocity $\cdot$ Electric Charge = Magnetic Flux

$\mu_0^2 \cdot$ Velocity$^2 \cdot$ Capacitance = Inductance

$\mu_0 \cdot$ Velocity$^3 \cdot$ Capacitance = Kinematic Viscosity

$\mu_0 \cdot$ Velocity$^2 \cdot$ Mass Density = Magnetic Flux Density^2

$\mu_0 \cdot$ Kinematic Viscosity = Velocity $\cdot$ Inductance

$\mu_0 \cdot$ Specific Volume $\cdot$ Electric Charge^2 = Area^2

$\mu_0 \cdot$ Current Density^2 = Area$^3 \cdot$ Pressure,Stress

$\mu_0^2 \cdot$ Current Density^4 = Area$^3 \cdot$ Energy,Work,Heat Quantity^2

$\mu_0^2 \cdot$ Current Density^2 = Area$^3 \cdot$ Magnetic Flux Density^2

$\mu_0^2 \cdot$ Current Density^2 = Area $\cdot$ Magnetic Flux^2

$\mu_0^2 \cdot$ Area$^3 \cdot$ Magnetic Field Strength^4 = Energy,Work,Heat Quantity^2

$\mu_0 \cdot$ Area $\cdot$ Magnetic Field Strength = Magnetic Flux

$\mu_0^2 \cdot$ Area $\cdot$ Frequency$^4 \cdot$ Capacitance^2 = A Constant

$\mu_0^2 \cdot$ Area $\cdot$ Frequency^2 = Electric Resistance^2

$\mu_0 \cdot$ Area$^2 \cdot$ Pressure,Stress = Magnetic Flux^2

$\mu_0^2 \cdot$ Energy,Work,Heat Quantity^2 = Area$^3 \cdot$ Magnetic Flux Density^4

$\mu_0^2 \cdot$ Area $\cdot$ Energy,Work,Heat Quantity^2 = Magnetic Flux^4

$\mu_0 \cdot$ Electric Charge^2 = Area$^2 \cdot$ Mass Density

$\mu_0^2 \cdot$ Area = Capacitance$^2 \cdot$ Electric Resistance^4

$\mu_0^2 \cdot$ Capacitance$^2 \cdot$ Absorbed Dose^2 = Area

$\mu_0^2 \cdot$ Capacitance$^2 \cdot$ Kinematic Viscosity^4 = Area^3

$\mu_0^2 \cdot$ Kinematic Viscosity^2 = Area $\cdot$ Electric Resistance^2

$\mu_0^3 \cdot$ Specific Volume$^3 \cdot$ Electric Charge^6 = Volume^4

$\mu_0 \cdot$ Current Density^2 = Volume$^2 \cdot$ Pressure,Stress

$\mu_0 \cdot$ Current Density^2 = Volume $\cdot$ Energy,Work,Heat Quantity

$\mu_0 \cdot$ Current Density = Volume $\cdot$ Magnetic Flux Density

$\mu_0^3 \cdot$ Current Density^3 = Volume $\cdot$ Magnetic Flux^3

$\mu_0 \cdot$ Volume $\cdot$ Magnetic Field Strength^2 = Energy,Work,Heat Quantity

$\mu_0^3 \cdot$ Volume$^2 \cdot$ Magnetic Field Strength^3 = Magnetic Flux^3

$\mu_0^3 \cdot$ Volume $\cdot$ Frequency$^6 \cdot$ Capacitance^3 = A Constant

$\mu_0^3 \cdot$ Volume $\cdot$ Frequency^3 = Electric Resistance^3

$\mu_0{}^3 \cdot \text{Volume}^4 \cdot \text{Pressure,Stress}^3 = \text{Magnetic Flux}^6$

$\mu_0 \cdot \text{Energy,Work,Heat Quantity} = \text{Volume} \cdot \text{Magnetic Flux Density}^2$

$\mu_0{}^3 \cdot \text{Volume} \cdot \text{Energy,Work,Heat Quantity}^3 = \text{Magnetic Flux}^6$

$\mu_0{}^3 \cdot \text{Electric Charge}^6 = \text{Volume}^4 \cdot \text{Mass Density}^3$

$\mu_0{}^3 \cdot \text{Volume} = \text{Capacitance}^3 \cdot \text{Electric Resistance}^6$

$\mu_0{}^3 \cdot \text{Capacitance}^3 \cdot \text{Absorbed Dose}^3 = \text{Volume}$

$\mu_0 \cdot \text{Capacitance} \cdot \text{Kinematic Viscosity}^2 = \text{Volume}$

$\mu_0{}^3 \cdot \text{Kinematic Viscosity}^3 = \text{Volume} \cdot \text{Electric Resistance}^3$

$\mu_0 \cdot \text{Wavenumber}^4 \cdot \text{Specific Volume} \cdot \text{Electric Charge}^2 = \text{A Constant}$

$\mu_0 \cdot \text{Wavenumber}^6 \cdot \text{Current Density}^2 = \text{Pressure,Stress}$

$\mu_0 \cdot \text{Wavenumber}^3 \cdot \text{Current Density}^2 = \text{Energy,Work,Heat Quantity}$

$\mu_0 \cdot \text{Wavenumber}^3 \cdot \text{Current Density} = \text{Magnetic Flux Density}$

$\mu_0 \cdot \text{Wavenumber} \cdot \text{Current Density} = \text{Magnetic Flux}$

$\mu_0 \cdot \text{Magnetic Field Strength}^2 = \text{Wavenumber}^3 \cdot \text{Energy,Work,Heat Quantity}$

$\mu_0 \cdot \text{Magnetic Field Strength} = \text{Wavenumber}^2 \cdot \text{Magnetic Flux}$

$\mu_0 \cdot \text{Frequency}^2 \cdot \text{Capacitance} = \text{Wavenumber}$

$\mu_0 \cdot \text{Frequency} = \text{Wavenumber} \cdot \text{Electric Resistance}$

$\mu_0 \cdot \text{Pressure,Stress} = \text{Wavenumber}^4 \cdot \text{Magnetic Flux}^2$

$\mu_0 \cdot \text{Wavenumber}^3 \cdot \text{Energy,Work,Heat Quantity} = \text{Magnetic Flux Density}^2$

$\mu_0 \cdot \text{Energy,Work,Heat Quantity} = \text{Wavenumber} \cdot \text{Magnetic Flux}^2$

$\mu_0 \cdot \text{Wavenumber}^4 \cdot \text{Electric Charge}^2 = \text{Mass Density}$

$\mu_0 = \text{Wavenumber} \cdot \text{Capacitance} \cdot \text{Electric Resistance}^2$

$\mu_0 \cdot$ Wavenumber $\cdot$ Capacitance $\cdot$ Absorbed Dose = A Constant

$\mu_0 \cdot$ Wavenumber$^3 \cdot$ Capacitance $\cdot$ Kinematic Viscosity^2 = A Constant

$\mu_0 \cdot$ Wavenumber $\cdot$ Kinematic Viscosity = Electric Resistance

$\mu_0{}^3 \cdot$ Specific Volume $\cdot$ Magnetic Field Strength^2 = Electric Resistance^2

$\mu_0 \cdot$ Specific Volume $\cdot$ Magnetic Field Strength^2 = Absorbed Dose

$\mu_0{}^2 \cdot$ Specific Volume $\cdot$ Pressure,Stress = Electric Resistance^2

$\mu_0{}^5 \cdot$ Specific Volume $\cdot$ Electric Charge^2 = Inductance^4

$\mu_0 \cdot$ Specific Volume $\cdot$ Magnetic Flux Density^2 = Electric Resistance^2

$\mu_0 \cdot$ Absorbed Dose = Specific Volume $\cdot$ Magnetic Flux Density^2

$\mu_0 \cdot$ Current Density $\cdot$ Magnetic Field Strength = Energy,Work,Heat Quantity

$\mu_0{}^3 \cdot$ Current Density = Magnetic Field Strength $\cdot$ Inductance^3

$\mu_0{}^3 \cdot$ Current Density$^2 \cdot$ Magnetic Field Strength = Magnetic Flux^3

$\mu_0 \cdot$ Current Density$^2 \cdot$ Pressure,Stress = Energy,Work,Heat Quantity^2

$\mu_0{}^7 \cdot$ Current Density^2 = Pressure,Stress $\cdot$ Inductance^6

$\mu_0{}^5 \cdot$ Current Density$^4 \cdot$ Pressure,Stress = Magnetic Flux^6

$\mu_0{}^4 \cdot$ Current Density^2 = Energy,Work,Heat Quantity $\cdot$ Inductance^3

$\mu_0{}^2 \cdot$ Current Density $\cdot$ Energy,Work,Heat Quantity = Magnetic Flux^3

$\mu_0{}^4 \cdot$ Current Density = Magnetic Flux Density $\cdot$ Inductance^3

$\mu_0{}^2 \cdot$ Current Density$^2 \cdot$ Magnetic Flux Density = Magnetic Flux^3

$\mu_0{}^2 \cdot$ Current Density = Inductance $\cdot$ Magnetic Flux

$\mu_0 \cdot$ Magnetic Field Strength^2 = Frequency $\cdot$ Dynamic Viscosity

$\mu_0{}^2 \cdot$ Energy,Work,Heat Quantity = Magnetic Field Strength$^2 \cdot$ Inductance^3

$\mu_0 \cdot$ Energy,Work,Heat Quantity^2 = Magnetic Field Strength $\cdot$ Magnetic Flux^3

$\mu_0 \cdot$ Magnetic Field Strength $\cdot$ Kinematic Viscosity = Electric Potential Difference

$\mu_0^3 \cdot$ Magnetic Field Strength^2 = Electric Resistance$^2 \cdot$ Mass Density

$\mu_0 \cdot$ Magnetic Flux = Magnetic Field Strength $\cdot$ Inductance^2

$\mu_0 \cdot$ Magnetic Field Strength^2 = Absorbed Dose $\cdot$ Mass Density

$\mu_0^2 \cdot$ Frequency$^2 \cdot$ Capacitance$^2 \cdot$ Absorbed Dose = A Constant

$\mu_0^2 \cdot$ Frequency$^3 \cdot$ Capacitance$^2 \cdot$ Kinematic Viscosity = A Constant

$\mu_0^2 \cdot$ Frequency $\cdot$ Kinematic Viscosity = Electric Resistance^2

$\mu_0 \cdot$ Frequency $\cdot$ Dynamic Viscosity = Magnetic Flux Density^2

$\mu_0^2 \cdot$ Absorbed Dose = Frequency$^2 \cdot$ Inductance^2

$\mu_0^2 \cdot$ Kinematic Viscosity = Frequency $\cdot$ Inductance^2

$\mu_0^3 \cdot$ Energy,Work,Heat Quantity = Pressure,Stress $\cdot$ Inductance^3

$\mu_0^3 \cdot$ Energy,Work,Heat Quantity^4 = Pressure,Stress $\cdot$ Magnetic Flux^6

$\mu_0 \cdot$ Pressure,Stress $\cdot$ Kinematic Viscosity^2 = Electric Potential Difference^2

$\mu_0^2 \cdot$ Pressure,Stress = Electric Resistance$^2 \cdot$ Mass Density

$\mu_0^3 \cdot$ Magnetic Flux^2 = Pressure,Stress $\cdot$ Inductance^4

$\mu_0^4 \cdot$ Energy,Work,Heat Quantity = Magnetic Flux Density$^2 \cdot$ Inductance^3

$\mu_0^2 \cdot$ Energy,Work,Heat Quantity^2 = Magnetic Flux Density $\cdot$ Magnetic Flux^3

$\mu_0^2 \cdot$ Power, Radiance Flux$^2 \cdot$ Absorbed Dose = Electric Potential Difference^4

$\mu_0^5 \cdot$ Electric Charge^2 = Inductance$^4 \cdot$ Mass Density

$\mu_0^2 \cdot$ Electric Charge$^2 \cdot$ Absorbed Dose = Magnetic Flux^2

$\mu_0^2 \cdot$ Kinematic Viscosity = Capacitance $\cdot$ Electric Resistance^3

$\mu_0^2 \cdot$ Capacitance $\cdot$ Absorbed Dose = Inductance

$\mu_0{}^4 \cdot \text{Capacitance} \cdot \text{Kinematic Viscosity}^2 = \text{Inductance}^3$

$\mu_0{}^2 \cdot \text{Capacitance}^2 \cdot \text{Absorbed Dose}^3 = \text{Kinematic Viscosity}^2$

$\mu_0 \cdot \text{Magnetic Flux Density}^2 = \text{Electric Resistance}^2 \cdot \text{Mass Density}$

$\mu_0{}^2 \cdot \text{Kinematic Viscosity} = \text{Electric Resistance} \cdot \text{Inductance}$

$\mu_0{}^2 \cdot \text{Magnetic Flux} = \text{Magnetic Flux Density} \cdot \text{Inductance}^2$

$\mu_0 \cdot \text{Absorbed Dose} \cdot \text{Mass Density} = \text{Magnetic Flux Density}^2$

$\mu_0{}^2 \cdot \text{Kinematic Viscosity}^2 = \text{Inductance}^2 \cdot \text{Absorbed Dose}$

$\varepsilon_0 \cdot c \cdot k^2 = h$

$\varepsilon_0 \cdot c \cdot k = \text{Electric Charge}$

$\varepsilon_0 \cdot c \cdot h = \text{Electric Charge}^2$

$\varepsilon_0 \cdot c \cdot \text{Magnetic Flux}^2 = h$

$\varepsilon_0 \cdot c^6 = G_c \cdot \text{Electric Current}^2$

$\varepsilon_0 \cdot G_c \cdot \text{Electric Potential Difference}^2 = c^4$

$\varepsilon_0 \cdot c^2 \cdot \text{Force} = \text{Electric Current}^2$

$\varepsilon_0 \cdot c \cdot \text{Power, Radiance Flux} = \text{Electric Current}^2$

$\varepsilon_0 \cdot c \cdot \text{Electric Potential Difference} = \text{Electric Current}$

$\varepsilon_0 \cdot c \cdot \text{Time} = \text{Capacitance}$

$\varepsilon_0 \cdot c \cdot \text{Inductance} = \text{Time}$

$\varepsilon_0 \cdot c^2 \cdot \text{Inductance} = \text{Length}$

$\varepsilon_0{}^2 \cdot \text{Volumetric Velocity} = c \cdot \text{Capacitance}^2$

$\varepsilon_0{}^2 \cdot c^5 \cdot \text{Inductance}^2 = \text{Volumetric Velocity}$

$\varepsilon_0 \cdot c^2 = \text{Acceleration} \cdot \text{Capacitance}$

$\varepsilon_0{}^2 \cdot c^4 \cdot \text{Inductance}^2 = \text{Area}$

$\varepsilon_0{}^3 \cdot c^6 \cdot \text{Inductance}^3 = \text{Volume}$

$\varepsilon_0 \cdot c^2 \cdot \text{Wavenumber} \cdot \text{Inductance} = \text{A Constant}$

$\varepsilon_0 \cdot c^4 = \text{Specific Volume} \cdot \text{Magnetic Field Strength}^2$

$\varepsilon_0 \cdot c^2 \cdot \text{Pressure,Stress} = \text{Magnetic Field Strength}^2$

$\varepsilon_0 \cdot c^2 \cdot \text{Magnetic Flux Density} = \text{Magnetic Field Strength}$

$\varepsilon_0 \cdot c^4 \cdot \text{Mass Density} = \text{Magnetic Field Strength}^2$

$\varepsilon_0 \cdot c = \text{Frequency} \cdot \text{Capacitance}$

$\varepsilon_0 \cdot c \cdot \text{Frequency} \cdot \text{Inductance} = \text{A Constant}$

$\varepsilon_0 \cdot c^2 \cdot \text{Magnetic Flux Density}^2 = \text{Pressure,Stress}$

$\varepsilon_0 \cdot c \cdot \text{Electric Potential Difference}^2 = \text{Power, Radiance Flux}$

$\varepsilon_0 \cdot c \cdot \text{Magnetic Flux} = \text{Electric Charge}$

$\varepsilon_0{}^2 \cdot c^2 \cdot \text{Inductance} = \text{Capacitance}$

$\varepsilon_0 \cdot \text{Kinematic Viscosity} = c \cdot \text{Capacitance}$

$\varepsilon_0 \cdot c^3 \cdot \text{Inductance} = \text{Kinematic Viscosity}$

$\varepsilon_0 \cdot k^2 \cdot \text{Velocity} = h$

$\varepsilon_0{}^2 \cdot k^4 \cdot \text{Absorbed Dose} = h^2$

$\varepsilon_0 \cdot k^2 \cdot G_c = \text{Kinematic Viscosity}^2$

$\varepsilon_0 \cdot k \cdot \text{Acceleration} = \text{Electric Current}$

$\varepsilon_0 \cdot k^2 = \text{Time}^2 \cdot \text{Force}$

$\varepsilon_0 \cdot k = \text{Time}^2 \cdot \text{Magnetic Field Strength}$

$\varepsilon_0 \cdot k^2 = \text{Length} \cdot \text{Mass}$

$\varepsilon_0 \cdot k^2 \cdot \text{Specific Volume} = \text{Length}^4$

$\varepsilon_0 \cdot k^2 = \text{Length}^4 \cdot \text{Mass Density}$

$\varepsilon_0 \cdot k^2 \cdot \text{Frequency}^2 = \text{Force}$

$\varepsilon_0 \cdot k^2 \cdot \text{Acceleration} = \text{Energy}$

$\varepsilon_0 \cdot k^2 \cdot \text{Acceleration} = \text{Energy, Work, Heat Quantity}$

$\varepsilon_0^2 \cdot k^4 = \text{Mass}^2 \cdot \text{Area}$

$\varepsilon_0^3 \cdot k^6 = \text{Mass}^3 \cdot \text{Volume}$

$\varepsilon_0 \cdot k^2 \cdot \text{Wavenumber} = \text{Mass}$

$\varepsilon_0^3 \cdot k^6 = \text{Mass}^4 \cdot \text{Specific Volume}$

$\varepsilon_0^2 \cdot k^2 = \text{Mass} \cdot \text{Capacitance}$

$\varepsilon_0^2 \cdot k^3 \cdot \text{Magnetic Flux Density} = \text{Mass}^2$

$\varepsilon_0^3 \cdot k^6 \cdot \text{Mass Density} = \text{Mass}^4$

$\varepsilon_0 \cdot k \cdot \text{Velocity} = \text{Electric Charge}$

$\varepsilon_0 \cdot k^2 \cdot \text{Specific Volume} = \text{Area}^2$

$\varepsilon_0 \cdot k^2 = \text{Area}^2 \cdot \text{Mass Density}$

$\varepsilon_0^3 \cdot k^6 \cdot \text{Specific Volume}^3 = \text{Volume}^4$

$\varepsilon_0^3 \cdot k^6 = \text{Volume}^4 \cdot \text{Mass Density}^3$

$\varepsilon_0 \cdot k^2 \cdot \text{Wavenumber}^4 \cdot \text{Specific Volume} = \text{A Constant}$

$\varepsilon_0 \cdot k^2 \cdot \text{Wavenumber}^4 = \text{Mass Density}$

$\varepsilon_0^5 \cdot k^2 \cdot \text{Specific Volume} = \text{Capacitance}^4$

$\varepsilon_0 \cdot k \cdot \text{Frequency}^2 = \text{Magnetic Field Strength}$

$\varepsilon_0 \cdot \text{Electric Potential Difference}^2 = k \cdot \text{Magnetic Field Strength}$

$\varepsilon_0^2 \cdot k^2 \cdot \text{Absorbed Dose} = \text{Electric Charge}^2$

$\varepsilon_0^2 \cdot k = \text{Capacitance}^2 \cdot \text{Magnetic Flux Density}$

$\varepsilon_0^5 \cdot k^2 = \text{Capacitance}^4 \cdot \text{Mass Density}$

$\varepsilon_0 \cdot h^2 \cdot G_c = \text{Current Density}^2$

$\varepsilon_0 \cdot h \cdot \text{Velocity} = \text{Electric Charge}^2$

$\varepsilon_0 \cdot \text{Velocity} \cdot \text{Magnetic Flux}^2 = h$

$\varepsilon_0^2 \cdot h^2 \cdot \text{Absorbed Dose} = \text{Electric Charge}^4$

$\varepsilon_0^3 \cdot h = \text{Capacitance}^3 \cdot \text{Dynamic Viscosity}$

$\varepsilon_0^2 \cdot \text{Absorbed Dose} \cdot \text{Magnetic Flux}^4 = h^2$

$\varepsilon_0^2 \cdot G_c \cdot \text{Force}^3 = \text{Electric Current}^4$

$\varepsilon_0 \cdot \text{Velocity}^6 = G_c \cdot \text{Electric Current}^2$

$\varepsilon_0^5 \cdot G_c \cdot \text{Electric Potential Difference}^6 = \text{Electric Current}^4$

$\varepsilon_0^5 \cdot G_c \cdot \text{Electric Current}^2 \cdot \text{Electric Resistance}^6 = \text{A Constant}$

$\varepsilon_0 \cdot \text{Absorbed Dose}^3 = G_c \cdot \text{Electric Current}^2$

$\varepsilon_0 \cdot G_c \cdot \text{Time}^2 \cdot \text{Magnetic Flux Density}^2 = \text{A Constant}$

$\varepsilon_0^4 \cdot G_c \cdot \text{Force} \cdot \text{Electric Resistance}^4 = \text{A Constant}$

$\varepsilon_0 \cdot G_c \cdot \text{Mass}^2 = \text{Electric Charge}^2$

$\varepsilon_0 \cdot G_c \cdot \text{Electric Potential Difference}^2 = \text{Velocity}^4$

$\varepsilon_0 \cdot G_c \cdot \text{Dynamic Viscosity}^2 = \text{Magnetic Field Strength}^2$

$\varepsilon_0 \cdot G_c \cdot \text{Magnetic Flux Density}^2 = \text{Frequency}^2$

$\varepsilon_0^2 \cdot G_c \cdot \text{Pressure,Stress} \cdot \text{Inductance}^2 = \text{A Constant}$

$\varepsilon_0^5 \cdot G_c \cdot \text{Power, Radiance Flux} \cdot \text{Electric Resistance}^5 = \text{A Constant}$

$\varepsilon_0^5 \cdot G_c \cdot \text{Electric Potential Difference}^2 \cdot \text{Electric Resistance}^4 = \text{A Constant}$

$\varepsilon_0 \cdot G_c \cdot \text{Electric Potential Difference}^2 = \text{Absorbed Dose}^2$

$\varepsilon_0 \cdot G_c \cdot \text{Magnetic Flux}^2 = \text{Kinematic Viscosity}^2$

$\varepsilon_0^2 \cdot \text{Luminous Intensity} = \text{Capacitance}^2 \cdot \text{Illuminance}$

$\varepsilon_0 \cdot \text{Force} \cdot \text{Velocity}^2 = \text{Electric Current}^2$

$\varepsilon_0 \cdot \text{Power, Radiance Flux}^2 = \text{Electric Current}^2 \cdot \text{Force}$

$\varepsilon_0 \cdot \text{Electric Current}^2 \cdot \text{Electric Resistance}^2 = \text{Force}$

$\varepsilon_0 \cdot \text{Force} \cdot \text{Absorbed Dose} = \text{Electric Current}^2$

$\varepsilon_0 \cdot \text{Energy} \cdot \text{Acceleration} = \text{Electric Current}^2$

$\varepsilon_0 \cdot \text{Acceleration} \cdot \text{Energy,Work,Heat Quantity} = \text{Electric Current}^2$

$\varepsilon_0 \cdot \text{Acceleration} \cdot \text{Magnetic Flux} = \text{Electric Current}$

$\varepsilon_0 \cdot \text{Velocity} \cdot \text{Power, Radiance Flux} = \text{Electric Current}^2$

$\varepsilon_0 \cdot \text{Velocity} \cdot \text{Electric Potential Difference} = \text{Electric Current}$

$\varepsilon_0^2 \cdot \text{Current Density} = \text{Electric Current} \cdot \text{Capacitance}^2$

$\varepsilon_0 \cdot \text{Electric Current} = \text{Magnetic Field Strength} \cdot \text{Capacitance}$

$\varepsilon_0 \cdot \text{Pressure,Stress} \cdot \text{Kinematic Viscosity}^2 = \text{Electric Current}^2$

$\varepsilon_0^2 \cdot \text{Power, Radiance Flux}^2 \cdot \text{Absorbed Dose} = \text{Electric Current}^4$

$\varepsilon_0^2 \cdot \text{Electric Potential Difference}^2 \cdot \text{Absorbed Dose} = \text{Electric Current}^2$

$\varepsilon_0 \cdot \text{Length} \cdot \text{Electric Resistance} = \text{Time}$

$\varepsilon_0 \cdot \text{Length} \cdot \text{Inductance} = \text{Time}^2$

$\varepsilon_0^3 \cdot \text{Time} \cdot \text{Volumetric Velocity} = \text{Capacitance}^3$

$\varepsilon_0^3 \cdot \text{Volumetric Velocity} \cdot \text{Electric Resistance}^3 = \text{Time}^2$

$\varepsilon_0^3 \cdot \text{Volumetric Velocity} \cdot \text{Inductance}^3 = \text{Time}^5$

$\varepsilon_0 \cdot \text{Force} = \text{Time}^2 \cdot \text{Magnetic Field Strength}^2$

$\varepsilon_0 \cdot \text{Magnetic Flux}^2 = \text{Time}^2 \cdot \text{Force}$

$\varepsilon_0 \cdot \text{Time}^2 \cdot \text{Acceleration} = \text{Capacitance}$

$\varepsilon_0 \cdot \text{Time} \cdot \text{Acceleration} \cdot \text{Electric Resistance} = \text{A Constant}$

$\varepsilon_0 \cdot \text{Time} \cdot \text{Velocity} = \text{Capacitance}$

$\varepsilon_0 \cdot \text{Velocity} \cdot \text{Inductance} = \text{Time}$

$\varepsilon_0^2 \cdot \text{Area} \cdot \text{Electric Resistance}^2 = \text{Time}^2$

$\varepsilon_0^2 \cdot \text{Area} \cdot \text{Inductance}^2 = \text{Time}^4$

$\varepsilon_0^3 \cdot \text{Volume} \cdot \text{Electric Resistance}^3 = \text{Time}^3$

$\varepsilon_0^3 \cdot \text{Volume} \cdot \text{Inductance}^3 = \text{Time}^6$

$\varepsilon_0 \cdot \text{Electric Resistance} = \text{Time} \cdot \text{Wavenumber}$

$\varepsilon_0 \cdot \text{Inductance} = \text{Time}^2 \cdot \text{Wavenumber}$

$\varepsilon_0 \cdot \text{Electric Potential Difference} = \text{Time} \cdot \text{Magnetic Field Strength}$

$\varepsilon_0 \cdot \text{Magnetic Flux} = \text{Time}^2 \cdot \text{Magnetic Field Strength}$

$\varepsilon_0^2 \cdot \text{Time}^2 \cdot \text{Absorbed Dose} = \text{Capacitance}^2$

$\varepsilon_0^2 \cdot \text{Time} \cdot \text{Kinematic Viscosity} = \text{Capacitance}^2$

$\varepsilon_0^2 \cdot \text{Electric Resistance}^2 \cdot \text{Kinematic Viscosity} = \text{Time}$

$\varepsilon_0^2 \cdot \text{Inductance}^2 \cdot \text{Absorbed Dose} = \text{Time}^2$

$\varepsilon_0^2 \cdot \text{Inductance}^2 \cdot \text{Kinematic Viscosity} = \text{Time}^3$

$\varepsilon_0 \cdot \text{Volumetric Velocity} \cdot \text{Electric Resistance} = \text{Length}^2$

$\varepsilon_0 \cdot \text{Volumetric Velocity}^2 \cdot \text{Inductance} = \text{Length}^5$

$\varepsilon_0 \cdot \text{Length}^2 \cdot \text{Force} = \text{Electric Charge}^2$

$\varepsilon_0 \cdot \text{Length} \cdot \text{Energy} = \text{Electric Charge}^2$

$\varepsilon_0 \cdot \text{Length} \cdot \text{Electric Potential Difference}^2 = \text{Energy}$

$\varepsilon_0{}^2 \cdot$ Length $\cdot$ Acceleration $\cdot$ Electric Resistance^2 = A Constant

$\varepsilon_0 \cdot$ Length$^3 \cdot$ Magnetic Flux Density^2 = Mass

$\varepsilon_0 \cdot$ Magnetic Flux^2 = Length $\cdot$ Mass

$\varepsilon_0 \cdot$ Velocity$^2 \cdot$ Inductance = Length

$\varepsilon_0 \cdot$ Specific Volume $\cdot$ Magnetic Flux^2 = Length^4

$\varepsilon_0 \cdot$ Length $\cdot$ Frequency $\cdot$ Electric Resistance = A Constant

$\varepsilon_0 \cdot$ Length $\cdot$ Frequency$^2 \cdot$ Inductance = A Constant

$\varepsilon_0 \cdot$ Length$^4 \cdot$ Pressure,Stress = Electric Charge^2

$\varepsilon_0 \cdot$ Electric Potential Difference^2 = Length$^2 \cdot$ Pressure,Stress

$\varepsilon_0 \cdot$ Length $\cdot$ Energy,Work,Heat Quantity = Electric Charge^2

$\varepsilon_0 \cdot$ Length $\cdot$ Electric Potential Difference^2 = Energy,Work,Heat Quantity

$\varepsilon_0 \cdot$ Length $\cdot$ Electric Potential Difference = Electric Charge

$\varepsilon_0 \cdot$ Length $\cdot$ Electric Resistance^2 = Inductance

$\varepsilon_0 \cdot$ Electric Resistance $\cdot$ Kinematic Viscosity = Length

$\varepsilon_0 \cdot$ Inductance $\cdot$ Absorbed Dose = Length

$\varepsilon_0 \cdot$ Inductance $\cdot$ Kinematic Viscosity^2 = Length^3

$\varepsilon_0 \cdot$ Magnetic Flux^2 = Length$^4 \cdot$ Mass Density

$\varepsilon_0 \cdot$ Volumetric Velocity$^2 \cdot$ Force = Current Density^2

$\varepsilon_0{}^5 \cdot$ Volumetric Velocity^2 = Acceleration $\cdot$ Capacitance^5

$\varepsilon_0{}^5 \cdot$ Volumetric Velocity $\cdot$ Acceleration$^2 \cdot$ Electric Resistance^5 = A Constant

$\varepsilon_0{}^2 \cdot$ Volumetric Velocity = Velocity $\cdot$ Capacitance^2

$\varepsilon_0{}^2 \cdot$ Velocity$^5 \cdot$ Inductance^2 = Volumetric Velocity

$\varepsilon_0 \cdot$ Volumetric Velocity $\cdot$ Electric Resistance = Area

$\varepsilon_0{}^2 \cdot$ Volumetric Velocity$^4 \cdot$ Inductance$^2 =$ Area^5

$\varepsilon_0{}^3 \cdot$ Volumetric Velocity$^3 \cdot$ Electric Resistance$^3 =$ Volume^2

$\varepsilon_0{}^3 \cdot$ Volumetric Velocity$^6 \cdot$ Inductance$^3 =$ Volume^5

$\varepsilon_0 \cdot$ Volumetric Velocity $\cdot$ Wavenumber$^2 \cdot$ Electric Resistance $=$ A Constant

$\varepsilon_0 \cdot$ Volumetric Velocity$^2 \cdot$ Wavenumber$^5 \cdot$ Inductance $=$ A Constant

$\varepsilon_0 \cdot$ Volumetric Velocity$^2 =$ Specific Volume $\cdot$ Electric Charge^2

$\varepsilon_0 \cdot$ Volumetric Velocity $\cdot$ Electric Potential Difference $=$ Current Density

$\varepsilon_0{}^3 \cdot$ Volumetric Velocity $=$ Frequency $\cdot$ Capacitance^3

$\varepsilon_0{}^3 \cdot$ Volumetric Velocity $\cdot$ Frequency$^2 \cdot$ Electric Resistance$^3 =$ A Constant

$\varepsilon_0{}^3 \cdot$ Volumetric Velocity $\cdot$ Frequency$^5 \cdot$ Inductance$^3 =$ A Constant

$\varepsilon_0 \cdot$ Volumetric Velocity $\cdot$ Magnetic Flux Density $=$ Electric Charge

$\varepsilon_0 \cdot$ Volumetric Velocity$^2 \cdot$ Mass Density $=$ Electric Charge^2

$\varepsilon_0{}^3 \cdot$ Volumetric Velocity $\cdot$ Electric Resistance $=$ Capacitance^2

$\varepsilon_0{}^6 \cdot$ Volumetric Velocity$^2 \cdot$ Inductance $=$ Capacitance^5

$\varepsilon_0{}^4 \cdot$ Volumetric Velocity$^2 =$ Capacitance$^4 \cdot$ Absorbed Dose

$\varepsilon_0 \cdot$ Volumetric Velocity $=$ Capacitance $\cdot$ Kinematic Viscosity

$\varepsilon_0{}^3 \cdot$ Volumetric Velocity $\cdot$ Electric Resistance$^5 =$ Inductance^2

$\varepsilon_0 \cdot$ Electric Resistance $\cdot$ Kinematic Viscosity$^2 =$ Volumetric Velocity

$\varepsilon_0{}^4 \cdot$ Inductance$^4 \cdot$ Absorbed Dose$^5 =$ Volumetric Velocity^2

$\varepsilon_0 \cdot$ Inductance $\cdot$ Kinematic Viscosity$^5 =$ Volumetric Velocity^3

$\varepsilon_0 \cdot$ Energy$^2 =$ Force $\cdot$ Electric Charge^2

$\varepsilon_0 \cdot$ Energy $=$ Force $\cdot$ Capacitance

$\epsilon_0 \cdot$ Force $\cdot$ Inductance = Mass

$\epsilon_0 \cdot$ Force $\cdot$ Area = Electric Charge^2

$\epsilon_0{}^3 \cdot$ Force$^3 \cdot$ Volume^2 = Electric Charge^6

$\epsilon_0 \cdot$ Force = Wavenumber$^2 \cdot$ Electric Charge^2

$\epsilon_0 \cdot$ Force $\cdot$ Frequency^2 = Magnetic Field Strength^2

$\epsilon_0 \cdot$ Frequency$^2 \cdot$ Magnetic Flux^2 = Force

$\epsilon_0 \cdot$ Force^2 = Pressure,Stress $\cdot$ Electric Charge^2

$\epsilon_0{}^2 \cdot$ Force = Pressure,Stress $\cdot$ Capacitance^2

$\epsilon_0 \cdot$ Energy,Work,Heat Quantity^2 = Force $\cdot$ Electric Charge^2

$\epsilon_0 \cdot$ Energy,Work,Heat Quantity = Force $\cdot$ Capacitance

$\epsilon_0 \cdot$ Power, Radiance Flux $\cdot$ Electric Resistance = Force

$\epsilon_0 \cdot$ Electric Charge^2 = Force $\cdot$ Capacitance^2

$\epsilon_0 \cdot$ Force $\cdot$ Magnetic Flux Density^2 = Dynamic Viscosity^2

$\epsilon_0 \cdot$ Magnetic Flux Density$^2 \cdot$ Kinematic Viscosity^2 = Force

$\epsilon_0 \cdot$ Acceleration $\cdot$ Magnetic Flux^2 = Energy

$\epsilon_0{}^2 \cdot$ Energy $\cdot$ Electric Resistance^2 = Mass

$\epsilon_0{}^2 \cdot$ Energy$^2 \cdot$ Area = Electric Charge^4

$\epsilon_0{}^2 \cdot$ Area $\cdot$ Electric Potential Difference^4 = Energy^2

$\epsilon_0{}^3 \cdot$ Energy$^3 \cdot$ Volume = Electric Charge^6

$\epsilon_0{}^3 \cdot$ Volume $\cdot$ Electric Potential Difference^6 = Energy^3

$\epsilon_0 \cdot$ Energy = Wavenumber $\cdot$ Electric Charge^2

$\epsilon_0 \cdot$ Electric Potential Difference^2 = Energy $\cdot$ Wavenumber

$\epsilon_0 \cdot$ Energy$^2 \cdot$ Specific Volume = Current Density^2

$\varepsilon_0 \cdot \text{Energy}^2 = \text{Current Density}^2 \cdot \text{Mass Density}$

$\varepsilon_0{}^3 \cdot \text{Energy}^4 = \text{Pressure,Stress} \cdot \text{Electric Charge}^6$

$\varepsilon_0{}^3 \cdot \text{Electric Potential Difference}^6 = \text{Energy}^2 \cdot \text{Pressure,Stress}$

$\varepsilon_0{}^3 \cdot \text{Energy} = \text{Pressure,Stress} \cdot \text{Capacitance}^3$

$\varepsilon_0 \cdot \text{Electric Potential Difference}^2 = \text{Acceleration} \cdot \text{Mass}$

$\varepsilon_0 \cdot \text{Velocity}^2 = \text{Acceleration} \cdot \text{Capacitance}$

$\varepsilon_0{}^4 \cdot \text{Acceleration}^2 \cdot \text{Area} \cdot \text{Electric Resistance}^4 = \text{A Constant}$

$\varepsilon_0{}^6 \cdot \text{Acceleration}^3 \cdot \text{Volume} \cdot \text{Electric Resistance}^6 = \text{A Constant}$

$\varepsilon_0{}^2 \cdot \text{Acceleration} \cdot \text{Electric Resistance}^2 = \text{Wavenumber}$

$\varepsilon_0 \cdot \text{Acceleration} = \text{Frequency}^2 \cdot \text{Capacitance}$

$\varepsilon_0 \cdot \text{Acceleration} \cdot \text{Electric Resistance} = \text{Frequency}$

$\varepsilon_0 \cdot \text{Acceleration} \cdot \text{Magnetic Flux}^2 = \text{Energy,Work,Heat Quantity}$

$\varepsilon_0 \cdot \text{Acceleration} \cdot \text{Capacitance} \cdot \text{Electric Resistance}^2 = \text{A Constant}$

$\varepsilon_0 \cdot \text{Absorbed Dose} = \text{Acceleration} \cdot \text{Capacitance}$

$\varepsilon_0{}^3 \cdot \text{Kinematic Viscosity}^2 = \text{Acceleration} \cdot \text{Capacitance}^3$

$\varepsilon_0{}^3 \cdot \text{Acceleration} \cdot \text{Electric Resistance}^3 \cdot \text{Kinematic Viscosity} = \text{A Constant}$

$\varepsilon_0{}^2 \cdot \text{Area}^3 \cdot \text{Magnetic Flux Density}^4 = \text{Mass}^2$

$\varepsilon_0{}^2 \cdot \text{Magnetic Flux}^4 = \text{Mass}^2 \cdot \text{Area}$

$\varepsilon_0 \cdot \text{Volume} \cdot \text{Magnetic Flux Density}^2 = \text{Mass}$

$\varepsilon_0{}^3 \cdot \text{Magnetic Flux}^6 = \text{Mass}^3 \cdot \text{Volume}$

$\varepsilon_0 \cdot \text{Magnetic Flux Density}^2 = \text{Mass} \cdot \text{Wavenumber}^3$

$\varepsilon_0 \cdot \text{Wavenumber} \cdot \text{Magnetic Flux}^2 = \text{Mass}$

$\varepsilon_0{}^3 \cdot$ Mass $\cdot$ Specific Volume = Capacitance^3

$\varepsilon_0{}^3 \cdot$ Magnetic Flux^6 = Mass$^4 \cdot$ Specific Volume

$\varepsilon_0{}^2 \cdot$ Energy,Work,Heat Quantity $\cdot$ Electric Resistance^2 = Mass

$\varepsilon_0{}^2 \cdot$ Electric Potential Difference$^2 \cdot$ Inductance = Mass

$\varepsilon_0{}^2 \cdot$ Mass = Capacitance$^3 \cdot$ Magnetic Flux Density^2

$\varepsilon_0{}^2 \cdot$ Magnetic Flux^2 = Mass $\cdot$ Capacitance

$\varepsilon_0{}^3 \cdot$ Mass = Capacitance$^3 \cdot$ Mass Density

$\varepsilon_0{}^2 \cdot$ Magnetic Flux Density $\cdot$ Magnetic Flux^3 = Mass^2

$\varepsilon_0{}^3 \cdot$ Magnetic Flux$^6 \cdot$ Mass Density = Mass^4

$\varepsilon_0{}^2 \cdot$ Velocity$^4 \cdot$ Inductance^2 = Area

$\varepsilon_0{}^3 \cdot$ Velocity$^6 \cdot$ Inductance^3 = Volume

$\varepsilon_0 \cdot$ Velocity$^2 \cdot$ Wavenumber $\cdot$ Inductance = A Constant

$\varepsilon_0 \cdot$ Velocity^4 = Specific Volume $\cdot$ Magnetic Field Strength^2

$\varepsilon_0 \cdot$ Velocity$^2 \cdot$ Pressure,Stress = Magnetic Field Strength^2

$\varepsilon_0 \cdot$ Velocity$^2 \cdot$ Magnetic Flux Density = Magnetic Field Strength

$\varepsilon_0 \cdot$ Velocity$^4 \cdot$ Mass Density = Magnetic Field Strength^2

$\varepsilon_0 \cdot$ Velocity = Frequency $\cdot$ Capacitance

$\varepsilon_0 \cdot$ Velocity $\cdot$ Frequency $\cdot$ Inductance = A Constant

$\varepsilon_0 \cdot$ Velocity$^2 \cdot$ Magnetic Flux Density^2 = Pressure,Stress

$\varepsilon_0 \cdot$ Velocity $\cdot$ Electric Potential Difference^2 = Power, Radiance Flux

$\varepsilon_0 \cdot$ Velocity $\cdot$ Magnetic Flux = Electric Charge

$\varepsilon_0{}^2 \cdot$ Velocity$^2 \cdot$ Inductance = Capacitance

$\varepsilon_0 \cdot$ Kinematic Viscosity = Velocity $\cdot$ Capacitance

$\varepsilon_0 \cdot \text{Velocity}^3 \cdot \text{Inductance} = \text{Kinematic Viscosity}$

$\varepsilon_0 \cdot \text{Specific Volume} \cdot \text{Magnetic Flux}^2 = \text{Area}^2$

$\varepsilon_0^2 \cdot \text{Area} \cdot \text{Frequency}^2 \cdot \text{Electric Resistance}^2 = \text{A Constant}$

$\varepsilon_0^2 \cdot \text{Area} \cdot \text{Frequency}^4 \cdot \text{Inductance}^2 = \text{A Constant}$

$\varepsilon_0 \cdot \text{Area}^2 \cdot \text{Pressure,Stress} = \text{Electric Charge}^2$

$\varepsilon_0 \cdot \text{Electric Potential Difference}^2 = \text{Area} \cdot \text{Pressure,Stress}$

$\varepsilon_0^2 \cdot \text{Area} \cdot \text{Energy,Work,Heat Quantity}^2 = \text{Electric Charge}^4$

$\varepsilon_0^2 \cdot \text{Area} \cdot \text{Electric Potential Difference}^4 = \text{Energy,Work,Heat Quantity}^2$

$\varepsilon_0^2 \cdot \text{Area} \cdot \text{Electric Potential Difference}^2 = \text{Electric Charge}^2$

$\varepsilon_0^2 \cdot \text{Area} \cdot \text{Electric Resistance}^4 = \text{Inductance}^2$

$\varepsilon_0^2 \cdot \text{Electric Resistance}^2 \cdot \text{Kinematic Viscosity}^2 = \text{Area}$

$\varepsilon_0^2 \cdot \text{Inductance}^2 \cdot \text{Absorbed Dose}^2 = \text{Area}$

$\varepsilon_0^2 \cdot \text{Inductance}^2 \cdot \text{Kinematic Viscosity}^4 = \text{Area}^3$

$\varepsilon_0 \cdot \text{Magnetic Flux}^2 = \text{Area}^2 \cdot \text{Mass Density}$

$\varepsilon_0^3 \cdot \text{Specific Volume}^3 \cdot \text{Magnetic Flux}^6 = \text{Volume}^4$

$\varepsilon_0^3 \cdot \text{Volume} \cdot \text{Frequency}^3 \cdot \text{Electric Resistance}^3 = \text{A Constant}$

$\varepsilon_0^3 \cdot \text{Volume} \cdot \text{Frequency}^6 \cdot \text{Inductance}^3 = \text{A Constant}$

$\varepsilon_0^3 \cdot \text{Volume}^4 \cdot \text{Pressure,Stress}^3 = \text{Electric Charge}^6$

$\varepsilon_0^3 \cdot \text{Electric Potential Difference}^6 = \text{Volume}^2 \cdot \text{Pressure,Stress}^3$

$\varepsilon_0^3 \cdot \text{Volume} \cdot \text{Energy,Work,Heat Quantity}^3 = \text{Electric Charge}^6$

$\varepsilon_0^3 \cdot \text{Volume} \cdot \text{Electric Potential Difference}^6 = \text{Energy,Work,Heat Quantity}^3$

$\varepsilon_0^3 \cdot \text{Volume} \cdot \text{Electric Potential Difference}^3 = \text{Electric Charge}^3$

$\varepsilon_0{}^3 \cdot$ Volume $\cdot$ Electric Resistance^6 = Inductance^3

$\varepsilon_0{}^3 \cdot$ Electric Resistance$^3 \cdot$ Kinematic Viscosity^3 = Volume

$\varepsilon_0{}^3 \cdot$ Inductance$^3 \cdot$ Absorbed Dose^3 = Volume

$\varepsilon_0 \cdot$ Inductance $\cdot$ Kinematic Viscosity^2 = Volume

$\varepsilon_0{}^3 \cdot$ Magnetic Flux^6 = Volume$^4 \cdot$ Mass Density^3

$\varepsilon_0 \cdot$ Wavenumber$^4 \cdot$ Specific Volume $\cdot$ Magnetic Flux^2 = A Constant

$\varepsilon_0 \cdot$ Frequency $\cdot$ Electric Resistance = Wavenumber

$\varepsilon_0 \cdot$ Frequency$^2 \cdot$ Inductance = Wavenumber

$\varepsilon_0 \cdot$ Pressure,Stress = Wavenumber$^4 \cdot$ Electric Charge^2

$\varepsilon_0 \cdot$ Wavenumber$^2 \cdot$ Electric Potential Difference^2 = Pressure,Stress

$\varepsilon_0 \cdot$ Energy,Work,Heat Quantity = Wavenumber $\cdot$ Electric Charge^2

$\varepsilon_0 \cdot$ Electric Potential Difference^2 = Wavenumber $\cdot$ Energy,Work,Heat Quantity

$\varepsilon_0 \cdot$ Electric Potential Difference = Wavenumber $\cdot$ Electric Charge

$\varepsilon_0 \cdot$ Electric Resistance^2 = Wavenumber $\cdot$ Inductance

$\varepsilon_0 \cdot$ Wavenumber $\cdot$ Electric Resistance $\cdot$ Kinematic Viscosity = A Constant

$\varepsilon_0 \cdot$ Wavenumber $\cdot$ Inductance $\cdot$ Absorbed Dose = A Constant

$\varepsilon_0 \cdot$ Wavenumber$^3 \cdot$ Inductance $\cdot$ Kinematic Viscosity^2 = A Constant

$\varepsilon_0 \cdot$ Wavenumber$^4 \cdot$ Magnetic Flux^2 = Mass Density

$\varepsilon_0 \cdot$ Specific Volume $\cdot$ Energy,Work,Heat Quantity^2 = Current Density^2

$\varepsilon_0 \cdot$ Specific Volume $\cdot$ Pressure,Stress^2 = Magnetic Field Strength^2

$\varepsilon_0{}^3 \cdot$ Specific Volume $\cdot$ Magnetic Field Strength$^2 \cdot$ Electric Resistance^4 = A Constant

$\varepsilon_0 \cdot$ Absorbed Dose^2 = Specific Volume $\cdot$ Magnetic Field Strength^2

$\varepsilon_0{}^2 \cdot$ Specific Volume $\cdot$ Pressure,Stress $\cdot$ Electric Resistance^2 = A Constant

$\varepsilon_0 \cdot$ Electric Potential Difference^2 = Specific Volume $\cdot$ Dynamic Viscosity^2

$\varepsilon_0 \cdot$ Specific Volume $\cdot$ Electric Potential Difference^2 = Kinematic Viscosity^2

$\varepsilon_0^5 \cdot$ Specific Volume $\cdot$ Magnetic Flux^2 = Capacitance^4

$\varepsilon_0^3 \cdot$ Current Density = Magnetic Field Strength $\cdot$ Capacitance^3

$\varepsilon_0 \cdot$ Energy,Work,Heat Quantity^2 = Current Density^2 $\cdot$ Mass Density

$\varepsilon_0 \cdot$ Frequency $\cdot$ Electric Potential Difference = Magnetic Field Strength

$\varepsilon_0 \cdot$ Frequency^2 $\cdot$ Magnetic Flux = Magnetic Field Strength

$\varepsilon_0 \cdot$ Magnetic Field Strength^2 $\cdot$ Electric Resistance^2 = Pressure,Stress

$\varepsilon_0 \cdot$ Pressure,Stress $\cdot$ Absorbed Dose = Magnetic Field Strength^2

$\varepsilon_0 \cdot$ Pressure,Stress^2 = Magnetic Field Strength^2 $\cdot$ Mass Density

$\varepsilon_0 \cdot$ Power, Radiance Flux = Magnetic Field Strength $\cdot$ Electric Charge

$\varepsilon_0 \cdot$ Electric Potential Difference^2 = Magnetic Field Strength $\cdot$ Magnetic Flux

$\varepsilon_0 \cdot$ Magnetic Field Strength $\cdot$ Electric Resistance^2 = Magnetic Flux Density

$\varepsilon_0^3 \cdot$ Magnetic Field Strength^2 $\cdot$ Electric Resistance^4 = Mass Density

$\varepsilon_0 \cdot$ Magnetic Flux Density $\cdot$ Absorbed Dose = Magnetic Field Strength

$\varepsilon_0 \cdot$ Absorbed Dose^2 $\cdot$ Mass Density = Magnetic Field Strength^2

$\varepsilon_0^2 \cdot$ Absorbed Dose = Frequency^2 $\cdot$ Capacitance^2

$\varepsilon_0^2 \cdot$ Kinematic Viscosity = Frequency $\cdot$ Capacitance^2

$\varepsilon_0^2 \cdot$ Frequency $\cdot$ Electric Resistance^2 $\cdot$ Kinematic Viscosity = A Constant

$\varepsilon_0^2 \cdot$ Frequency^2 $\cdot$ Inductance^2 $\cdot$ Absorbed Dose = A Constant

$\varepsilon_0^2 \cdot$ Frequency^3 $\cdot$ Inductance^2 $\cdot$ Kinematic Viscosity = A Constant

$\varepsilon_0^3 \cdot$ Energy,Work,Heat Quantity^4 = Pressure,Stress $\cdot$ Electric Charge^6

$\varepsilon_0{}^3 \cdot$ Electric Potential Difference^6 = Pressure,Stress $\cdot$ Energy,Work,Heat Quantity^2

$\varepsilon_0{}^3 \cdot$ Energy,Work,Heat Quantity = Pressure,Stress $\cdot$ Capacitance^3

$\varepsilon_0{}^3 \cdot$ Electric Potential Difference^4 = Pressure,Stress $\cdot$ Electric Charge^2

$\varepsilon_0{}^3 \cdot$ Electric Charge^2 = Pressure,Stress $\cdot$ Capacitance^4

$\varepsilon_0{}^3 \cdot$ Electric Potential Difference^2 = Pressure,Stress $\cdot$ Capacitance^2

$\varepsilon_0 \cdot$ Pressure,Stress $\cdot$ Electric Resistance^2 = Magnetic Flux Density^2

$\varepsilon_0{}^2 \cdot$ Pressure,Stress $\cdot$ Electric Resistance^2 = Mass Density

$\varepsilon_0 \cdot$ Magnetic Flux Density^2 $\cdot$ Absorbed Dose = Pressure,Stress

$\varepsilon_0{}^2 \cdot$ Electric Potential Difference^4 $\cdot$ Absorbed Dose = Power, Radiance Flux^2

$\varepsilon_0{}^2 \cdot$ Absorbed Dose $\cdot$ Magnetic Flux^2 = Electric Charge^2

$\varepsilon_0 \cdot$ Electric Potential Difference $\cdot$ Magnetic Flux Density = Dynamic Viscosity

Luminous Intensity $\cdot$ Electric Current = Illuminance $\cdot$ Current Density

Luminous Intensity $\cdot$ Magnetic Field Strength^2 = Illuminance $\cdot$ Electric Current^2

Luminous Intensity^3 = Illuminance^3 $\cdot$ Time^2 $\cdot$ Volume Velocity^2

Luminous Intensity = Illuminance $\cdot$ Time^4 $\cdot$ Acceleration^2

Luminous Intensity = Illuminance $\cdot$ Time^2 $\cdot$ Velocity^2

Luminous Intensity = Illuminance $\cdot$ Time^2 $\cdot$ Absorbed Dose

Luminous Intensity = Illuminance $\cdot$ Time $\cdot$ Kinematic Viscosity

Luminous Intensity^5 $\cdot$ Acceleration^2 = Illuminance^5 $\cdot$ Volume Velocity^4

Luminous Intensity $\cdot$ Velocity = Illuminance $\cdot$ Volume Velocity

Luminous Intensity^3 $\cdot$ Frequency^2 = Illuminance^3 $\cdot$ Volume Velocity^2

Luminous Intensity^2 $\cdot$ Absorbed Dose = Illuminance^2 $\cdot$ Volume Velocity^2

Luminous Intensity $\cdot$ Kinematic Viscosity^2 = Illuminance $\cdot$ Volume Velocity^2

Luminous Intensity · Force^2 = Illuminance · Energy^2

Luminous Intensity · Pressure,Stress = Illuminance · Force

Luminous Intensity · Force^2 = Illuminance · Energy,Work,Heat Quantity^2

Luminous Intensity^3 · Pressure,Stress^2 = Illuminance^3 · Energy^2

Luminous Intensity · Acceleration^2 = Illuminance · Velocity^4

Luminous Intensity · Frequency^4 = Illuminance · Acceleration^2

Luminous Intensity · Acceleration^2 = Illuminance · Absorbed Dose^2

Luminous Intensity^3 · Acceleration^2 = Illuminance^3 · Kinematic Viscosity^4

Luminous Intensity^3 = Illuminance^3 · Mass^2 · Specific Volume^2

Luminous Intensity^3 · Mass Density^2 = Illuminance^3 · Mass^2

Luminous Intensity · Frequency^2 = Illuminance · Velocity^2

Luminous Intensity · Velocity^2 = Illuminance · Kinematic Viscosity^2

Luminous Intensity^3 · Magnetic Field Strength^2 = Illuminance^3 · Current Density^2

Luminous Intensity · Frequency^2 = Illuminance · Absorbed Dose

Luminous Intensity · Frequency = Illuminance · Kinematic Viscosity

Luminous Intensity^3 · Pressure,Stress^2 = Illuminance^3 · Energy,Work,Heat Quantity^2

Luminous Intensity · Magnetic Flux Density = Illuminance · Magnetic Flux

Luminous Intensity · Absorbed Dose = Illuminance · Kinematic Viscosity^2

Amount of a Substance · Electric Current = Electric Charge · Catalytic Activity

Amount of a Substance · Electric Current = Catalytic Activity · Electric Charge

Amount of a Substance · Volume Velocity = Catalytic Activity · Length^3

Amount of a Substance^2 · Acceleration = Catalytic Activity^2 · Length

Amount of a Substance $\cdot$ Velocity = Catalytic Activity $\cdot$ Length

Amount of a Substance^2 $\cdot$ Absorbed Dose = Catalytic Activity^2 $\cdot$ Length^2

Amount of a Substance $\cdot$ Kinematic Viscosity = Catalytic Activity $\cdot$ Length^2

Amount of a Substance^5 $\cdot$ Acceleration^3 = Catalytic Activity^5 $\cdot$ Volume Velocity

Amount of a Substance^2 $\cdot$ Velocity^3 = Catalytic Activity^2 $\cdot$ Volume Velocity

Amount of a Substance^2 $\cdot$ Volume Velocity^2 = Catalytic Activity^2 $\cdot$ Area^3

Amount of a Substance $\cdot$ Volume Velocity = Catalytic Activity $\cdot$ Volume

Amount of a Substance $\cdot$ Volume Velocity $\cdot$ Wavenumber^3 = Catalytic Activity

Amount of a Substance^4 $\cdot$ Absorbed Dose^3 = Catalytic Activity^4 $\cdot$ Volume Velocity^2

Amount of a Substance $\cdot$ Kinematic Viscosity^3 = Catalytic Activity $\cdot$ Volume Velocity^2

Amount of a Substance $\cdot$ Power, Radiance Flux = Catalytic Activity $\cdot$ Energy

Amount of a Substance $\cdot$ Acceleration = Catalytic Activity $\cdot$ Velocity

Amount of a Substance^4 $\cdot$ Acceleration^2 = Catalytic Activity^4 $\cdot$ Area

Amount of a Substance^6 $\cdot$ Acceleration^3 = Catalytic Activity^6 $\cdot$ Volume

Amount of a Substance^2 $\cdot$ Acceleration $\cdot$ Wavenumber = Catalytic Activity^2

Amount of a Substance^2 $\cdot$ Acceleration^2 = Catalytic Activity^2 $\cdot$ Absorbed Dose

Amount of a Substance^3 $\cdot$ Acceleration^2 = Catalytic Activity^3 $\cdot$ Kinematic Viscosity

Amount of a Substance^2 $\cdot$ Velocity^2 = Catalytic Activity^2 $\cdot$ Area

Amount of a Substance^3 $\cdot$ Velocity^3 = Catalytic Activity^3 $\cdot$ Volume

Amount of a Substance $\cdot$ Velocity $\cdot$ Wavenumber = Catalytic Activity

Amount of a Substance $\cdot$ Velocity^2 = Catalytic Activity $\cdot$ Kinematic Viscosity

Amount of a Substance^2 $\cdot$ Absorbed Dose = Catalytic Activity^2 $\cdot$ Area

Amount of a Substance $\cdot$ Kinematic Viscosity = Catalytic Activity $\cdot$ Area

Amount of a Substance^6 · Absorbed Dose^3 = Catalytic Activity^6 · Volume^2

Amount of a Substance^3 · Kinematic Viscosity^3 = Catalytic Activity^3 · Volume^2

Amount of a Substance^2 · Wavenumber^2 · Absorbed Dose = Catalytic Activity^2

Amount of a Substance · Wavenumber^2 · Kinematic Viscosity = Catalytic Activity

Amount of a Substance · Pressure,Stress = Catalytic Activity · Dynamic Viscosity

Amount of a Substance · Power, Radiance Flux = Catalytic Activity · Energy,Work,Heat Quantity

Amount of a Substance · Electric Potential Difference = Catalytic Activity · Magnetic Flux

Amount of a Substance = Catalytic Activity · Capacitance · Electric Resistance

Amount of a Substance^2 = Catalytic Activity^2 · Capacitance · Inductance

Amount of a Substance · Electric Resistance = Catalytic Activity · Inductance

Electric Current · Electric Charge^2 · Volume Velocity^2 = Current Density^3

Electric Current^3 · Current Density = Electric Charge^4 · Acceleration^2

Electric Current · Current Density = Electric Charge^2 · Velocity^2

Electric Current · Current Density = Electric Charge^2 · Absorbed Dose

Electric Current^2 · Inductance = Current Density · Magnetic Flux Density

Electric Current · Magnetic Flux = Current Density · Magnetic Flux Density

Electric Current · Current Density · Magnetic Flux Density^2 = Force^2

Electric Current^3 · Magnetic Flux Density^2 = Current Density · Pressure,Stress^2

Electric Current^5 · Inductance^2 = Current Density · Force^2

Electric Current^7 · Inductance^2 = Current Density^3 · Pressure,Stress^2

Electric Current^3 · Magnetic Flux^2 = Current Density · Force^2

Electric Current^5 · Magnetic Flux^2 = Current Density^3 · Pressure,Stress^2

Electric Current^3 · Time^2 · Volume Velocity^2 = Current Density^3

Electric Current · Time^4 · Acceleration^2 = Current Density

Electric Current · Time^2 · Velocity^2 = Current Density

Electric Current · Time^2 · Absorbed Dose = Current Density

Electric Current · Time · Kinematic Viscosity = Current Density

Electric Current^5 · Volume Velocity^4 = Current Density^5 · Acceleration^2

Electric Current · Volume Velocity = Current Density · Velocity

Electric Current^3 · Volume Velocity^2 = Current Density^3 · Frequency^2

Electric Current^2 · Volume Velocity^2 = Current Density^2 · Absorbed Dose

Electric Current · Volume Velocity^2 = Current Density · Kinematic Viscosity^2

Electric Current · Energy^2 = Current Density · Force^2

Electric Current · Force = Current Density · Pressure,Stress

Electric Current · Energy,Work,Heat Quantity^2 = Current Density · Force^2

Electric Current^3 · Energy^2 = Current Density^3 · Pressure,Stress^2

Electric Current · Velocity^4 = Current Density · Acceleration^2

Electric Current · Acceleration^2 = Current Density · Frequency^4

Electric Current · Absorbed Dose^2 = Current Density · Acceleration^2

Electric Current^3 · Kinematic Viscosity^4 = Current Density^3 · Acceleration^2

Electric Current^3 · Mass^2 · Specific Volume^2 = Current Density^3

Electric Current^3 · Mass^2 = Current Density^3 · Mass Density^2

Electric Current · Velocity^2 = Current Density · Frequency^2

Electric Current · Kinematic Viscosity^2 = Current Density · Velocity^2

70

Electric Current $\cdot$ Absorbed Dose = Current Density $\cdot$ Frequency^2

Electric Current $\cdot$ Kinematic Viscosity = Current Density $\cdot$ Frequency

Electric Current^3 $\cdot$ Energy,Work,Heat Quantity^2 = Current Density^3 $\cdot$ Pressure,Stress^2

Electric Current $\cdot$ Kinematic Viscosity^2 = Current Density $\cdot$ Absorbed Dose

Electric Current^4 = Magnetic Field Strength^3 $\cdot$ Electric Charge $\cdot$ Volume Velocity

Electric Current^3 = Magnetic Field Strength $\cdot$ Electric Charge^2 $\cdot$ Acceleration

Electric Current^2 = Magnetic Field Strength $\cdot$ Electric Charge $\cdot$ Velocity

Electric Current^4 = Magnetic Field Strength^2 $\cdot$ Electric Charge^2 $\cdot$ Absorbed Dose

Electric Current^3 = Magnetic Field Strength^2 $\cdot$ Electric Charge $\cdot$ Kinematic Viscosity

Electric Current $\cdot$ Magnetic Flux Density = Magnetic Field Strength^2 $\cdot$ Inductance

Electric Current^2 $\cdot$ Magnetic Flux Density = Magnetic Field Strength^2 $\cdot$ Magnetic Flux

Electric Current^2 $\cdot$ Magnetic Flux Density = Magnetic Field Strength $\cdot$ Force

Electric Current^3 $\cdot$ Magnetic Flux Density = Magnetic Field Strength^2 $\cdot$ Energy

Electric Current^3 $\cdot$ Magnetic Flux Density = Magnetic Field Strength^2 $\cdot$ Energy,Work,Heat Quantity

ε_0 $\cdot$ Electric Potential Difference^2 = Dynamic Viscosity $\cdot$ Kinematic Viscosity

ε_0 $\cdot$ Electric Potential Difference^2 $\cdot$ Mass Density = Dynamic Viscosity^2

ε_0 $\cdot$ Electric Potential Difference^2 = Kinematic Viscosity^2 $\cdot$ Mass Density

$\varepsilon_0{}^2$ $\cdot$ Electric Resistance $\cdot$ Kinematic Viscosity = Capacitance

$\varepsilon_0{}^2$ $\cdot$ Magnetic Flux = Capacitance^2 $\cdot$ Magnetic Flux Density

$\varepsilon_0{}^2$ $\cdot$ Inductance $\cdot$ Absorbed Dose = Capacitance

$\varepsilon_0{}^4$ $\cdot$ Inductance $\cdot$ Kinematic Viscosity^2 = Capacitance^3

$\varepsilon_0{}^2$ $\cdot$ Kinematic Viscosity^2 = Capacitance^2 $\cdot$ Absorbed Dose

71

$\varepsilon_0{}^5 \cdot$ Magnetic Flux^2 = Capacitance$^4 \cdot$ Mass Density

$\varepsilon_0{}^2 \cdot$ Electric Resistance$^3 \cdot$ Kinematic Viscosity = Inductance

$\varepsilon_0 \cdot$ Magnetic Flux Density$^2 \cdot$ Kinematic Viscosity = Dynamic Viscosity

$\varepsilon_0{}^2 \cdot$ Inductance$^2 \cdot$ Absorbed Dose^3 = Kinematic Viscosity^2

$c^4 = k \cdot G_c \cdot$ Magnetic Field Strength

$c^2 \cdot$ Mass $= k \cdot$ Electric Current

$c^2 \cdot$ Time$^2 \cdot$ Magnetic Flux Density $= k$

$c \cdot k =$ Length $\cdot$ Electric Potential Difference

$c^3 \cdot k^2 =$ Volume Velocity $\cdot$ Electric Potential Difference^2

$c \cdot k =$ Volume Velocity $\cdot$ Magnetic Flux Density

$c \cdot$ Electric Potential Difference $= k \cdot$ Acceleration

$c^4 \cdot$ Magnetic Flux Density $= k \cdot$ Acceleration^2

$c^2 \cdot$ Mass $\cdot$ Inductance $= k^2$

$c^2 \cdot k^2 =$ Area $\cdot$ Electric Potential Difference^2

$c^3 \cdot k^3 =$ Volume $\cdot$ Electric Potential Difference^3

$c \cdot k \cdot$ Wavenumber = Electric Potential Difference

$c^2 \cdot k \cdot$ Capacitance = Current Density

$c \cdot k \cdot$ Magnetic Field Strength = Power, Radiance Flux

$c^2 \cdot$ Magnetic Flux Density $= k \cdot$ Frequency^2

$c^2 \cdot k \cdot$ Magnetic Flux Density = Electric Potential Difference^2

$c^2 \cdot k =$ Electric Potential Difference $\cdot$ Kinematic Viscosity

$c^2 \cdot k =$ Magnetic Flux Density $\cdot$ Kinematic Viscosity^2

$c^5 \cdot$ Time$^2 = h \cdot G_c$

$$c^3 \cdot \text{Length}^2 = h \cdot G_c$$

$$c^2 \cdot \text{Volume Velocity} = h \cdot G_c$$

$$c^5 \cdot h = G_c \cdot \text{Energy}^2$$

$$c^7 = h \cdot G_c \cdot \text{Acceleration}^2$$

$$c \cdot h = G_c \cdot \text{Mass}^2$$

$$c^3 \cdot \text{Area} = h \cdot G_c$$

$$c^9 \cdot \text{Volume}^2 = h^3 \cdot G_c^{\,3}$$

$$c^3 = h \cdot G_c \cdot \text{Wavenumber}^2$$

$$c^5 \cdot \text{Specific Volume} = h \cdot G_c^{\,2}$$

$$c^5 = h \cdot G_c \cdot \text{Frequency}^2$$

$$c^7 = h \cdot G_c^{\,2} \cdot \text{Pressure,Stress}$$

$$c^5 \cdot h = G_c \cdot \text{Energy,Work,Heat Quantity}^2$$

$$c^9 = h \cdot G_c^{\,3} \cdot \text{Dynamic Viscosity}^2$$

$$c \cdot \text{Kinematic Viscosity}^2 = h \cdot G_c$$

$$c^5 = h \cdot G_c^{\,2} \cdot \text{Mass Density}$$

$$c \cdot \text{Time}^2 \cdot \text{Force} = h$$

$$c^2 \cdot \text{Time} \cdot \text{Mass} = h$$

$$c^5 \cdot \text{Time}^4 = h \cdot \text{Specific Volume}$$

$$c^3 \cdot \text{Time}^4 \cdot \text{Pressure,Stress} = h$$

$$c^3 \cdot \text{Time}^3 \cdot \text{Dynamic Viscosity} = h$$

$$c^5 \cdot \text{Time}^4 \cdot \text{Mass Density} = h$$

$$c \cdot h = \text{Length}^2 \cdot \text{Force}$$

$c \cdot h = \text{Length} \cdot \text{Energy}$

$c \cdot \text{Length} \cdot \text{Mass} = h$

$c \cdot \text{Length}^4 = h \cdot \text{Specific Volume}$

$c \cdot h = \text{Length}^4 \cdot \text{Pressure,Stress}$

$c \cdot h = \text{Length} \cdot \text{Energy,Work,Heat Quantity}$

$c^2 \cdot h = \text{Length}^2 \cdot \text{Power, Radiance Flux}$

$c \cdot \text{Length}^4 \cdot \text{Mass Density} = h$

$c^2 \cdot h = \text{Volume Velocity} \cdot \text{Force}$

$c^3 \cdot h^2 = \text{Volume Velocity} \cdot \text{Energy}^2$

$c \cdot \text{Volume Velocity} \cdot \text{Mass}^2 = h^2$

$c \cdot h \cdot \text{Specific Volume} = \text{Volume Velocity}^2$

$c^3 \cdot h = \text{Volume Velocity}^2 \cdot \text{Pressure,Stress}$

$c^3 \cdot h^2 = \text{Volume Velocity} \cdot \text{Energy,Work,Heat Quantity}^2$

$c^3 \cdot h = \text{Volume Velocity} \cdot \text{Power, Radiance Flux}$

$c^3 \cdot h^2 = \text{Volume Velocity}^3 \cdot \text{Dynamic Viscosity}^2$

$c \cdot h = \text{Volume Velocity}^2 \cdot \text{Mass Density}$

$c \cdot h \cdot \text{Force} = \text{Energy}^2$

$c^3 \cdot \text{Force} = h \cdot \text{Acceleration}^2$

$c^3 \cdot \text{Mass}^2 = h \cdot \text{Force}$

$c \cdot h = \text{Force} \cdot \text{Area}$

$c^3 \cdot h^3 = \text{Force}^3 \cdot \text{Volume}^2$

$c \cdot h \cdot \text{Wavenumber}^2 = \text{Force}$

$c^3 \cdot h = \text{Force}^2 \cdot \text{Specific Volume}$

$c \cdot \text{Force} = h \cdot \text{Frequency}^2$

$c \cdot h \cdot \text{Pressure,Stress} = \text{Force}^2$

$c \cdot h \cdot \text{Force} = \text{Energy,Work,Heat Quantity}^2$

$c^3 \cdot h \cdot \text{Dynamic Viscosity}^2 = \text{Force}^3$

$c^3 \cdot h = \text{Force} \cdot \text{Kinematic Viscosity}^2$

$c^3 \cdot h \cdot \text{Mass Density} = \text{Force}^2$

$c \cdot \text{Energy} = h \cdot \text{Acceleration}$

$c^2 \cdot h^2 = \text{Energy}^2 \cdot \text{Area}$

$c^3 \cdot h^3 = \text{Energy}^3 \cdot \text{Volume}$

$c \cdot h \cdot \text{Wavenumber} = \text{Energy}$

$c^5 \cdot h^3 = \text{Energy}^4 \cdot \text{Specific Volume}$

$c^3 \cdot h^3 \cdot \text{Pressure,Stress} = \text{Energy}^4$

$c^3 \cdot h^2 \cdot \text{Dynamic Viscosity} = \text{Energy}^3$

$c^2 \cdot h = \text{Energy} \cdot \text{Kinematic Viscosity}$

$c^5 \cdot h^3 \cdot \text{Mass Density} = \text{Energy}^4$

$c^3 \cdot \text{Mass} = h \cdot \text{Acceleration}$

$c^9 = h \cdot \text{Acceleration}^4 \cdot \text{Specific Volume}$

$c^7 \cdot \text{Pressure,Stress} = h \cdot \text{Acceleration}^4$

$c \cdot \text{Energy,Work,Heat Quantity} = h \cdot \text{Acceleration}$

$c^2 \cdot \text{Power, Radiance Flux} = h \cdot \text{Acceleration}^2$

$c^6 \cdot \text{Dynamic Viscosity} = h \cdot \text{Acceleration}^3$

$c^9 \cdot \text{Mass Density} = h \cdot \text{Acceleration}^4$

$c^2 \cdot \text{Mass}^2 \cdot \text{Area} = h^2$

$c^3 \cdot \text{Mass}^3 \cdot \text{Volume} = h^3$

$c \cdot \text{Mass} = h \cdot \text{Wavenumber}$

$c^3 \cdot \text{Mass}^4 \cdot \text{Specific Volume} = h^3$

$c^2 \cdot \text{Mass} = h \cdot \text{Frequency}$

$c^5 \cdot \text{Mass}^4 = h^3 \cdot \text{Pressure,Stress}$

$c^4 \cdot \text{Mass}^2 = h \cdot \text{Power, Radiance Flux}$

$c^3 \cdot \text{Mass}^3 = h^2 \cdot \text{Dynamic Viscosity}$

$c^3 \cdot \text{Mass}^4 = h^3 \cdot \text{Mass Density}$

$c \cdot \text{Area}^2 = h \cdot \text{Specific Volume}$

$c \cdot h = \text{Area}^2 \cdot \text{Pressure,Stress}$

$c^2 \cdot h^2 = \text{Area} \cdot \text{Energy,Work,Heat Quantity}^2$

$c^2 \cdot h = \text{Area} \cdot \text{Power, Radiance Flux}$

$c \cdot \text{Area}^2 \cdot \text{Mass Density} = h$

$c^3 \cdot \text{Volume}^4 = h^3 \cdot \text{Specific Volume}^3$

$c^3 \cdot h^3 = \text{Volume}^4 \cdot \text{Pressure,Stress}^3$

$c^3 \cdot h^3 = \text{Volume} \cdot \text{Energy,Work,Heat Quantity}^3$

$c^6 \cdot h^3 = \text{Volume}^2 \cdot \text{Power, Radiance Flux}^3$

$c^3 \cdot \text{Volume}^4 \cdot \text{Mass Density}^3 = h^3$

$c = h \cdot \text{Wavenumber}^4 \cdot \text{Specific Volume}$

$c \cdot h \cdot \text{Wavenumber}^4 = \text{Pressure,Stress}$

$c \cdot h \cdot \text{Wavenumber} = \text{Energy,Work,Heat Quantity}$

$c^2 \cdot h \cdot \text{Wavenumber}^2 = \text{Power, Radiance Flux}$

$c \cdot$ Mass Density $= h \cdot$ Wavenumber^4

$c^5 = h \cdot$ Specific Volume $\cdot$ Frequency^4

$c^5 \cdot h^3 =$ Specific Volume $\cdot$ Energy,Work,Heat Quantity^4

$c^5 \cdot h =$ Specific Volume $\cdot$ Power, Radiance Flux^2

$c^3 \cdot h =$ Specific Volume$^3 \cdot$ Dynamic Viscosity^4

$c^3 \cdot h \cdot$ Specific Volume $=$ Kinematic Viscosity^4

$c^2 \cdot h =$ Current Density $\cdot$ Electric Potential Difference

$c^3 \cdot$ Pressure,Stress $= h \cdot$ Frequency^4

$c^3 \cdot$ Dynamic Viscosity $= h \cdot$ Frequency^3

$c^5 \cdot$ Mass Density $= h \cdot$ Frequency^4

$c^3 \cdot h^3 \cdot$ Pressure,Stress $=$ Energy,Work,Heat Quantity^4

$c^3 \cdot h \cdot$ Pressure,Stress $=$ Power, Radiance Flux^2

$c^3 \cdot$ Dynamic Viscosity$^4 = h \cdot$ Pressure,Stress^3

$c^5 \cdot h =$ Pressure,Stress $\cdot$ Kinematic Viscosity^4

$c^3 \cdot h^2 \cdot$ Dynamic Viscosity $=$ Energy,Work,Heat Quantity^3

$c^2 \cdot h =$ Energy,Work,Heat Quantity $\cdot$ Kinematic Viscosity

$c^5 \cdot h^3 \cdot$ Mass Density $=$ Energy,Work,Heat Quantity^4

$c^6 \cdot h \cdot$ Dynamic Viscosity$^2 =$ Power, Radiance Flux^3

$c^4 \cdot h =$ Power, Radiance Flux $\cdot$ Kinematic Viscosity^2

$c^5 \cdot h \cdot$ Mass Density $=$ Power, Radiance Flux^2

$c^3 \cdot h =$ Dynamic Viscosity $\cdot$ Kinematic Viscosity^3

$c^3 \cdot h \cdot$ Mass Density$^3 =$ Dynamic Viscosity^4

$c^3 \cdot h = $ Kinematic Viscosity$^4 \cdot$ Mass Density

$c^5 = G_c \cdot$ Electric Current $\cdot$ Electric Potential Difference

$c^5 = G_c \cdot$ Electric Current$^2 \cdot$ Electric Resistance

$c^5 \cdot$ Time $= G_c \cdot$ Energy

$c^3 \cdot$ Time $= G_c \cdot$ Mass

$c^2 = G_c \cdot$ Time$^2 \cdot$ Pressure,Stress

$c^5 \cdot$ Time $= G_c \cdot$ Energy,Work,Heat Quantity

$c^2 = G_c \cdot$ Time $\cdot$ Dynamic Viscosity

$c^4 \cdot$ Length $= G_c \cdot$ Energy

$c^2 \cdot$ Length $= G_c \cdot$ Mass

$c^2 \cdot$ Specific Volume $= G_c \cdot$ Length^2

$c^4 = G_c \cdot$ Length$^2 \cdot$ Pressure,Stress

$c^4 \cdot$ Length $= G_c \cdot$ Energy,Work,Heat Quantity

$c^3 = G_c \cdot$ Length $\cdot$ Dynamic Viscosity

$c^2 = G_c \cdot$ Length$^2 \cdot$ Mass Density

$c^7 \cdot$ Volume Velocity $= G_c^2 \cdot$ Energy^2

$c^3 \cdot$ Volume Velocity $= G_c^2 \cdot$ Mass^2

$c^3 \cdot$ Specific Volume $= G_c \cdot$ Volume Velocity

$c^5 = G_c \cdot$ Volume Velocity $\cdot$ Pressure,Stress

$c^7 \cdot$ Volume Velocity $= G_c^2 \cdot$ Energy,Work,Heat Quantity^2

$c^7 = G_c^2 \cdot$ Volume Velocity $\cdot$ Dynamic Viscosity^2

$c^3 = G_c \cdot$ Volume Velocity $\cdot$ Mass Density

$c^6 = G_c \cdot$ Energy $\cdot$ Acceleration

$$c^8 \cdot \text{Area} = G_c{}^2 \cdot \text{Energy}^2$$

$$c^4 = G_c \cdot \text{Energy} \cdot \text{Wavenumber}$$

$$c^5 = G_c \cdot \text{Energy} \cdot \text{Frequency}$$

$$c^7 = G_c{}^2 \cdot \text{Energy} \cdot \text{Dynamic Viscosity}$$

$$c^3 \cdot \text{Kinematic Viscosity} = G_c \cdot \text{Energy}$$

$$c^4 = G_c \cdot \text{Acceleration} \cdot \text{Mass}$$

$$c^2 \cdot G_c = \text{Acceleration}^2 \cdot \text{Specific Volume}$$

$$c^6 = G_c \cdot \text{Acceleration} \cdot \text{Energy,Work,Heat Quantity}$$

$$c \cdot \text{Acceleration} = G_c \cdot \text{Dynamic Viscosity}$$

$$c^2 \cdot G_c \cdot \text{Mass Density} = \text{Acceleration}^2$$

$$c^4 \cdot \text{Area} = G_c{}^2 \cdot \text{Mass}^2$$

$$c^6 \cdot \text{Volume} = G_c{}^3 \cdot \text{Mass}^3$$

$$c^2 = G_c \cdot \text{Mass} \cdot \text{Wavenumber}$$

$$c^6 \cdot \text{Specific Volume} = G_c{}^3 \cdot \text{Mass}^2$$

$$c^3 = G_c \cdot \text{Mass} \cdot \text{Frequency}$$

$$c^8 = G_c{}^3 \cdot \text{Mass}^2 \cdot \text{Pressure,Stress}$$

$$c^5 = G_c{}^2 \cdot \text{Mass} \cdot \text{Dynamic Viscosity}$$

$$c \cdot \text{Kinematic Viscosity} = G_c \cdot \text{Mass}$$

$$c^6 = G_c{}^3 \cdot \text{Mass}^2 \cdot \text{Mass Density}$$

$$c^2 \cdot \text{Specific Volume} = G_c \cdot \text{Area}$$

$$c^4 = G_c \cdot \text{Area} \cdot \text{Pressure,Stress}$$

$$c^8 \cdot \text{Area} = G_c{}^2 \cdot \text{Energy,Work,Heat Quantity}^2$$

$$c^6 = G_c^2 \cdot \text{Area} \cdot \text{Dynamic Viscosity}^2$$

$$c^2 = G_c \cdot \text{Area} \cdot \text{Mass Density}$$

$$c^6 \cdot \text{Specific Volume}^3 = G_c^3 \cdot \text{Volume}^2$$

$$c^9 = G_c^3 \cdot \text{Volume} \cdot \text{Dynamic Viscosity}^3$$

$$c^6 = G_c^3 \cdot \text{Volume}^2 \cdot \text{Mass Density}^3$$

$$c^2 \cdot \text{Wavenumber}^2 \cdot \text{Specific Volume} = G_c$$

$$c^4 \cdot \text{Wavenumber}^2 = G_c \cdot \text{Pressure,Stress}$$

$$c^4 = G_c \cdot \text{Wavenumber} \cdot \text{Energy,Work,Heat Quantity}$$

$$c^3 \cdot \text{Wavenumber} = G_c \cdot \text{Dynamic Viscosity}$$

$$c^2 \cdot \text{Wavenumber}^2 = G_c \cdot \text{Mass Density}$$

$$c^4 = G_c \cdot \text{Specific Volume} \cdot \text{Dynamic Viscosity}^2$$

$$c^4 \cdot \text{Specific Volume} = G_c \cdot \text{Kinematic Viscosity}^2$$

$$c^4 = G_c \cdot \text{Magnetic Field Strength} \cdot \text{Magnetic Flux}$$

$$c^2 \cdot \text{Frequency}^2 = G_c \cdot \text{Pressure,Stress}$$

$$c^5 = G_c \cdot \text{Frequency} \cdot \text{Energy,Work,Heat Quantity}$$

$$c^2 \cdot \text{Frequency} = G_c \cdot \text{Dynamic Viscosity}$$

$$c^2 \cdot \text{Pressure,Stress} = G_c \cdot \text{Dynamic Viscosity}^2$$

$$c^6 = G_c \cdot \text{Pressure,Stress} \cdot \text{Kinematic Viscosity}^2$$

$$c^7 = G_c^2 \cdot \text{Energy,Work,Heat Quantity} \cdot \text{Dynamic Viscosity}$$

$$c^3 \cdot \text{Kinematic Viscosity} = G_c \cdot \text{Energy,Work,Heat Quantity}$$

$$c^3 = G_c \cdot \text{Electric Charge} \cdot \text{Magnetic Flux Density}$$

$$c^5 \cdot \text{Electric Resistance} = G_c \cdot \text{Electric Potential Difference}^2$$

$$c^4 = G_c \cdot \text{Dynamic Viscosity} \cdot \text{Kinematic Viscosity}$$

$c^4 \cdot$ Mass Density $= G_c \cdot$ Dynamic Viscosity^2

$c^4 = G_c \cdot$ Kinematic Viscosity$^2 \cdot$ Mass Density

$c^2 \cdot$ Time$^2 \cdot$ Illuminance $=$ Luminous Intensity

$c \cdot$ Luminous Intensity $=$ Volume Velocity $\cdot$ Illuminance

$c^4 \cdot$ Illuminance $=$ Luminous Intensity $\cdot$ Acceleration^2

$c^2 \cdot$ Illuminance $=$ Luminous Intensity $\cdot$ Frequency^2

$c^2 \cdot$ Luminous Intensity $=$ Illuminance $\cdot$ Kinematic Viscosity^2

$c \cdot$ Amount of a Substance $=$ Length $\cdot$ Catalytic Activity

$c^3 \cdot$ Amount of a Substance$^2 =$ Volume Velocity $\cdot$ Catalytic Activity^2

$c \cdot$ Catalytic Activity $=$ Amount of a Substance $\cdot$ Acceleration

$c^2 \cdot$ Amount of a Substance$^2 =$ Area $\cdot$ Catalytic Activity^2

$c^3 \cdot$ Amount of a Substance$^3 =$ Volume $\cdot$ Catalytic Activity^3

$c \cdot$ Amount of a Substance $\cdot$ Wavenumber $=$ Catalytic Activity

$c^2 \cdot$ Amount of a Substance $=$ Catalytic Activity $\cdot$ Kinematic Viscosity

$c^2 \cdot$ Electric Current $\cdot$ Time$^2 =$ Current Density

$c \cdot$ Time $\cdot$ Magnetic Field Strength $=$ Electric Current

$c \cdot$ Electric Charge $=$ Electric Current $\cdot$ Length

$c \cdot$ Current Density $=$ Electric Current $\cdot$ Volume Velocity

$c \cdot$ Electric Current$^2 =$ Volume Velocity $\cdot$ Magnetic Field Strength^2

$c^3 \cdot$ Electric Charge$^2 =$ Electric Current$^2 \cdot$ Volume Velocity

$c \cdot$ Force $=$ Electric Current $\cdot$ Electric Potential Difference

$c \cdot$ Force $=$ Electric Current$^2 \cdot$ Electric Resistance

$c^4 \cdot$ Electric Current = Acceleration$^2 \cdot$ Current Density

$c^2 \cdot$ Magnetic Field Strength = Electric Current $\cdot$ Acceleration

$c \cdot$ Electric Current = Acceleration $\cdot$ Electric Charge

$c^2 \cdot$ Mass = Electric Current$^2 \cdot$ Inductance

$c^2 \cdot$ Mass = Electric Current $\cdot$ Magnetic Flux

$c^2 \cdot$ Electric Charge^2 = Electric Current$^2 \cdot$ Area

$c^3 \cdot$ Electric Charge^3 = Electric Current$^3 \cdot$ Volume

$c \cdot$ Wavenumber $\cdot$ Electric Charge = Electric Current

$c^2 \cdot$ Electric Current = Current Density $\cdot$ Frequency^2

$c^2 \cdot$ Electric Charge^2 = Electric Current $\cdot$ Current Density

$c^2 \cdot$ Current Density = Electric Current $\cdot$ Kinematic Viscosity^2

$c \cdot$ Magnetic Field Strength = Electric Current $\cdot$ Frequency

$c \cdot$ Magnetic Field Strength $\cdot$ Electric Charge = Electric Current^2

$c \cdot$ Electric Current = Magnetic Field Strength $\cdot$ Kinematic Viscosity

$c^2 \cdot$ Electric Charge = Electric Current $\cdot$ Kinematic Viscosity

$c^2 \cdot$ Capacitance $\cdot$ Magnetic Flux Density = Electric Current

$c^3 \cdot$ Capacitance $\cdot$ Dynamic Viscosity = Electric Current^2

$c \cdot$ Dynamic Viscosity = Electric Current $\cdot$ Magnetic Flux Density

$c \cdot$ Time $\cdot$ Force = Energy

$c \cdot$ Mass = Time $\cdot$ Force

$c^4 \cdot$ Time^2 = Force $\cdot$ Specific Volume

$c^2 \cdot$ Time$^2 \cdot$ Pressure,Stress = Force

$c \cdot$ Time $\cdot$ Force = Energy,Work,Heat Quantity

$c^2 \cdot$ Time $\cdot$ Dynamic Viscosity = Force

$c^4 \cdot$ Time$^2 \cdot$ Mass Density = Force

$c^5 \cdot$ Time^3 = Energy $\cdot$ Specific Volume

$c^3 \cdot$ Time$^3 \cdot$ Pressure,Stress = Energy

$c^3 \cdot$ Time$^2 \cdot$ Dynamic Viscosity = Energy

$c^5 \cdot$ Time$^3 \cdot$ Mass Density = Energy

$c^3 \cdot$ Time^3 = Mass $\cdot$ Specific Volume

$c \cdot$ Time$^3 \cdot$ Pressure,Stress = Mass

$c^2 \cdot$ Mass = Time $\cdot$ Power, Radiance Flux

$c \cdot$ Time$^2 \cdot$ Dynamic Viscosity = Mass

$c^3 \cdot$ Time$^3 \cdot$ Mass Density = Mass

$c^5 \cdot$ Time^3 = Specific Volume $\cdot$ Energy,Work,Heat Quantity

$c^5 \cdot$ Time^2 = Specific Volume $\cdot$ Power, Radiance Flux

$c^2 \cdot$ Time = Specific Volume $\cdot$ Dynamic Viscosity

$c^3 \cdot$ Time$^3 \cdot$ Magnetic Field Strength = Current Density

$c^2 \cdot$ Time $\cdot$ Electric Charge = Current Density

$c \cdot$ Time$^2 \cdot$ Magnetic Field Strength = Electric Charge

$c^3 \cdot$ Time$^3 \cdot$ Pressure,Stress = Energy,Work,Heat Quantity

$c^3 \cdot$ Time$^2 \cdot$ Pressure,Stress = Power, Radiance Flux

$c^3 \cdot$ Time$^2 \cdot$ Dynamic Viscosity = Energy,Work,Heat Quantity

$c^5 \cdot$ Time$^3 \cdot$ Mass Density = Energy,Work,Heat Quantity

$c^3 \cdot$ Time $\cdot$ Dynamic Viscosity = Power, Radiance Flux

$c^5 \cdot \text{Time}^2 \cdot \text{Mass Density} = \text{Power, Radiance Flux}$

$c^2 \cdot \text{Time} \cdot \text{Magnetic Flux Density} = \text{Electric Potential Difference}$

$c^2 \cdot \text{Time}^2 \cdot \text{Magnetic Flux Density} = \text{Magnetic Flux}$

$c^2 \cdot \text{Time} \cdot \text{Mass Density} = \text{Dynamic Viscosity}$

$c^2 \cdot \text{Mass} = \text{Length} \cdot \text{Force}$

$c^2 \cdot \text{Length}^2 = \text{Force} \cdot \text{Specific Volume}$

$c \cdot \text{Length} \cdot \text{Dynamic Viscosity} = \text{Force}$

$c^2 \cdot \text{Length}^2 \cdot \text{Mass Density} = \text{Force}$

$c^2 \cdot \text{Length}^3 = \text{Energy} \cdot \text{Specific Volume}$

$c \cdot \text{Energy} = \text{Length} \cdot \text{Power, Radiance Flux}$

$c \cdot \text{Length}^2 \cdot \text{Dynamic Viscosity} = \text{Energy}$

$c^2 \cdot \text{Length}^3 \cdot \text{Mass Density} = \text{Energy}$

$c^2 \cdot \text{Mass} = \text{Length}^3 \cdot \text{Pressure,Stress}$

$c^3 \cdot \text{Mass} = \text{Length} \cdot \text{Power, Radiance Flux}$

$c \cdot \text{Mass} = \text{Length}^2 \cdot \text{Dynamic Viscosity}$

$c^2 \cdot \text{Length}^3 = \text{Specific Volume} \cdot \text{Energy,Work,Heat Quantity}$

$c^3 \cdot \text{Length}^2 = \text{Specific Volume} \cdot \text{Power, Radiance Flux}$

$c \cdot \text{Length} = \text{Specific Volume} \cdot \text{Dynamic Viscosity}$

$c \cdot \text{Length} \cdot \text{Electric Charge} = \text{Current Density}$

$c \cdot \text{Electric Charge} = \text{Length}^2 \cdot \text{Magnetic Field Strength}$

$c \cdot \text{Length}^2 \cdot \text{Pressure,Stress} = \text{Power, Radiance Flux}$

$c \cdot \text{Dynamic Viscosity} = \text{Length} \cdot \text{Pressure,Stress}$

$c \cdot \text{Energy,Work,Heat Quantity} = \text{Length} \cdot \text{Power, Radiance Flux}$

$c \cdot \text{Length}^2 \cdot \text{Dynamic Viscosity} = \text{Energy,Work,Heat Quantity}$

$c^2 \cdot \text{Length}^3 \cdot \text{Mass Density} = \text{Energy,Work,Heat Quantity}$

$c^2 \cdot \text{Length} \cdot \text{Dynamic Viscosity} = \text{Power, Radiance Flux}$

$c^3 \cdot \text{Length}^2 \cdot \text{Mass Density} = \text{Power, Radiance Flux}$

$c \cdot \text{Length} \cdot \text{Magnetic Flux Density} = \text{Electric Potential Difference}$

$c \cdot \text{Magnetic Flux} = \text{Length} \cdot \text{Electric Potential Difference}$

$c \cdot \text{Capacitance} \cdot \text{Electric Resistance} = \text{Length}$

$c^2 \cdot \text{Capacitance} \cdot \text{Inductance} = \text{Length}^2$

$c \cdot \text{Inductance} = \text{Length} \cdot \text{Electric Resistance}$

$c \cdot \text{Length} \cdot \text{Mass Density} = \text{Dynamic Viscosity}$

$c \cdot \text{Energy}^2 = \text{Volume Velocity} \cdot \text{Force}^2$

$c^5 \cdot \text{Mass}^2 = \text{Volume Velocity} \cdot \text{Force}^2$

$c \cdot \text{Volume Velocity} = \text{Force} \cdot \text{Specific Volume}$

$c \cdot \text{Force} = \text{Volume Velocity} \cdot \text{Pressure,Stress}$

$c \cdot \text{Energy,Work,Heat Quantity}^2 = \text{Volume Velocity} \cdot \text{Force}^2$

$c \cdot \text{Volume Velocity} \cdot \text{Dynamic Viscosity}^2 = \text{Force}^2$

$c \cdot \text{Volume Velocity} \cdot \text{Mass Density} = \text{Force}$

$c \cdot \text{Volume Velocity}^3 = \text{Energy}^2 \cdot \text{Specific Volume}^2$

$c^3 \cdot \text{Energy}^2 = \text{Volume Velocity}^3 \cdot \text{Pressure,Stress}^2$

$c^3 \cdot \text{Energy}^2 = \text{Volume Velocity} \cdot \text{Power, Radiance Flux}^2$

$c \cdot \text{Volume Velocity}^3 \cdot \text{Mass Density}^2 = \text{Energy}^2$

$c^3 \cdot \text{Mass}^2 \cdot \text{Specific Volume}^2 = \text{Volume Velocity}^3$

$c^7 \cdot \text{Mass}^2 = \text{Volume Velocity}^3 \cdot \text{Pressure,Stress}^2$

$c^7 \cdot \text{Mass}^2 = \text{Volume Velocity} \cdot \text{Power, Radiance Flux}^2$

$c^2 \cdot \text{Mass} = \text{Volume Velocity} \cdot \text{Dynamic Viscosity}$

$c^3 \cdot \text{Mass}^2 = \text{Volume Velocity}^3 \cdot \text{Mass Density}^2$

$c \cdot \text{Volume Velocity}^3 = \text{Specific Volume}^2 \cdot \text{Energy,Work,Heat Quantity}^2$

$c^2 \cdot \text{Volume Velocity} = \text{Specific Volume} \cdot \text{Power, Radiance Flux}$

$c \cdot \text{Volume Velocity} = \text{Specific Volume}^2 \cdot \text{Dynamic Viscosity}^2$

$c^3 \cdot \text{Current Density}^2 = \text{Volume Velocity}^3 \cdot \text{Magnetic Field Strength}^2$

$c \cdot \text{Volume Velocity} \cdot \text{Electric Charge}^2 = \text{Current Density}^2$

$c^2 \cdot \text{Electric Charge} = \text{Volume Velocity} \cdot \text{Magnetic Field Strength}$

$c^3 \cdot \text{Energy,Work,Heat Quantity}^2 = \text{Volume Velocity}^3 \cdot \text{Pressure,Stress}^2$

$c^3 \cdot \text{Dynamic Viscosity}^2 = \text{Volume Velocity} \cdot \text{Pressure,Stress}^2$

$c^3 \cdot \text{Energy,Work,Heat Quantity}^2 = \text{Volume Velocity} \cdot \text{Power, Radiance Flux}^2$

$c \cdot \text{Volume Velocity}^3 \cdot \text{Mass Density}^2 = \text{Energy,Work,Heat Quantity}^2$

$c^3 \cdot \text{Volume Velocity} \cdot \text{Dynamic Viscosity}^2 = \text{Power, Radiance Flux}^2$

$c^2 \cdot \text{Volume Velocity} \cdot \text{Mass Density} = \text{Power, Radiance Flux}$

$c \cdot \text{Volume Velocity} \cdot \text{Magnetic Flux Density}^2 = \text{Electric Potential Difference}^2$

$c^3 \cdot \text{Magnetic Flux}^2 = \text{Volume Velocity} \cdot \text{Electric Potential Difference}^2$

$c^3 \cdot \text{Capacitance}^2 \cdot \text{Electric Resistance}^2 = \text{Volume Velocity}$

$c^3 \cdot \text{Capacitance} \cdot \text{Inductance} = \text{Volume Velocity}$

$c^3 \cdot \text{Inductance}^2 = \text{Volume Velocity} \cdot \text{Electric Resistance}^2$

$c \cdot \text{Magnetic Flux} = \text{Volume Velocity} \cdot \text{Magnetic Flux Density}$

$c \cdot \text{Volume Velocity} \cdot \text{Mass Density}^2 = \text{Dynamic Viscosity}^2$

$c^2 \cdot$ Force $=$ Energy $\cdot$ Acceleration

$c^2 \cdot$ Energy$^2 =$ Force$^3 \cdot$ Specific Volume

$c \cdot$ Force $=$ Energy $\cdot$ Frequency

$c \cdot$ Energy $\cdot$ Dynamic Viscosity $=$ Force^2

$c \cdot$ Energy $=$ Force $\cdot$ Kinematic Viscosity

$c^2 \cdot$ Energy$^2 \cdot$ Mass Density $=$ Force^3

$c^6 =$ Force $\cdot$ Acceleration$^2 \cdot$ Specific Volume

$c^4 \cdot$ Pressure,Stress $=$ Force $\cdot$ Acceleration^2

$c^2 \cdot$ Force $=$ Acceleration $\cdot$ Energy,Work,Heat Quantity

$c^3 \cdot$ Dynamic Viscosity $=$ Force $\cdot$ Acceleration

$c^6 \cdot$ Mass Density $=$ Force $\cdot$ Acceleration^2

$c^4 \cdot$ Mass$^2 =$ Force$^2 \cdot$ Area

$c^6 \cdot$ Mass$^3 =$ Force$^3 \cdot$ Volume

$c^2 \cdot$ Mass $\cdot$ Wavenumber $=$ Force

$c^6 \cdot$ Mass$^2 =$ Force$^3 \cdot$ Specific Volume

$c \cdot$ Mass $\cdot$ Frequency $=$ Force

$c^4 \cdot$ Mass$^2 \cdot$ Pressure,Stress $=$ Force^3

$c^3 \cdot$ Mass $\cdot$ Dynamic Viscosity $=$ Force^2

$c^3 \cdot$ Mass $=$ Force $\cdot$ Kinematic Viscosity

$c^6 \cdot$ Mass$^2 \cdot$ Mass Density $=$ Force^3

$c^2 \cdot$ Area $=$ Force $\cdot$ Specific Volume

$c^2 \cdot$ Area $\cdot$ Dynamic Viscosity$^2 =$ Force^2

$c^2 \cdot$ Area $\cdot$ Mass Density $=$ Force

$c^6 \cdot$ Volume$^2 =$ Force$^3 \cdot$ Specific Volume^3

$c^3 \cdot$ Volume $\cdot$ Dynamic Viscosity$^3 =$ Force^3

$c^6 \cdot$ Volume$^2 \cdot$ Mass Density$^3 =$ Force^3

$c^2 =$ Force $\cdot$ Wavenumber$^2 \cdot$ Specific Volume

$c \cdot$ Dynamic Viscosity $=$ Force $\cdot$ Wavenumber

$c^2 \cdot$ Mass Density $=$ Force $\cdot$ Wavenumber^2

$c^4 =$ Force $\cdot$ Specific Volume $\cdot$ Frequency^2

$c^2 \cdot$ Energy,Work,Heat Quantity$^2 =$ Force$^3 \cdot$ Specific Volume

$c^2 \cdot$ Pressure,Stress $=$ Force $\cdot$ Frequency^2

$c \cdot$ Force $=$ Frequency $\cdot$ Energy,Work,Heat Quantity

$c^2 \cdot$ Dynamic Viscosity $=$ Force $\cdot$ Frequency

$c^4 \cdot$ Mass Density $=$ Force $\cdot$ Frequency^2

$c^2 \cdot$ Dynamic Viscosity$^2 =$ Force $\cdot$ Pressure,Stress

$c^2 \cdot$ Force $=$ Pressure,Stress $\cdot$ Kinematic Viscosity^2

$c \cdot$ Energy,Work,Heat Quantity $\cdot$ Dynamic Viscosity $=$ Force^2

$c \cdot$ Energy,Work,Heat Quantity $=$ Force $\cdot$ Kinematic Viscosity

$c^2 \cdot$ Energy,Work,Heat Quantity$^2 \cdot$ Mass Density $=$ Force^3

$c \cdot$ Electric Charge $\cdot$ Magnetic Flux Density $=$ Force

$c \cdot$ Force $\cdot$ Electric Resistance $=$ Electric Potential Difference^2

$c^8 =$ Energy $\cdot$ Acceleration$^3 \cdot$ Specific Volume

$c^6 \cdot$ Pressure,Stress $=$ Energy $\cdot$ Acceleration^3

$c \cdot$ Power, Radiance Flux $=$ Energy $\cdot$ Acceleration

$c^5 \cdot$ Dynamic Viscosity = Energy $\cdot$ Acceleration^2

$c^8 \cdot$ Mass Density = Energy $\cdot$ Acceleration^3

$c^4 \cdot$ Area^3 = Energy$^2 \cdot$ Specific Volume^2

$c^2 \cdot$ Energy^2 = Area $\cdot$ Power, Radiance Flux^2

$c \cdot$ Area $\cdot$ Dynamic Viscosity = Energy

$c^4 \cdot$ Area$^3 \cdot$ Mass Density^2 = Energy^2

$c^2 \cdot$ Volume = Energy $\cdot$ Specific Volume

$c^3 \cdot$ Energy^3 = Volume $\cdot$ Power, Radiance Flux^3

$c^3 \cdot$ Volume$^2 \cdot$ Dynamic Viscosity^3 = Energy^3

$c^2 \cdot$ Volume $\cdot$ Mass Density = Energy

c^2 = Energy $\cdot$ Wavenumber$^3 \cdot$ Specific Volume

$c \cdot$ Energy $\cdot$ Wavenumber = Power, Radiance Flux

$c \cdot$ Dynamic Viscosity = Energy $\cdot$ Wavenumber^2

$c^2 \cdot$ Mass Density = Energy $\cdot$ Wavenumber^3

c^5 = Energy $\cdot$ Specific Volume $\cdot$ Frequency^3

$c^5 \cdot$ Energy^2 = Specific Volume $\cdot$ Power, Radiance Flux^3

$c \cdot$ Energy = Specific Volume$^2 \cdot$ Dynamic Viscosity^3

$c \cdot$ Energy $\cdot$ Specific Volume = Kinematic Viscosity^3

$c^3 \cdot$ Pressure,Stress = Energy $\cdot$ Frequency^3

$c^3 \cdot$ Dynamic Viscosity = Energy $\cdot$ Frequency^2

$c^5 \cdot$ Mass Density = Energy $\cdot$ Frequency^3

$c^3 \cdot$ Energy$^2 \cdot$ Pressure,Stress = Power, Radiance Flux^3

$c^3 \cdot \text{Dynamic Viscosity}^3 = \text{Energy} \cdot \text{Pressure,Stress}^2$

$c^3 \cdot \text{Energy} = \text{Pressure,Stress} \cdot \text{Kinematic Viscosity}^3$

$c^3 \cdot \text{Energy} \cdot \text{Dynamic Viscosity} = \text{Power, Radiance Flux}^2$

$c^2 \cdot \text{Energy} = \text{Power, Radiance Flux} \cdot \text{Kinematic Viscosity}$

$c^5 \cdot \text{Energy}^2 \cdot \text{Mass Density} = \text{Power, Radiance Flux}^3$

$c \cdot \text{Energy} = \text{Dynamic Viscosity} \cdot \text{Kinematic Viscosity}^2$

$c \cdot \text{Energy} \cdot \text{Mass Density}^2 = \text{Dynamic Viscosity}^3$

$c \cdot \text{Energy} = \text{Kinematic Viscosity}^3 \cdot \text{Mass Density}$

$c^6 = \text{Acceleration}^3 \cdot \text{Mass} \cdot \text{Specific Volume}$

$c^4 \cdot \text{Pressure,Stress} = \text{Acceleration}^3 \cdot \text{Mass}$

$c \cdot \text{Acceleration} \cdot \text{Mass} = \text{Power, Radiance Flux}$

$c^3 \cdot \text{Dynamic Viscosity} = \text{Acceleration}^2 \cdot \text{Mass}$

$c^6 \cdot \text{Mass Density} = \text{Acceleration}^3 \cdot \text{Mass}$

$c^8 = \text{Acceleration}^3 \cdot \text{Specific Volume} \cdot \text{Energy,Work,Heat Quantity}$

$c^7 = \text{Acceleration}^2 \cdot \text{Specific Volume} \cdot \text{Power, Radiance Flux}$

$c^3 = \text{Acceleration} \cdot \text{Specific Volume} \cdot \text{Dynamic Viscosity}$

$c^6 \cdot \text{Magnetic Field Strength} = \text{Acceleration}^3 \cdot \text{Current Density}$

$c^3 \cdot \text{Electric Charge} = \text{Acceleration} \cdot \text{Current Density}$

$c^3 \cdot \text{Magnetic Field Strength} = \text{Acceleration}^2 \cdot \text{Electric Charge}$

$c^6 \cdot \text{Pressure,Stress} = \text{Acceleration}^3 \cdot \text{Energy,Work,Heat Quantity}$

$c^5 \cdot \text{Pressure,Stress} = \text{Acceleration}^2 \cdot \text{Power, Radiance Flux}$

$c \cdot \text{Pressure,Stress} = \text{Acceleration} \cdot \text{Dynamic Viscosity}$

$c \cdot \text{Power, Radiance Flux} = \text{Acceleration} \cdot \text{Energy,Work,Heat Quantity}$

$c^5 \cdot$ Dynamic Viscosity = Acceleration$^2 \cdot$ Energy,Work,Heat Quantity

$c^8 \cdot$ Mass Density = Acceleration$^3 \cdot$ Energy,Work,Heat Quantity

$c^4 \cdot$ Dynamic Viscosity = Acceleration $\cdot$ Power, Radiance Flux

$c^7 \cdot$ Mass Density = Acceleration$^2 \cdot$ Power, Radiance Flux

$c^3 \cdot$ Magnetic Flux Density = Acceleration $\cdot$ Electric Potential Difference

$c \cdot$ Electric Potential Difference = Acceleration $\cdot$ Magnetic Flux

c = Acceleration $\cdot$ Capacitance $\cdot$ Electric Resistance

c^2 = Acceleration$^2 \cdot$ Capacitance $\cdot$ Inductance

$c \cdot$ Electric Resistance = Acceleration $\cdot$ Inductance

$c^4 \cdot$ Magnetic Flux Density = Acceleration$^2 \cdot$ Magnetic Flux

$c^3 \cdot$ Mass Density = Acceleration $\cdot$ Dynamic Viscosity

$c^4 \cdot$ Mass^2 = Area$^3 \cdot$ Pressure,Stress^2

$c^6 \cdot$ Mass^2 = Area $\cdot$ Power, Radiance Flux^2

$c \cdot$ Mass = Area $\cdot$ Dynamic Viscosity

$c^2 \cdot$ Mass = Volume $\cdot$ Pressure,Stress

$c^9 \cdot$ Mass^3 = Volume $\cdot$ Power, Radiance Flux^3

$c^3 \cdot$ Mass^3 = Volume$^2 \cdot$ Dynamic Viscosity^3

$c^2 \cdot$ Mass $\cdot$ Wavenumber^3 = Pressure,Stress

$c^3 \cdot$ Mass $\cdot$ Wavenumber = Power, Radiance Flux

$c \cdot$ Mass $\cdot$ Wavenumber^2 = Dynamic Viscosity

c^3 = Mass $\cdot$ Specific Volume $\cdot$ Frequency^3

$c^9 \cdot$ Mass^2 = Specific Volume $\cdot$ Power, Radiance Flux^3

$c^3 \cdot$ Mass = Specific Volume$^2 \cdot$ Dynamic Viscosity^3

$c^3 \cdot$ Mass $\cdot$ Specific Volume = Kinematic Viscosity^3

$c^2 \cdot$ Mass = Current Density $\cdot$ Magnetic Flux Density

$c \cdot$ Pressure,Stress = Mass $\cdot$ Frequency^3

$c^2 \cdot$ Mass $\cdot$ Frequency = Power, Radiance Flux

$c \cdot$ Dynamic Viscosity = Mass $\cdot$ Frequency^2

$c^3 \cdot$ Mass Density = Mass $\cdot$ Frequency^3

$c^7 \cdot$ Mass$^2 \cdot$ Pressure,Stress = Power, Radiance Flux^3

$c \cdot$ Dynamic Viscosity^3 = Mass $\cdot$ Pressure,Stress^2

$c^5 \cdot$ Mass = Pressure,Stress $\cdot$ Kinematic Viscosity^3

$c^5 \cdot$ Mass $\cdot$ Dynamic Viscosity = Power, Radiance Flux^2

$c^4 \cdot$ Mass = Power, Radiance Flux $\cdot$ Kinematic Viscosity

$c^9 \cdot$ Mass$^2 \cdot$ Mass Density = Power, Radiance Flux^3

$c^2 \cdot$ Mass = Electric Charge $\cdot$ Electric Potential Difference

$c^2 \cdot$ Mass $\cdot$ Capacitance = Electric Charge^2

$c^2 \cdot$ Mass = Electric Potential Difference$^2 \cdot$ Capacitance

$c^2 \cdot$ Mass $\cdot$ Inductance = Magnetic Flux^2

$c^3 \cdot$ Mass = Dynamic Viscosity $\cdot$ Kinematic Viscosity^2

$c^3 \cdot$ Mass $\cdot$ Mass Density^2 = Dynamic Viscosity^3

$c^3 \cdot$ Mass = Kinematic Viscosity$^3 \cdot$ Mass Density

$c^4 \cdot$ Area^3 = Specific Volume$^2 \cdot$ Energy,Work,Heat Quantity^2

$c^3 \cdot$ Area = Specific Volume $\cdot$ Power, Radiance Flux

$c^2 \cdot$ Area = Specific Volume$^2 \cdot$ Dynamic Viscosity^2

$$c^2 \cdot \text{Area} \cdot \text{Electric Charge}^2 = \text{Current Density}^2$$

$$c \cdot \text{Electric Charge} = \text{Area} \cdot \text{Magnetic Field Strength}$$

$$c \cdot \text{Area} \cdot \text{Pressure,Stress} = \text{Power, Radiance Flux}$$

$$c^2 \cdot \text{Dynamic Viscosity}^2 = \text{Area} \cdot \text{Pressure,Stress}^2$$

$$c^2 \cdot \text{Energy,Work,Heat Quantity}^2 = \text{Area} \cdot \text{Power, Radiance Flux}^2$$

$$c \cdot \text{Area} \cdot \text{Dynamic Viscosity} = \text{Energy,Work,Heat Quantity}$$

$$c^4 \cdot \text{Area}^3 \cdot \text{Mass Density}^2 = \text{Energy,Work,Heat Quantity}^2$$

$$c^4 \cdot \text{Area} \cdot \text{Dynamic Viscosity}^2 = \text{Power, Radiance Flux}^2$$

$$c^3 \cdot \text{Area} \cdot \text{Mass Density} = \text{Power, Radiance Flux}$$

$$c^2 \cdot \text{Area} \cdot \text{Magnetic Flux Density}^2 = \text{Electric Potential Difference}^2$$

$$c^2 \cdot \text{Magnetic Flux}^2 = \text{Area} \cdot \text{Electric Potential Difference}^2$$

$$c^2 \cdot \text{Capacitance}^2 \cdot \text{Electric Resistance}^2 = \text{Area}$$

$$c^2 \cdot \text{Capacitance} \cdot \text{Inductance} = \text{Area}$$

$$c^2 \cdot \text{Inductance}^2 = \text{Area} \cdot \text{Electric Resistance}^2$$

$$c^2 \cdot \text{Area} \cdot \text{Mass Density}^2 = \text{Dynamic Viscosity}^2$$

$$c^2 \cdot \text{Volume} = \text{Specific Volume} \cdot \text{Energy,Work,Heat Quantity}$$

$$c^9 \cdot \text{Volume}^2 = \text{Specific Volume}^3 \cdot \text{Power, Radiance Flux}^3$$

$$c^3 \cdot \text{Volume} = \text{Specific Volume}^3 \cdot \text{Dynamic Viscosity}^3$$

$$c^3 \cdot \text{Volume} \cdot \text{Electric Charge}^3 = \text{Current Density}^3$$

$$c^3 \cdot \text{Electric Charge}^3 = \text{Volume}^2 \cdot \text{Magnetic Field Strength}^3$$

$$c^3 \cdot \text{Volume}^2 \cdot \text{Pressure,Stress}^3 = \text{Power, Radiance Flux}^3$$

$$c^3 \cdot \text{Dynamic Viscosity}^3 = \text{Volume} \cdot \text{Pressure,Stress}^3$$

$c^3 \cdot$ Energy,Work,Heat Quantity^3 = Volume $\cdot$ Power, Radiance Flux^3

$c^3 \cdot$ Volume$^2 \cdot$ Dynamic Viscosity^3 = Energy,Work,Heat Quantity^3

$c^2 \cdot$ Volume $\cdot$ Mass Density = Energy,Work,Heat Quantity

$c^6 \cdot$ Volume $\cdot$ Dynamic Viscosity^3 = Power, Radiance Flux^3

$c^9 \cdot$ Volume$^2 \cdot$ Mass Density^3 = Power, Radiance Flux^3

$c^3 \cdot$ Volume $\cdot$ Magnetic Flux Density^3 = Electric Potential Difference^3

$c^3 \cdot$ Magnetic Flux^3 = Volume $\cdot$ Electric Potential Difference^3

$c^3 \cdot$ Capacitance$^3 \cdot$ Electric Resistance^3 = Volume

$c^6 \cdot$ Capacitance$^3 \cdot$ Inductance^3 = Volume^2

$c^3 \cdot$ Inductance^3 = Volume $\cdot$ Electric Resistance^3

$c^3 \cdot$ Volume $\cdot$ Mass Density^3 = Dynamic Viscosity^3

c^2 = Wavenumber$^3 \cdot$ Specific Volume $\cdot$ Energy,Work,Heat Quantity

c^3 = Wavenumber$^2 \cdot$ Specific Volume $\cdot$ Power, Radiance Flux

c = Wavenumber $\cdot$ Specific Volume $\cdot$ Dynamic Viscosity

c $\cdot$ Electric Charge = Wavenumber $\cdot$ Current Density

c $\cdot$ Wavenumber$^2 \cdot$ Electric Charge = Magnetic Field Strength

c $\cdot$ Pressure,Stress = Wavenumber$^2 \cdot$ Power, Radiance Flux

c $\cdot$ Wavenumber $\cdot$ Dynamic Viscosity = Pressure,Stress

c $\cdot$ Wavenumber $\cdot$ Energy,Work,Heat Quantity = Power, Radiance Flux

c $\cdot$ Dynamic Viscosity = Wavenumber$^2 \cdot$ Energy,Work,Heat Quantity

$c^2 \cdot$ Mass Density = Wavenumber$^3 \cdot$ Energy,Work,Heat Quantity

$c^2 \cdot$ Dynamic Viscosity = Wavenumber $\cdot$ Power, Radiance Flux

$c^3 \cdot$ Mass Density = Wavenumber$^2 \cdot$ Power, Radiance Flux

$c \cdot$ Magnetic Flux Density = Wavenumber $\cdot$ Electric Potential Difference

$c \cdot$ Wavenumber $\cdot$ Magnetic Flux = Electric Potential Difference

$c \cdot$ Wavenumber $\cdot$ Capacitance $\cdot$ Electric Resistance = A Constant

$c^2 \cdot$ Wavenumber$^2 \cdot$ Capacitance $\cdot$ Inductance = A Constant

$c \cdot$ Wavenumber $\cdot$ Inductance = Electric Resistance

$c \cdot$ Mass Density = Wavenumber $\cdot$ Dynamic Viscosity

c^3 = Specific Volume $\cdot$ Magnetic Field Strength$^2 \cdot$ Electric Resistance

c^2 = Specific Volume $\cdot$ Magnetic Field Strength $\cdot$ Magnetic Flux Density

c^5 = Specific Volume $\cdot$ Frequency$^3 \cdot$ Energy,Work,Heat Quantity

c^5 = Specific Volume $\cdot$ Frequency$^2 \cdot$ Power, Radiance Flux

c^2 = Specific Volume $\cdot$ Frequency $\cdot$ Dynamic Viscosity

$c^5 \cdot$ Energy,Work,Heat Quantity^2 = Specific Volume $\cdot$ Power, Radiance Flux^3

$c \cdot$ Energy,Work,Heat Quantity = Specific Volume$^2 \cdot$ Dynamic Viscosity^3

$c \cdot$ Specific Volume $\cdot$ Energy,Work,Heat Quantity = Kinematic Viscosity^3

$c \cdot$ Specific Volume $\cdot$ Dynamic Viscosity^2 = Power, Radiance Flux

$c \cdot$ Kinematic Viscosity^2 = Specific Volume $\cdot$ Power, Radiance Flux

$c \cdot$ Electric Resistance = Specific Volume $\cdot$ Magnetic Flux Density^2

$c^3 \cdot$ Magnetic Field Strength = Current Density $\cdot$ Frequency^3

$c^3 \cdot$ Electric Charge^3 = Current Density$^2 \cdot$ Magnetic Field Strength

$c^3 \cdot$ Current Density = Magnetic Field Strength $\cdot$ Kinematic Viscosity^3

$c^2 \cdot$ Electric Charge = Current Density $\cdot$ Frequency

$c^2 \cdot$ Capacitance $\cdot$ Magnetic Flux = Current Density

$c \cdot$ Magnetic Field Strength = Frequency$^2 \cdot$ Electric Charge

$c \cdot$ Pressure,Stress = Magnetic Field Strength$^2 \cdot$ Electric Resistance

$c \cdot$ Magnetic Field Strength $\cdot$ Magnetic Flux = Power, Radiance Flux

$c^3 \cdot$ Electric Charge = Magnetic Field Strength $\cdot$ Kinematic Viscosity^2

$c \cdot$ Magnetic Field Strength $\cdot$ Inductance = Electric Potential Difference

$c^2 \cdot$ Dynamic Viscosity = Magnetic Field Strength $\cdot$ Electric Potential Difference

$c \cdot$ Magnetic Flux Density = Magnetic Field Strength $\cdot$ Electric Resistance

$c^3 \cdot$ Mass Density = Magnetic Field Strength$^2 \cdot$ Electric Resistance

$c^2 \cdot$ Mass Density = Magnetic Field Strength $\cdot$ Magnetic Flux Density

$c \cdot$ Dynamic Viscosity = Magnetic Field Strength$^2 \cdot$ Inductance

$c^3 \cdot$ Pressure,Stress = Frequency$^3 \cdot$ Energy,Work,Heat Quantity

$c^3 \cdot$ Pressure,Stress = Frequency$^2 \cdot$ Power, Radiance Flux

$c^3 \cdot$ Dynamic Viscosity = Frequency$^2 \cdot$ Energy,Work,Heat Quantity

$c^5 \cdot$ Mass Density = Frequency$^3 \cdot$ Energy,Work,Heat Quantity

$c^3 \cdot$ Dynamic Viscosity = Frequency $\cdot$ Power, Radiance Flux

$c^5 \cdot$ Mass Density = Frequency$^2 \cdot$ Power, Radiance Flux

$c^2 \cdot$ Magnetic Flux Density = Frequency $\cdot$ Electric Potential Difference

$c^2 \cdot$ Magnetic Flux Density = Frequency$^2 \cdot$ Magnetic Flux

$c^2 \cdot$ Mass Density = Frequency $\cdot$ Dynamic Viscosity

$c^3 \cdot$ Pressure,Stress $\cdot$ Energy,Work,Heat Quantity^2 = Power, Radiance Flux^3

$c^3 \cdot$ Dynamic Viscosity^3 = Pressure,Stress$^2 \cdot$ Energy,Work,Heat Quantity

$c^3 \cdot$ Energy,Work,Heat Quantity = Pressure,Stress $\cdot$ Kinematic Viscosity^3

$c^3 \cdot$ Dynamic Viscosity^2 = Pressure,Stress $\cdot$ Power, Radiance Flux

$c \cdot$ Power, Radiance Flux = Pressure,Stress $\cdot$ Kinematic Viscosity^2

$c \cdot$ Magnetic Flux Density^2 = Pressure,Stress $\cdot$ Electric Resistance

$c^2 \cdot$ Dynamic Viscosity = Pressure,Stress $\cdot$ Kinematic Viscosity

$c^3 \cdot$ Energy,Work,Heat Quantity $\cdot$ Dynamic Viscosity = Power, Radiance Flux^2

$c^2 \cdot$ Energy,Work,Heat Quantity = Power, Radiance Flux $\cdot$ Kinematic Viscosity

$c^5 \cdot$ Energy,Work,Heat Quantity$^2 \cdot$ Mass Density = Power, Radiance Flux^3

$c \cdot$ Energy,Work,Heat Quantity = Dynamic Viscosity $\cdot$ Kinematic Viscosity^2

$c \cdot$ Energy,Work,Heat Quantity $\cdot$ Mass Density^2 = Dynamic Viscosity^3

$c \cdot$ Energy,Work,Heat Quantity = Kinematic Viscosity$^3 \cdot$ Mass Density

$c^2 \cdot$ Electric Charge $\cdot$ Magnetic Flux Density = Power, Radiance Flux

$c \cdot$ Dynamic Viscosity $\cdot$ Kinematic Viscosity = Power, Radiance Flux

$c \cdot$ Dynamic Viscosity^2 = Power, Radiance Flux $\cdot$ Mass Density

$c \cdot$ Kinematic Viscosity$^2 \cdot$ Mass Density = Power, Radiance Flux

$c^2 \cdot$ Magnetic Flux Density $\cdot$ Magnetic Flux = Electric Potential Difference^2

$c^3 \cdot$ Inductance $\cdot$ Dynamic Viscosity = Electric Potential Difference^2

$c^2 \cdot$ Magnetic Flux = Electric Potential Difference $\cdot$ Kinematic Viscosity

$c^2 \cdot$ Capacitance $\cdot$ Electric Resistance = Kinematic Viscosity

$c \cdot$ Capacitance $\cdot$ Magnetic Flux Density^2 = Dynamic Viscosity

$c^4 \cdot$ Capacitance $\cdot$ Inductance = Kinematic Viscosity^2

$c \cdot$ Electric Resistance $\cdot$ Mass Density = Magnetic Flux Density^2

$c^2 \cdot$ Inductance = Electric Resistance $\cdot$ Kinematic Viscosity

$c^2 \cdot$ Magnetic Flux = Magnetic Flux Density $\cdot$ Kinematic Viscosity^2

$k \cdot$ Electric Current $\cdot$ Time $= h$

$k \cdot$ Electric Current $= h \cdot$ Frequency

$k^2 \cdot$ Electric Current$^2 = h \cdot$ Power, Radiance Flux

$k^2 \cdot$ Electric Current $= h \cdot$ Electric Potential Difference

$k^3 \cdot$ Electric Current $\cdot$ Capacitance $= h^2$

$k^2 \cdot$ Capacitance $= h \cdot$ Time

$k^2 \cdot$ Time $= h \cdot$ Inductance

$k \cdot$ Energy $= h \cdot$ Electric Potential Difference

$k^2 \cdot$ Energy $\cdot$ Capacitance $= h^2$

$k \cdot$ Mass $\cdot$ Current Density $= h^2$

$k \cdot$ Current Density $= h \cdot$ Kinematic Viscosity

$k^2 \cdot$ Frequency $\cdot$ Capacitance $= h$

$k^2 = h \cdot$ Frequency $\cdot$ Inductance

$k \cdot$ Energy,Work,Heat Quantity $= h \cdot$ Electric Potential Difference

$k^2 \cdot$ Energy,Work,Heat Quantity $\cdot$ Capacitance $= h^2$

$k^2 \cdot$ Power, Radiance Flux $= h \cdot$ Electric Potential Difference^2

$k^4 \cdot$ Power, Radiance Flux $\cdot$ Capacitance$^2 = h^3$

$k^4 = h \cdot$ Power, Radiance Flux $\cdot$ Inductance^2

$k \cdot$ Electric Potential Difference $\cdot$ Capacitance $= h$

$k^3 = h \cdot$ Electric Potential Difference $\cdot$ Inductance

$k^4 \cdot$ Capacitance $= h^2 \cdot$ Inductance

$k^3 \cdot$ Dynamic Viscosity$^2 = h^2 \cdot$ Magnetic Flux Density^3

$k \cdot G_c \cdot$ Magnetic Field Strength $=$ Velocity^4

$k^2 \cdot G_c$ = Specific Volume $\cdot$ Electric Potential Difference^2

$k^5 \cdot G_c \cdot$ Magnetic Field Strength^5 = Power, Radiance Flux^4

$k \cdot G_c \cdot$ Magnetic Field Strength = Absorbed Dose^2

$k^2 \cdot G_c \cdot$ Mass Density = Electric Potential Difference^2

$k \cdot$ Illuminance = Luminous Intensity $\cdot$ Magnetic Flux Density

$k \cdot$ Catalytic Activity = Amount of a Substance $\cdot$ Electric Potential Difference

$k \cdot$ Electric Current = Time $\cdot$ Power, Radiance Flux

$k \cdot$ Capacitance = Electric Current $\cdot$ Time^2

k = Electric Current $\cdot$ Time $\cdot$ Electric Resistance

$k \cdot$ Electric Current = Length $\cdot$ Force

$k \cdot$ Electric Current = Length$^3 \cdot$ Pressure,Stress

$k \cdot$ Electric Current = Volumetric Velocity $\cdot$ Dynamic Viscosity

$k^2 \cdot$ Electric Current^2 = Force$^2 \cdot$ Area

$k^3 \cdot$ Electric Current^3 = Force$^3 \cdot$ Volume

$k \cdot$ Electric Current $\cdot$ Wavenumber = Force

$k^2 \cdot$ Electric Current^3 = Force$^2 \cdot$ Current Density

$k^2 \cdot$ Electric Current$^2 \cdot$ Pressure,Stress = Force^3

$k \cdot$ Electric Current$^2 \cdot$ Magnetic Flux Density = Force^2

$k \cdot$ Electric Current = Mass $\cdot$ Velocity^2

$k \cdot$ Electric Current = Mass $\cdot$ Absorbed Dose

$k^2 \cdot$ Electric Current^2 = Area$^3 \cdot$ Pressure,Stress^2

$k \cdot$ Electric Current = Volume $\cdot$ Pressure,Stress

$k \cdot$ Electric Current $\cdot$ Wavenumber^3 = Pressure,Stress

$k^2 \cdot$ Electric Current^5 = Current Density$^3 \cdot$ Pressure,Stress^2

$k \cdot$ Electric Current = Current Density $\cdot$ Magnetic Flux Density

$k \cdot$ Magnetic Field Strength^3 = Electric Current$^2 \cdot$ Pressure,Stress

$k \cdot$ Magnetic Field Strength^2 = Electric Current$^2 \cdot$ Magnetic Flux Density

$k \cdot$ Electric Current $\cdot$ Frequency = Power, Radiance Flux

$k \cdot$ Frequency$^2 \cdot$ Capacitance = Electric Current

$k \cdot$ Frequency = Electric Current $\cdot$ Electric Resistance

$k \cdot$ Pressure,Stress^2 = Electric Current$^2 \cdot$ Magnetic Flux Density^3

$k \cdot$ Electric Current^2 = Power, Radiance Flux $\cdot$ Electric Charge

$k \cdot$ Electric Current^3 = Power, Radiance Flux$^2 \cdot$ Capacitance

$k \cdot$ Electric Current = Electric Charge $\cdot$ Electric Potential Difference

$k \cdot$ Electric Current $\cdot$ Capacitance = Electric Charge^2

$k \cdot$ Electric Current = Electric Potential Difference$^2 \cdot$ Capacitance

k = Electric Current $\cdot$ Capacitance $\cdot$ Electric Resistance^2

k^3 = Time$^2 \cdot$ Volumetric Velocity$^2 \cdot$ Magnetic Flux Density^3

$k \cdot$ Electric Charge = Time $\cdot$ Energy

$k^2 \cdot$ Capacitance = Time$^2 \cdot$ Energy

k^2 = Time $\cdot$ Energy $\cdot$ Electric Resistance

k = Time$^4 \cdot$ Acceleration$^2 \cdot$ Magnetic Flux Density

k = Time$^2 \cdot$ Velocity$^2 \cdot$ Magnetic Flux Density

$k \cdot$ Electric Charge = Time $\cdot$ Energy,Work,Heat Quantity

$k^2 \cdot$ Capacitance = Time$^2 \cdot$ Energy,Work,Heat Quantity

k^2 = Time · Energy,Work,Heat Quantity · Electric Resistance

k · Electric Charge = Time^2 · Power, Radiance Flux

k^2 · Capacitance = Time^3 · Power, Radiance Flux

k^2 = Time^2 · Power, Radiance Flux · Electric Resistance

k^2 = Time · Power, Radiance Flux · Inductance

k · Capacitance = Time · Electric Charge

k · Time = Electric Charge · Inductance

k = Time^2 · Magnetic Flux Density · Absorbed Dose

k = Time · Magnetic Flux Density · Kinematic Viscosity

k · Volumetric Velocity = Length^3 · Electric Potential Difference

k · Current Density = Length^3 · Force

k^2 = Length · Force · Inductance

k · Current Density = Length^2 · Energy

k · Length · Magnetic Field Strength = Energy

k^2 · Acceleration = Length · Electric Potential Difference^2

k^2 · Capacitance = Length^2 · Mass

k · Velocity = Length · Electric Potential Difference

k^2 · Specific Volume · Capacitance = Length^5

k · Current Density = Length^5 · Pressure,Stress

k · Current Density = Length^2 · Energy,Work,Heat Quantity

k · Length^2 = Current Density · Inductance

k · Magnetic Field Strength = Length^2 · Pressure,Stress

$k \cdot$ Length $\cdot$ Magnetic Field Strength = Energy,Work,Heat Quantity

$k =$ Length $\cdot$ Magnetic Field Strength $\cdot$ Inductance

$k^2 =$ Length$^3 \cdot$ Pressure,Stress $\cdot$ Inductance

$k \cdot$ Electric Charge = Length$^3 \cdot$ Dynamic Viscosity

$k^2 \cdot$ Absorbed Dose = Length$^2 \cdot$ Electric Potential Difference^2

$k \cdot$ Kinematic Viscosity = Length$^2 \cdot$ Electric Potential Difference

$k^2 \cdot$ Capacitance = Length$^5 \cdot$ Mass Density

$k^2 =$ Length$^3 \cdot$ Electric Resistance $\cdot$ Dynamic Viscosity

$k^5 \cdot$ Acceleration^3 = Volumetric Velocity $\cdot$ Electric Potential Difference^5

$k^5 \cdot$ Acceleration^2 = Volumetric Velocity$^4 \cdot$ Magnetic Flux Density^5

$k^2 \cdot$ Velocity^3 = Volumetric Velocity $\cdot$ Electric Potential Difference^2

$k \cdot$ Velocity = Volumetric Velocity $\cdot$ Magnetic Flux Density

$k^2 \cdot$ Volumetric Velocity^2 = Area$^3 \cdot$ Electric Potential Difference^2

$k \cdot$ Volumetric Velocity = Volume $\cdot$ Electric Potential Difference

$k \cdot$ Volumetric Velocity $\cdot$ Wavenumber^3 = Electric Potential Difference

$k^3 \cdot$ Frequency^2 = Volumetric Velocity$^2 \cdot$ Magnetic Flux Density^3

$k \cdot$ Electric Potential Difference^2 = Volumetric Velocity$^2 \cdot$ Magnetic Flux Density^3

$k^4 \cdot$ Absorbed Dose^3 = Volumetric Velocity$^2 \cdot$ Electric Potential Difference^4

$k \cdot$ Kinematic Viscosity^3 = Volumetric Velocity$^2 \cdot$ Electric Potential Difference

$k^2 \cdot$ Absorbed Dose = Volumetric Velocity$^2 \cdot$ Magnetic Flux Density^2

$k \cdot$ Kinematic Viscosity^2 = Volumetric Velocity$^2 \cdot$ Magnetic Flux Density

$k^2 =$ Volumetric Velocity $\cdot$ Inductance $\cdot$ Dynamic Viscosity

$k \cdot$ Force$^2 \cdot$ Current Density = Energy^3

$k \cdot Force^2 = Energy^2 \cdot Magnetic\ Flux\ Density$

$k^2 \cdot Current\ Density^2 = Force^2 \cdot Area^3$

$k^4 = Force^2 \cdot Area \cdot Inductance^2$

$k \cdot Current\ Density = Force \cdot Volume$

$k^6 = Force^3 \cdot Volume \cdot Inductance^3$

$k \cdot Wavenumber^3 \cdot Current\ Density = Force$

$k^2 \cdot Wavenumber = Force \cdot Inductance$

$k^2 \cdot Current\ Density^2 \cdot Pressure,Stress^3 = Force^5$

$k \cdot Force^2 \cdot Current\ Density = Energy,Work,Heat\ Quantity^3$

$k \cdot Force^2 = Current\ Density^2 \cdot Magnetic\ Flux\ Density^3$

$k^5 = Force^2 \cdot Current\ Density \cdot Inductance^3$

$k \cdot Pressure,Stress = Force \cdot Magnetic\ Flux\ Density$

$k^4 \cdot Pressure,Stress = Force^3 \cdot Inductance^2$

$k \cdot Force^2 = Energy,Work,Heat\ Quantity^2 \cdot Magnetic\ Flux\ Density$

$k^3 \cdot Magnetic\ Flux\ Density = Force^2 \cdot Inductance^2$

$k \cdot Current\ Density = Energy \cdot Area$

$k^2 \cdot Area \cdot Magnetic\ Field\ Strength^2 = Energy^2$

$k^3 \cdot Current\ Density^3 = Energy^3 \cdot Volume^2$

$k^3 \cdot Volume \cdot Magnetic\ Field\ Strength^3 = Energy^3$

$k \cdot Wavenumber^2 \cdot Current\ Density = Energy$

$k \cdot Magnetic\ Field\ Strength = Energy \cdot Wavenumber$

$k^3 \cdot Current\ Density \cdot Magnetic\ Field\ Strength^2 = Energy^3$

$k^3 \cdot \text{Current Density}^3 \cdot \text{Pressure,Stress}^2 = \text{Energy}^5$

$k^3 \cdot \text{Magnetic Field Strength}^3 = \text{Energy}^2 \cdot \text{Pressure,Stress}$

$k^3 \cdot \text{Magnetic Field Strength}^2 = \text{Energy}^2 \cdot \text{Magnetic Flux Density}$

$k \cdot \text{Frequency} \cdot \text{Electric Charge} = \text{Energy}$

$k^2 \cdot \text{Frequency}^2 \cdot \text{Capacitance} = \text{Energy}$

$k^2 \cdot \text{Frequency} = \text{Energy} \cdot \text{Electric Resistance}$

$k^3 \cdot \text{Pressure,Stress}^2 = \text{Energy}^2 \cdot \text{Magnetic Flux Density}^3$

$k \cdot \text{Power, Radiance Flux} \cdot \text{Electric Charge} = \text{Energy}^2$

$k \cdot \text{Power, Radiance Flux} = \text{Energy} \cdot \text{Electric Potential Difference}$

$k^2 \cdot \text{Power, Radiance Flux}^2 \cdot \text{Capacitance} = \text{Energy}^3$

$k^2 \cdot \text{Power, Radiance Flux} = \text{Energy}^2 \cdot \text{Electric Resistance}$

$k \cdot \text{Electric Potential Difference} = \text{Energy} \cdot \text{Electric Resistance}$

$k^2 = \text{Energy} \cdot \text{Capacitance} \cdot \text{Electric Resistance}^2$

$k \cdot \text{Magnetic Field Strength} = \text{Acceleration} \cdot \text{Mass}$

$k \cdot \text{Acceleration} = \text{Velocity} \cdot \text{Electric Potential Difference}$

$k \cdot \text{Acceleration}^2 = \text{Velocity}^4 \cdot \text{Magnetic Flux Density}$

$k^4 \cdot \text{Acceleration}^2 = \text{Area} \cdot \text{Electric Potential Difference}^4$

$k^6 \cdot \text{Acceleration}^3 = \text{Volume} \cdot \text{Electric Potential Difference}^6$

$k^2 \cdot \text{Acceleration} \cdot \text{Wavenumber} = \text{Electric Potential Difference}^2$

$k \cdot \text{Frequency}^4 = \text{Acceleration}^2 \cdot \text{Magnetic Flux Density}$

$k^3 \cdot \text{Acceleration}^2 \cdot \text{Magnetic Flux Density} = \text{Electric Potential Difference}^4$

$k^2 \cdot \text{Acceleration}^2 = \text{Electric Potential Difference}^2 \cdot \text{Absorbed Dose}$

$k^3 \cdot \text{Acceleration}^2 = \text{Electric Potential Difference}^3 \cdot \text{Kinematic Viscosity}$

$k \cdot \text{Acceleration}^2 = \text{Magnetic Flux Density} \cdot \text{Absorbed Dose}^2$

$k^3 \cdot \text{Acceleration}^2 = \text{Magnetic Flux Density}^3 \cdot \text{Kinematic Viscosity}^4$

$k^2 = \text{Mass} \cdot \text{Velocity}^2 \cdot \text{Inductance}$

$k^2 \cdot \text{Capacitance} = \text{Mass} \cdot \text{Area}$

$k^6 \cdot \text{Capacitance}^3 = \text{Mass}^3 \cdot \text{Volume}^2$

$k^2 \cdot \text{Wavenumber}^2 \cdot \text{Capacitance} = \text{Mass}$

$k^6 \cdot \text{Capacitance}^3 = \text{Mass}^5 \cdot \text{Specific Volume}^2$

$k^3 = \text{Mass}^2 \cdot \text{Specific Volume}^2 \cdot \text{Magnetic Flux Density}^3$

$k \cdot \text{Electric Charge}^2 = \text{Mass} \cdot \text{Current Density}$

$k^3 = \text{Mass} \cdot \text{Current Density} \cdot \text{Electric Resistance}^2$

$k \cdot \text{Current Density} = \text{Mass} \cdot \text{Kinematic Viscosity}^2$

$k \cdot \text{Electric Charge} = \text{Mass} \cdot \text{Kinematic Viscosity}$

$k \cdot \text{Capacitance} \cdot \text{Magnetic Flux Density} = \text{Mass}$

$k^6 \cdot \text{Capacitance}^3 \cdot \text{Mass Density}^2 = \text{Mass}^5$

$k^2 = \text{Mass} \cdot \text{Electric Resistance} \cdot \text{Kinematic Viscosity}$

$k^3 \cdot \text{Mass Density}^2 = \text{Mass}^2 \cdot \text{Magnetic Flux Density}^3$

$k^2 = \text{Mass} \cdot \text{Inductance} \cdot \text{Absorbed Dose}$

$k^2 \cdot \text{Velocity}^2 = \text{Area} \cdot \text{Electric Potential Difference}^2$

$k^3 \cdot \text{Velocity}^3 = \text{Volume} \cdot \text{Electric Potential Difference}^3$

$k \cdot \text{Velocity} \cdot \text{Wavenumber} = \text{Electric Potential Difference}$

$k \cdot \text{Velocity}^2 \cdot \text{Capacitance} = \text{Current Density}$

$k \cdot \text{Velocity} \cdot \text{Magnetic Field Strength} = \text{Power, Radiance Flux}$

$k \cdot \text{Frequency}^2 = \text{Velocity}^2 \cdot \text{Magnetic Flux Density}$

$k \cdot \text{Velocity}^2 \cdot \text{Magnetic Flux Density} = \text{Electric Potential Difference}^2$

$k \cdot \text{Velocity}^2 = \text{Electric Potential Difference} \cdot \text{Kinematic Viscosity}$

$k \cdot \text{Velocity}^2 = \text{Magnetic Flux Density} \cdot \text{Kinematic Viscosity}^2$

$k^4 \cdot \text{Specific Volume}^2 \cdot \text{Capacitance}^2 = \text{Area}^5$

$k^2 \cdot \text{Current Density}^2 = \text{Area}^5 \cdot \text{Pressure,Stress}^2$

$k \cdot \text{Current Density} = \text{Area} \cdot \text{Energy,Work,Heat Quantity}$

$k \cdot \text{Area} = \text{Current Density} \cdot \text{Inductance}$

$k \cdot \text{Magnetic Field Strength} = \text{Area} \cdot \text{Pressure,Stress}$

$k^2 \cdot \text{Area} \cdot \text{Magnetic Field Strength}^2 = \text{Energy,Work,Heat Quantity}^2$

$k^2 = \text{Area} \cdot \text{Magnetic Field Strength}^2 \cdot \text{Inductance}^2$

$k^4 = \text{Area}^3 \cdot \text{Pressure,Stress}^2 \cdot \text{Inductance}^2$

$k^2 \cdot \text{Electric Charge}^2 = \text{Area}^3 \cdot \text{Dynamic Viscosity}^2$

$k^2 \cdot \text{Absorbed Dose} = \text{Area} \cdot \text{Electric Potential Difference}^2$

$k \cdot \text{Kinematic Viscosity} = \text{Area} \cdot \text{Electric Potential Difference}$

$k^4 \cdot \text{Capacitance}^2 = \text{Area}^5 \cdot \text{Mass Density}^2$

$k^4 = \text{Area}^3 \cdot \text{Electric Resistance}^2 \cdot \text{Dynamic Viscosity}^2$

$k^6 \cdot \text{Specific Volume}^3 \cdot \text{Capacitance}^3 = \text{Volume}^5$

$k^3 \cdot \text{Current Density}^3 = \text{Volume}^5 \cdot \text{Pressure,Stress}^3$

$k^3 \cdot \text{Current Density}^3 = \text{Volume}^2 \cdot \text{Energy,Work,Heat Quantity}^3$

$k^3 \cdot \text{Volume}^2 = \text{Current Density}^3 \cdot \text{Inductance}^3$

$k^3 \cdot \text{Magnetic Field Strength}^3 = \text{Volume}^2 \cdot \text{Pressure,Stress}^3$

$k^3 \cdot \text{Volume} \cdot \text{Magnetic Field Strength}^3 = \text{Energy,Work,Heat Quantity}^3$

k^3 = Volume · Magnetic Field Strength^3 · Inductance^3

k^2 = Volume · Pressure,Stress · Inductance

k · Electric Charge = Volume · Dynamic Viscosity

k^6 · Absorbed Dose^3 = Volume^2 · Electric Potential Difference^6

k^3 · Kinematic Viscosity^3 = Volume^2 · Electric Potential Difference^3

k^6 · Capacitance^3 = Volume^5 · Mass Density^3

k^2 = Volume · Electric Resistance · Dynamic Viscosity

k^2 · Wavenumber^5 · Specific Volume · Capacitance = A Constant

k · Wavenumber^5 · Current Density = Pressure,Stress

k · Wavenumber^2 · Current Density = Energy,Work,Heat Quantity

k = Wavenumber^2 · Current Density · Inductance

k · Wavenumber^2 · Magnetic Field Strength = Pressure,Stress

k · Magnetic Field Strength = Wavenumber · Energy,Work,Heat Quantity

k · Wavenumber = Magnetic Field Strength · Inductance

k^2 · Wavenumber^3 = Pressure,Stress · Inductance

k · Wavenumber^3 · Electric Charge = Dynamic Viscosity

k^2 · Wavenumber^2 · Absorbed Dose = Electric Potential Difference^2

k · Wavenumber^2 · Kinematic Viscosity = Electric Potential Difference

k^2 · Wavenumber^5 · Capacitance = Mass Density

k^2 · Wavenumber^3 = Electric Resistance · Dynamic Viscosity

k · Magnetic Field Strength = Specific Volume · Dynamic Viscosity^2

k · Specific Volume · Magnetic Field Strength = Kinematic Viscosity^2

$k = \text{Specific Volume}^2 \cdot \text{Capacitance}^2 \cdot \text{Magnetic Flux Density}^5$

$k^3 \cdot \text{Magnetic Field Strength}^5 = \text{Current Density}^2 \cdot \text{Pressure,Stress}^3$

$k^3 \cdot \text{Current Density} \cdot \text{Magnetic Field Strength}^2 = \text{Energy,Work,Heat Quantity}^3$

$k^3 \cdot \text{Magnetic Field Strength}^2 = \text{Current Density}^2 \cdot \text{Magnetic Flux Density}^3$

$k^3 = \text{Current Density} \cdot \text{Magnetic Field Strength}^2 \cdot \text{Inductance}^3$

$k^3 \cdot \text{Current Density}^3 \cdot \text{Pressure,Stress}^2 = \text{Energy,Work,Heat Quantity}^5$

$k^3 \cdot \text{Pressure,Stress}^2 = \text{Current Density}^2 \cdot \text{Magnetic Flux Density}^5$

$k^7 = \text{Current Density}^3 \cdot \text{Pressure,Stress}^2 \cdot \text{Inductance}^5$

$k \cdot \text{Capacitance} \cdot \text{Absorbed Dose} = \text{Current Density}$

$k \cdot \text{Kinematic Viscosity} = \text{Current Density} \cdot \text{Electric Resistance}$

$k^2 = \text{Current Density} \cdot \text{Magnetic Flux Density} \cdot \text{Inductance}$

$k^3 \cdot \text{Magnetic Field Strength}^3 = \text{Pressure,Stress} \cdot \text{Energy,Work,Heat Quantity}^2$

$k \cdot \text{Pressure,Stress} = \text{Magnetic Field Strength}^3 \cdot \text{Inductance}^2$

$k^3 \cdot \text{Magnetic Field Strength}^2 = \text{Energy,Work,Heat Quantity}^2 \cdot \text{Magnetic Flux Density}$

$k^2 \cdot \text{Magnetic Field Strength}^2 \cdot \text{Absorbed Dose} = \text{Power, Radiance Flux}^2$

$k \cdot \text{Magnetic Flux Density} = \text{Magnetic Field Strength}^2 \cdot \text{Inductance}^2$

$k \cdot \text{Magnetic Field Strength} = \text{Dynamic Viscosity} \cdot \text{Kinematic Viscosity}$

$k \cdot \text{Magnetic Field Strength} \cdot \text{Mass Density} = \text{Dynamic Viscosity}^2$

$k \cdot \text{Magnetic Field Strength} = \text{Kinematic Viscosity}^2 \cdot \text{Mass Density}$

$k \cdot \text{Frequency} \cdot \text{Electric Charge} = \text{Energy,Work,Heat Quantity}$

$k^2 \cdot \text{Frequency}^2 \cdot \text{Capacitance} = \text{Energy,Work,Heat Quantity}$

$k^2 \cdot \text{Frequency} = \text{Energy,Work,Heat Quantity} \cdot \text{Electric Resistance}$

$k \cdot \text{Frequency}^2 \cdot \text{Electric Charge} = \text{Power, Radiance Flux}$

$k^2 \cdot \text{Frequency}^3 \cdot \text{Capacitance} = \text{Power, Radiance Flux}$

$k^2 \cdot \text{Frequency}^2 = \text{Power, Radiance Flux} \cdot \text{Electric Resistance}$

$k^2 \cdot \text{Frequency} = \text{Power, Radiance Flux} \cdot \text{Inductance}$

$k \cdot \text{Frequency} \cdot \text{Capacitance} = \text{Electric Charge}$

$k = \text{Frequency} \cdot \text{Electric Charge} \cdot \text{Inductance}$

$k \cdot \text{Frequency}^2 = \text{Magnetic Flux Density} \cdot \text{Absorbed Dose}$

$k \cdot \text{Frequency} = \text{Magnetic Flux Density} \cdot \text{Kinematic Viscosity}$

$k^3 \cdot \text{Pressure,Stress}^2 = \text{Energy,Work,Heat Quantity}^2 \cdot \text{Magnetic Flux Density}^3$

$k \cdot \text{Pressure,Stress} = \text{Electric Potential Difference} \cdot \text{Dynamic Viscosity}$

$k \cdot \text{Magnetic Flux Density}^3 = \text{Pressure,Stress}^2 \cdot \text{Inductance}^2$

$k \cdot \text{Power, Radiance Flux} \cdot \text{Electric Charge} = \text{Energy,Work,Heat Quantity}^2$

$k \cdot \text{Power, Radiance Flux} = \text{Energy,Work,Heat Quantity} \cdot \text{Electric Potential Difference}$

$k^2 \cdot \text{Power, Radiance Flux}^2 \cdot \text{Capacitance} = \text{Energy,Work,Heat Quantity}^3$

$k^2 \cdot \text{Power, Radiance Flux} = \text{Energy,Work,Heat Quantity}^2 \cdot \text{Electric Resistance}$

$k \cdot \text{Electric Potential Difference} = \text{Energy,Work,Heat Quantity} \cdot \text{Electric Resistance}$

$k^2 = \text{Energy,Work,Heat Quantity} \cdot \text{Capacitance} \cdot \text{Electric Resistance}^2$

$k \cdot \text{Power, Radiance Flux} = \text{Electric Charge} \cdot \text{Electric Potential Difference}^2$

$k \cdot \text{Power, Radiance Flux} \cdot \text{Capacitance}^2 = \text{Electric Charge}^3$

$k^3 = \text{Power, Radiance Flux} \cdot \text{Electric Charge} \cdot \text{Inductance}^2$

$k \cdot \text{Power, Radiance Flux} = \text{Electric Potential Difference}^3 \cdot \text{Capacitance}$

$k \cdot \text{Electric Potential Difference} = \text{Power, Radiance Flux} \cdot \text{Inductance}$

$k^2 = \text{Power, Radiance Flux} \cdot \text{Capacitance}^2 \cdot \text{Electric Resistance}^3$

k^4 = Power, Radiance Flux^2 · Capacitance · Inductance^3

k^2 · Electric Resistance = Power, Radiance Flux · Inductance^2

k^2 = Electric Charge · Electric Potential Difference · Inductance

k^2 · Capacitance = Electric Charge^2 · Inductance

k · Dynamic Viscosity^2 = Electric Charge^2 · Magnetic Flux Density^3

k = Electric Potential Difference · Capacitance · Electric Resistance

k^2 = Electric Potential Difference^2 · Capacitance · Inductance

k · Electric Resistance = Electric Potential Difference · Inductance

k · Magnetic Flux Density · Absorbed Dose = Electric Potential Difference^2

k · Absorbed Dose = Electric Potential Difference · Kinematic Viscosity

k · Mass Density^2 = Capacitance^2 · Magnetic Flux Density^5

k · Magnetic Flux Density^3 = Electric Resistance^2 · Dynamic Viscosity^2

k · Absorbed Dose = Magnetic Flux Density · Kinematic Viscosity^2

h · G_c · Time^3 = Length^5

Luminous Intensity · Electric Current = Illuminance · Current Density

Luminous Intensity · Magnetic Field Strength^2 = Illuminance · Electric Current^2

Luminous Intensity^3 = Illuminance^3 · Time^2 · Volumetric Velocity^2

Luminous Intensity = Illuminance · Time^4 · Acceleration^2

Luminous Intensity = Illuminance · Time^2 · Velocity^2

Luminous Intensity = Illuminance · Time^2 · Absorbed Dose

Luminous Intensity = Illuminance · Time · Kinematic Viscosity

Luminous Intensity^5 · Acceleration^2 = Illuminance^5 · Volumetric Velocity^4

Luminous Intensity · Velocity = Illuminance · Volumetric Velocity

Luminous Intensity^3 · Frequency^2 = Illuminance^3 · Volumetric Velocity^2

Luminous Intensity^2 · Absorbed Dose = Illuminance^2 · Volumetric Velocity^2

Luminous Intensity · Kinematic Viscosity^2 = Illuminance · Volumetric Velocity^2

Luminous Intensity · Force^2 = Illuminance · Energy^2

Luminous Intensity · Pressure,Stress = Illuminance · Force

Luminous Intensity · Force^2 = Illuminance · Energy,Work,Heat Quantity^2

Luminous Intensity^3 · Pressure,Stress^2 = Illuminance^3 · Energy^2

Luminous Intensity · Acceleration^2 = Illuminance · Velocity^4

Luminous Intensity · Frequency^4 = Illuminance · Acceleration^2

Luminous Intensity · Acceleration^2 = Illuminance · Absorbed Dose^2

Luminous Intensity^3 · Acceleration^2 = Illuminance^3 · Kinematic Viscosity^4

Luminous Intensity^3 = Illuminance^3 · Mass^2 · Specific Volume^2

Luminous Intensity^3 · Mass Density^2 = Illuminance^3 · Mass^2

Luminous Intensity · Frequency^2 = Illuminance · Velocity^2

Luminous Intensity · Velocity^2 = Illuminance · Kinematic Viscosity^2

Luminous Intensity^3 · Magnetic Field Strength^2 = Illuminance^3 · Current Density^2

Luminous Intensity · Frequency^2 = Illuminance · Absorbed Dose

Luminous Intensity · Frequency = Illuminance · Kinematic Viscosity

Luminous Intensity^3 · Pressure,Stress^2 = Illuminance^3 · Energy,Work,Heat Quantity^2

Luminous Intensity · Magnetic Flux Density = Illuminance · Magnetic Flux

Luminous Intensity · Absorbed Dose = Illuminance · Kinematic Viscosity^2

Amount of a Substance · Electric Current = Electric Charge · Catalytic Activity

Amount of a Substance · Electric Current = Catalytic Activity · Electric Charge

Amount of a Substance · Volumetric Velocity = Catalytic Activity · Length^3

Amount of a Substance^2 · Acceleration = Catalytic Activity^2 · Length

Amount of a Substance · Velocity = Catalytic Activity · Length

Amount of a Substance^2 · Absorbed Dose = Catalytic Activity^2 · Length^2

Amount of a Substance · Kinematic Viscosity = Catalytic Activity · Length^2

Amount of a Substance^5 · Acceleration^3 = Catalytic Activity^5 · Volumetric Velocity

Amount of a Substance^2 · Velocity^3 = Catalytic Activity^2 · Volumetric Velocity

Amount of a Substance^2 · Volumetric Velocity^2 = Catalytic Activity^2 · Area^3

Amount of a Substance · Volumetric Velocity = Catalytic Activity · Volume

Amount of a Substance · Volumetric Velocity · Wavenumber^3 = Catalytic Activity

Amount of a Substance^4 · Absorbed Dose^3 = Catalytic Activity^4 · Volumetric Velocity^2

Amount of a Substance · Kinematic Viscosity^3 = Catalytic Activity · Volumetric Velocity^2

Amount of a Substance · Power, Radiance Flux = Catalytic Activity · Energy

Amount of a Substance · Acceleration = Catalytic Activity · Velocity

Amount of a Substance^4 · Acceleration^2 = Catalytic Activity^4 · Area

Amount of a Substance^6 · Acceleration^3 = Catalytic Activity^6 · Volume

Amount of a Substance^2 · Acceleration · Wavenumber = Catalytic Activity^2

Amount of a Substance^2 · Acceleration^2 = Catalytic Activity^2 · Absorbed Dose

Amount of a Substance^3 · Acceleration^2 = Catalytic Activity^3 · Kinematic Viscosity

Amount of a Substance^2 · Velocity^2 = Catalytic Activity^2 · Area

Amount of a Substance^3 · Velocity^3 = Catalytic Activity^3 · Volume

Amount of a Substance · Velocity · Wavenumber = Catalytic Activity

Amount of a Substance $\cdot$ Velocity^2 = Catalytic Activity $\cdot$ Kinematic Viscosity

Amount of a Substance^2 $\cdot$ Absorbed Dose = Catalytic Activity^2 $\cdot$ Area

Amount of a Substance $\cdot$ Kinematic Viscosity = Catalytic Activity $\cdot$ Area

Amount of a Substance^6 $\cdot$ Absorbed Dose^3 = Catalytic Activity^6 $\cdot$ Volume^2

Amount of a Substance^3 $\cdot$ Kinematic Viscosity^3 = Catalytic Activity^3 $\cdot$ Volume^2

Amount of a Substance^2 $\cdot$ Wavenumber^2 $\cdot$ Absorbed Dose = Catalytic Activity^2

Amount of a Substance $\cdot$ Wavenumber^2 $\cdot$ Kinematic Viscosity = Catalytic Activity

Amount of a Substance $\cdot$ Pressure,Stress = Catalytic Activity $\cdot$ Dynamic Viscosity

Amount of a Substance $\cdot$ Power, Radiance Flux = Catalytic Activity $\cdot$ Energy,Work,Heat Quantity

Amount of a Substance $\cdot$ Electric Potential Difference = Catalytic Activity $\cdot$ Magnetic Flux

Amount of a Substance = Catalytic Activity $\cdot$ Capacitance $\cdot$ Electric Resistance

Amount of a Substance^2 = Catalytic Activity^2 $\cdot$ Capacitance $\cdot$ Inductance

Amount of a Substance $\cdot$ Electric Resistance = Catalytic Activity $\cdot$ Inductance

Electric Current $\cdot$ Electric Charge^2 $\cdot$ Volumetric Velocity^2 = Current Density^3

Electric Current^3 $\cdot$ Current Density = Electric Charge^4 $\cdot$ Acceleration^2

Electric Current $\cdot$ Current Density = Electric Charge^2 $\cdot$ Velocity^2

Electric Current $\cdot$ Current Density = Electric Charge^2 $\cdot$ Absorbed Dose

Electric Current^2 $\cdot$ Inductance = Current Density $\cdot$ Magnetic Flux Density

Electric Current $\cdot$ Magnetic Flux = Current Density $\cdot$ Magnetic Flux Density

Electric Current $\cdot$ Current Density $\cdot$ Magnetic Flux Density^2 = Force^2

Electric Current^3 $\cdot$ Magnetic Flux Density^2 = Current Density $\cdot$ Pressure,Stress^2

Electric Current^5 $\cdot$ Inductance^2 = Current Density $\cdot$ Force^2

Electric Current^7 · Inductance^2 = Current Density^3 · Pressure,Stress^2

Electric Current^3 · Magnetic Flux^2 = Current Density · Force^2

Electric Current^5 · Magnetic Flux^2 = Current Density^3 · Pressure,Stress^2

Electric Current^3 · Time^2 · Volumetric Velocity^2 = Current Density^3

Electric Current · Time^4 · Acceleration^2 = Current Density

Electric Current · Time^2 · Velocity^2 = Current Density

Electric Current · Time^2 · Absorbed Dose = Current Density

Electric Current · Time · Kinematic Viscosity = Current Density

Electric Current^5 · Volumetric Velocity^4 = Current Density^5 · Acceleration^2

Electric Current · Volumetric Velocity = Current Density · Velocity

Electric Current^3 · Volumetric Velocity^2 = Current Density^3 · Frequency^2

Electric Current^2 · Volumetric Velocity^2 = Current Density^2 · Absorbed Dose

Electric Current · Volumetric Velocity^2 = Current Density · Kinematic Viscosity^2

Electric Current · Energy^2 = Current Density · Force^2

Electric Current · Force = Current Density · Pressure,Stress

Electric Current · Energy,Work,Heat Quantity^2 = Current Density · Force^2

Electric Current^3 · Energy^2 = Current Density^3 · Pressure,Stress^2

Electric Current · Velocity^4 = Current Density · Acceleration^2

Electric Current · Acceleration^2 = Current Density · Frequency^4

Electric Current · Absorbed Dose^2 = Current Density · Acceleration^2

Electric Current^3 · Kinematic Viscosity^4 = Current Density^3 · Acceleration^2

Electric Current^3 · Mass^2 · Specific Volume^2 = Current Density^3

Electric Current^3 · Mass^2 = Current Density^3 · Mass Density^2

Electric Current · Velocity^2 = Current Density · Frequency^2

Electric Current · Kinematic Viscosity^2 = Current Density · Velocity^2

Electric Current · Absorbed Dose = Current Density · Frequency^2

Electric Current · Kinematic Viscosity = Current Density · Frequency

Electric Current^3 · Energy,Work,Heat Quantity^2 = Current Density^3 · Pressure,Stress^2

Electric Current · Kinematic Viscosity^2 = Current Density · Absorbed Dose

Electric Current^4 = Magnetic Field Strength^3 · Electric Charge · Volumetric Velocity

Electric Current^3 = Magnetic Field Strength · Electric Charge^2 · Acceleration

Electric Current^2 = Magnetic Field Strength · Electric Charge · Velocity

Electric Current^4 = Magnetic Field Strength^2 · Electric Charge^2 · Absorbed Dose

Electric Current^3 = Magnetic Field Strength^2 · Electric Charge · Kinematic Viscosity

Electric Current · Magnetic Flux Density = Magnetic Field Strength^2 · Inductance

Electric Current^2 · Magnetic Flux Density = Magnetic Field Strength^2 · Magnetic Flux

Electric Current^2 · Magnetic Flux Density = Magnetic Field Strength · Force

Electric Current^3 · Magnetic Flux Density = Magnetic Field Strength^2 · Energy

Electric Current^3 · Magnetic Flux Density = Magnetic Field Strength^2 · Energy,Work,Heat Quantity

Electric Current · Velocity = Magnetic Field Strength · Kinematic Viscosity

Electric Current^2 · Frequency^2 = Magnetic Field Strength^2 · Absorbed Dose

Electric Current^2 · Frequency = Magnetic Field Strength^2 · Kinematic Viscosity

Electric Current^3 · Pressure,Stress = Magnetic Field Strength^3 · Energy,Work,Heat Quantity

Electric Current^2 · Absorbed Dose = Magnetic Field Strength^2 · Kinematic Viscosity^2

Electric Current^2 · Inductance = Electric Charge · Electric Potential Difference

Electric Current · Magnetic Flux = Electric Charge · Electric Potential Difference

Electric Current · Capacitance · Electric Resistance = Electric Charge

Electric Current^2 · Capacitance · Inductance = Electric Charge^2

Electric Current · Capacitance · Magnetic Flux = Electric Charge^2

Electric Current · Electric Charge = Capacitance · Power, Radiance Flux

Electric Current · Inductance = Electric Charge · Electric Resistance

Electric Current · Electric Charge · Electric Resistance = Energy

Electric Current · Electric Charge · Electric Resistance = Energy,Work,Heat Quantity

Electric Current · Mass = Electric Charge^2 · Magnetic Flux Density

Electric Current^3 · Inductance = Electric Charge · Power, Radiance Flux

Electric Current^2 · Magnetic Flux = Electric Charge · Power, Radiance Flux

Electric Current · Length^3 = Electric Charge · Volumetric Velocity

Electric Current^2 · Length = Electric Charge^2 · Acceleration

Electric Current · Length = Electric Charge · Velocity

Electric Current^2 · Length^2 = Electric Charge^2 · Absorbed Dose

Electric Current · Length^2 = Electric Charge · Kinematic Viscosity

Electric Current^5 · Volumetric Velocity = Electric Charge^5 · Acceleration^3

Electric Current^2 · Volumetric Velocity = Electric Charge^2 · Velocity^3

Electric Current^2 · Area^3 = Electric Charge^2 · Volumetric Velocity^2

Electric Current · Volume = Electric Charge · Volumetric Velocity

Electric Current = Electric Charge · Volumetric Velocity · Wavenumber^3

Electric Current^4 · Volumetric Velocity^2 = Electric Charge^4 · Absorbed Dose^3

Electric Current $\cdot$ Volumetric Velocity^2 = Electric Charge $\cdot$ Kinematic Viscosity^3

Electric Current $\cdot$ Energy = Electric Charge $\cdot$ Power, Radiance Flux

Electric Current $\cdot$ Velocity = Electric Charge $\cdot$ Acceleration

Electric Current^4 $\cdot$ Area = Electric Charge^4 $\cdot$ Acceleration^2

Electric Current^6 $\cdot$ Volume = Electric Charge^6 $\cdot$ Acceleration^3

Electric Current^2 = Electric Charge^2 $\cdot$ Acceleration $\cdot$ Wavenumber

Electric Current^2 $\cdot$ Absorbed Dose = Electric Charge^2 $\cdot$ Acceleration^2

Electric Current^3 $\cdot$ Kinematic Viscosity = Electric Charge^3 $\cdot$ Acceleration^2

Electric Current^2 $\cdot$ Area = Electric Charge^2 $\cdot$ Velocity^2

Electric Current^3 $\cdot$ Volume = Electric Charge^3 $\cdot$ Velocity^3

Electric Current = Electric Charge $\cdot$ Velocity $\cdot$ Wavenumber

Electric Current $\cdot$ Kinematic Viscosity = Electric Charge $\cdot$ Velocity^2

Electric Current^2 $\cdot$ Area = Electric Charge^2 $\cdot$ Absorbed Dose

Electric Current $\cdot$ Area = Electric Charge $\cdot$ Kinematic Viscosity

Electric Current^6 $\cdot$ Volume^2 = Electric Charge^6 $\cdot$ Absorbed Dose^3

Electric Current^3 $\cdot$ Volume^2 = Electric Charge^3 $\cdot$ Kinematic Viscosity^3

Electric Current^2 = Electric Charge^2 $\cdot$ Wavenumber^2 $\cdot$ Absorbed Dose

Electric Current = Electric Charge $\cdot$ Wavenumber^2 $\cdot$ Kinematic Viscosity

Electric Current $\cdot$ Dynamic Viscosity = Electric Charge $\cdot$ Pressure,Stress

Electric Current $\cdot$ Energy,Work,Heat Quantity = Electric Charge $\cdot$ Power, Radiance Flux

Electric Current $\cdot$ Kinematic Viscosity = Electric Charge $\cdot$ Absorbed Dose

Electric Current^2 $\cdot$ Inductance = Electric Potential Difference^2 $\cdot$ Capacitance

Electric Current $\cdot$ Magnetic Flux = Electric Potential Difference^2 $\cdot$ Capacitance

Electric Current $\cdot$ Time = Electric Potential Difference $\cdot$ Capacitance

Electric Current = Electric Potential Difference $\cdot$ Capacitance $\cdot$ Frequency

Electric Current $\cdot$ Inductance = Electric Potential Difference $\cdot$ Time

Electric Current $\cdot$ Inductance $\cdot$ Frequency = Electric Potential Difference

Electric Current $\cdot$ Electric Potential Difference $\cdot$ Time = Energy

Electric Current $\cdot$ Electric Potential Difference $\cdot$ Time = Energy,Work,Heat Quantity

Electric Current $\cdot$ Electric Potential Difference = Volumetric Velocity $\cdot$ Pressure,Stress

Electric Current $\cdot$ Electric Potential Difference = Force $\cdot$ Velocity

Electric Current^2 $\cdot$ Electric Potential Difference^2 = Force^2 $\cdot$ Absorbed Dose

Electric Current $\cdot$ Electric Potential Difference = Energy $\cdot$ Frequency

Electric Current $\cdot$ Electric Potential Difference = Frequency $\cdot$ Energy,Work,Heat Quantity

Electric Current $\cdot$ Capacitance $\cdot$ Electric Resistance^2 = Magnetic Flux

Electric Current^2 $\cdot$ Capacitance $\cdot$ Electric Resistance^2 = Energy

Electric Current^2 $\cdot$ Capacitance $\cdot$ Electric Resistance^2 = Energy,Work,Heat Quantity

Electric Current = Capacitance $\cdot$ Magnetic Flux Density $\cdot$ Velocity^2

Electric Current = Capacitance $\cdot$ Magnetic Flux Density $\cdot$ Absorbed Dose

Electric Current $\cdot$ Capacitance $\cdot$ Magnetic Flux Density^3 = Dynamic Viscosity^2

Electric Current^4 $\cdot$ Inductance = Capacitance $\cdot$ Power, Radiance Flux^2

Electric Current $\cdot$ Time^2 = Capacitance $\cdot$ Magnetic Flux

Electric Current = Capacitance $\cdot$ Magnetic Flux $\cdot$ Frequency^2

Electric Current^3 $\cdot$ Magnetic Flux = Capacitance $\cdot$ Power, Radiance Flux^2

Electric Current^2 $\cdot$ Time^2 = Capacitance $\cdot$ Energy

Electric Current^2 · Time^2 = Capacitance · Energy,Work,Heat Quantity

Electric Current^2 · Time = Capacitance · Power, Radiance Flux

Electric Current^2 = Capacitance · Force · Acceleration

Electric Current^2 · Mass = Capacitance · Force^2

Electric Current^2 = Capacitance · Energy · Frequency^2

Electric Current^2 · Energy = Capacitance · Power, Radiance Flux^2

Electric Current^2 = Capacitance · Acceleration^2 · Mass

Electric Current^2 = Capacitance · Velocity^3 · Dynamic Viscosity

Electric Current^2 = Capacitance · Frequency^2 · Energy,Work,Heat Quantity

Electric Current^2 = Capacitance · Frequency · Power, Radiance Flux

Electric Current^2 · Energy,Work,Heat Quantity = Capacitance · Power, Radiance Flux^2

Electric Current^4 = Capacitance^2 · Absorbed Dose^3 · Dynamic Viscosity^2

Electric Current · Electric Resistance = Magnetic Flux Density · Kinematic Viscosity

Electric Current · Electric Resistance · Time = Magnetic Flux

Electric Current · Electric Resistance = Magnetic Flux · Frequency

Electric Current^2 · Electric Resistance · Time = Energy

Electric Current^2 · Electric Resistance · Time = Energy,Work,Heat Quantity

Electric Current^2 · Electric Resistance = Volumetric Velocity · Pressure,Stress

Electric Current^2 · Electric Resistance = Force · Velocity

Electric Current^4 · Electric Resistance^2 = Force^2 · Absorbed Dose

Electric Current^2 · Electric Resistance = Energy · Frequency

Electric Current^2 · Electric Resistance = Frequency · Energy,Work,Heat Quantity

Electric Current $\cdot$ Inductance = Magnetic Flux Density $\cdot$ Length^2

Electric Current^3 $\cdot$ Magnetic Flux Density $\cdot$ Inductance = Force^2

Electric Current $\cdot$ Inductance = Magnetic Flux Density $\cdot$ Area

Electric Current^3 $\cdot$ Inductance^3 = Magnetic Flux Density^3 $\cdot$ Volume^2

Electric Current $\cdot$ Inductance $\cdot$ Wavenumber^2 = Magnetic Flux Density

Electric Current $\cdot$ Magnetic Flux Density^3 = Inductance $\cdot$ Pressure,Stress^2

Electric Current^2 $\cdot$ Magnetic Flux Density $\cdot$ Magnetic Flux = Force^2

Electric Current^2 $\cdot$ Magnetic Flux Density^3 = Magnetic Flux $\cdot$ Pressure,Stress^2

Electric Current $\cdot$ Magnetic Flux Density $\cdot$ Time^2 = Mass

Electric Current $\cdot$ Magnetic Flux Density $\cdot$ Length = Force

Electric Current $\cdot$ Magnetic Flux Density $\cdot$ Length^2 = Energy

Electric Current $\cdot$ Magnetic Flux Density = Length $\cdot$ Pressure,Stress

Electric Current $\cdot$ Magnetic Flux Density $\cdot$ Length^2 = Energy,Work,Heat Quantity

Electric Current $\cdot$ Magnetic Flux Density $\cdot$ Energy = Force^2

Electric Current^2 $\cdot$ Magnetic Flux Density^2 $\cdot$ Area = Force^2

Electric Current^3 $\cdot$ Magnetic Flux Density^3 $\cdot$ Volume = Force^3

Electric Current $\cdot$ Magnetic Flux Density = Force $\cdot$ Wavenumber

Electric Current^2 $\cdot$ Magnetic Flux Density^2 = Force $\cdot$ Pressure,Stress

Electric Current $\cdot$ Magnetic Flux Density $\cdot$ Energy,Work,Heat Quantity = Force^2

Electric Current $\cdot$ Magnetic Flux Density $\cdot$ Area = Energy

Electric Current^3 $\cdot$ Magnetic Flux Density^3 $\cdot$ Volume^2 = Energy^3

Electric Current $\cdot$ Magnetic Flux Density = Energy $\cdot$ Wavenumber^2

Electric Current^3 $\cdot$ Magnetic Flux Density^3 = Energy $\cdot$ Pressure,Stress^2

Electric Current · Magnetic Flux Density = Mass · Frequency^2

Electric Current · Magnetic Flux Density = Velocity · Dynamic Viscosity

Electric Current^2 · Magnetic Flux Density^2 = Area · Pressure,Stress^2

Electric Current · Magnetic Flux Density · Area = Energy,Work,Heat Quantity

Electric Current^3 · Magnetic Flux Density^3 = Volume · Pressure,Stress^3

Electric Current^3 · Magnetic Flux Density^3 · Volume^2 = Energy,Work,Heat Quantity^3

Electric Current · Magnetic Flux Density · Wavenumber = Pressure,Stress

Electric Current · Magnetic Flux Density = Wavenumber^2 · Energy,Work,Heat Quantity

Electric Current^3 · Magnetic Flux Density^3 = Pressure,Stress^2 · Energy,Work,Heat Quantity

Electric Current · Magnetic Flux Density · Kinematic Viscosity = Power, Radiance Flux

Electric Current^2 · Magnetic Flux Density^2 = Absorbed Dose · Dynamic Viscosity^2

Electric Current^2 · Inductance = Time · Power, Radiance Flux

Electric Current^2 · Inductance = Length · Force

Electric Current^2 · Inductance = Length^3 · Pressure,Stress

Electric Current^2 · Inductance = Volumetric Velocity · Dynamic Viscosity

Electric Current^4 · Inductance^2 = Force^2 · Area

Electric Current^6 · Inductance^3 = Force^3 · Volume

Electric Current^2 · Inductance · Wavenumber = Force

Electric Current^4 · Inductance^2 · Pressure,Stress = Force^3

Electric Current^2 · Inductance = Mass · Velocity^2

Electric Current^2 · Inductance = Mass · Absorbed Dose

Time^2 · Force = Length · Mass

$\text{Time}^2 \cdot \text{Force} \cdot \text{Specific Volume} = \text{Length}^4$

$\text{Time} \cdot \text{Power, Radiance Flux} = \text{Length} \cdot \text{Force}$

$\text{Time} \cdot \text{Force} = \text{Length}^2 \cdot \text{Dynamic Viscosity}$

$\text{Time}^2 \cdot \text{Force} = \text{Length}^4 \cdot \text{Mass Density}$

$\text{Time}^2 \cdot \text{Energy} = \text{Length}^2 \cdot \text{Mass}$

$\text{Time}^2 \cdot \text{Energy} \cdot \text{Specific Volume} = \text{Length}^5$

$\text{Time} \cdot \text{Energy} = \text{Length}^3 \cdot \text{Dynamic Viscosity}$

$\text{Time}^2 \cdot \text{Energy} = \text{Length}^5 \cdot \text{Mass Density}$

$\text{Time}^2 \cdot \text{Length} \cdot \text{Pressure, Stress} = \text{Mass}$

$\text{Time}^2 \cdot \text{Energy, Work, Heat Quantity} = \text{Length}^2 \cdot \text{Mass}$

$\text{Time}^3 \cdot \text{Power, Radiance Flux} = \text{Length}^2 \cdot \text{Mass}$

$\text{Time} \cdot \text{Length} \cdot \text{Dynamic Viscosity} = \text{Mass}$

$\text{Time}^2 \cdot \text{Specific Volume} \cdot \text{Pressure, Stress} = \text{Length}^2$

$\text{Time}^2 \cdot \text{Specific Volume} \cdot \text{Energy, Work, Heat Quantity} = \text{Length}^5$

$\text{Time}^3 \cdot \text{Specific Volume} \cdot \text{Power, Radiance Flux} = \text{Length}^5$

$\text{Time} \cdot \text{Specific Volume} \cdot \text{Dynamic Viscosity} = \text{Length}^2$

$\text{Time} \cdot \text{Current Density} = \text{Length}^2 \cdot \text{Electric Charge}$

$\text{Time} \cdot \text{Length} \cdot \text{Magnetic Field Strength} = \text{Electric Charge}$

$\text{Time} \cdot \text{Power, Radiance Flux} = \text{Length}^3 \cdot \text{Pressure, Stress}$

$\text{Time}^2 \cdot \text{Pressure, Stress} = \text{Length}^2 \cdot \text{Mass Density}$

$\text{Time} \cdot \text{Energy, Work, Heat Quantity} = \text{Length}^3 \cdot \text{Dynamic Viscosity}$

$\text{Time}^2 \cdot \text{Energy, Work, Heat Quantity} = \text{Length}^5 \cdot \text{Mass Density}$

$\text{Time}^2 \cdot \text{Power, Radiance Flux} = \text{Length}^3 \cdot \text{Dynamic Viscosity}$

$Time^3 \cdot$ Power, Radiance Flux $= Length^5 \cdot$ Mass Density

$Time \cdot$ Electric Potential Difference $= Length^2 \cdot$ Magnetic Flux Density

$Time \cdot$ Dynamic Viscosity $= Length^2 \cdot$ Mass Density

$Time \cdot$ Volumetric Velocity $\cdot Force^3 = Energy^3$

$Time^5 \cdot Force^3 =$ Volumetric Velocity $\cdot Mass^3$

$Time^2 \cdot Force^3 \cdot$ Specific Volume$^3 =$ Volumetric Velocity^4

$Time^2 \cdot$ Volumetric Velocity$^2 \cdot$ Pressure,Stress$^3 = Force^3$

$Time \cdot$ Volumetric Velocity $\cdot Force^3 =$ Energy,Work,Heat Quantity^3

$Time^2 \cdot$ Power, Radiance Flux$^3 =$ Volumetric Velocity $\cdot Force^3$

$Time \cdot Force^3 =$ Volumetric Velocity$^2 \cdot$ Dynamic Viscosity^3

$Time^2 \cdot Force^3 =$ Volumetric Velocity$^4 \cdot$ Mass Density^3

$Time^4 \cdot Energy^3 =$ Volumetric Velocity$^2 \cdot Mass^3$

$Time \cdot Energy^3 \cdot$ Specific Volume$^3 =$ Volumetric Velocity^5

$Time \cdot$ Volumetric Velocity $\cdot$ Pressure,Stress $= Energy$

$Time \cdot Energy^3 =$ Volumetric Velocity$^5 \cdot$ Mass Density^3

$Time \cdot$ Volumetric Velocity $= Mass \cdot$ Specific Volume

$Time^7 \cdot$ Volumetric Velocity $\cdot$ Pressure,Stress$^3 = Mass^3$

$Time^4 \cdot$ Energy,Work,Heat Quantity$^3 =$ Volumetric Velocity$^2 \cdot Mass^3$

$Time^7 \cdot$ Power, Radiance Flux$^3 =$ Volumetric Velocity$^2 \cdot Mass^3$

$Time^4 \cdot$ Volumetric Velocity $\cdot$ Dynamic Viscosity$^3 = Mass^3$

$Time \cdot$ Volumetric Velocity $\cdot$ Mass Density $= Mass$

$Time^4 \cdot$ Specific Volume$^3 \cdot$ Pressure,Stress$^3 =$ Volumetric Velocity^2

$$\text{Time} \cdot \text{Specific Volume}^3 \cdot \text{Energy,Work,Heat Quantity}^3 = \text{Volumetric Velocity}^5$$

$$\text{Time}^4 \cdot \text{Specific Volume}^3 \cdot \text{Power, Radiance Flux}^3 = \text{Volumetric Velocity}^5$$

$$\text{Time} \cdot \text{Specific Volume}^3 \cdot \text{Dynamic Viscosity}^3 = \text{Volumetric Velocity}^2$$

$$\text{Time} \cdot \text{Volumetric Velocity} \cdot \text{Magnetic Field Strength} = \text{Current Density}$$

$$\text{Time} \cdot \text{Current Density}^3 = \text{Volumetric Velocity}^2 \cdot \text{Electric Charge}^3$$

$$\text{Time}^4 \cdot \text{Volumetric Velocity} \cdot \text{Magnetic Field Strength}^3 = \text{Electric Charge}^3$$

$$\text{Time} \cdot \text{Volumetric Velocity} \cdot \text{Pressure,Stress} = \text{Energy,Work,Heat Quantity}$$

$$\text{Time}^4 \cdot \text{Pressure,Stress}^3 = \text{Volumetric Velocity}^2 \cdot \text{Mass Density}^3$$

$$\text{Time} \cdot \text{Energy,Work,Heat Quantity}^3 = \text{Volumetric Velocity}^5 \cdot \text{Mass Density}^3$$

$$\text{Time} \cdot \text{Power, Radiance Flux} = \text{Volumetric Velocity} \cdot \text{Dynamic Viscosity}$$

$$\text{Time}^4 \cdot \text{Power, Radiance Flux}^3 = \text{Volumetric Velocity}^5 \cdot \text{Mass Density}^3$$

$$\text{Time} \cdot \text{Electric Potential Difference}^3 = \text{Volumetric Velocity}^2 \cdot \text{Magnetic Flux Density}^3$$

$$\text{Time}^2 \cdot \text{Volumetric Velocity}^2 \cdot \text{Magnetic Flux Density}^3 = \text{Magnetic Flux}^3$$

$$\text{Time} \cdot \text{Dynamic Viscosity}^3 = \text{Volumetric Velocity}^2 \cdot \text{Mass Density}^3$$

$$\text{Time}^2 \cdot \text{Force} \cdot \text{Acceleration} = \text{Energy}$$

$$\text{Time}^2 \cdot \text{Force}^2 = \text{Energy} \cdot \text{Mass}$$

$$\text{Time} \cdot \text{Force} \cdot \text{Velocity} = \text{Energy}$$

$$\text{Time}^2 \cdot \text{Force}^5 \cdot \text{Specific Volume} = \text{Energy}^4$$

$$\text{Time}^2 \cdot \text{Force}^2 \cdot \text{Absorbed Dose} = \text{Energy}^2$$

$$\text{Time} \cdot \text{Force}^3 = \text{Energy}^2 \cdot \text{Dynamic Viscosity}$$

$$\text{Time} \cdot \text{Force}^2 \cdot \text{Kinematic Viscosity} = \text{Energy}^2$$

$$\text{Time}^2 \cdot \text{Force}^5 = \text{Energy}^4 \cdot \text{Mass Density}$$

$$\text{Time}^6 \cdot \text{Acceleration}^4 = \text{Force} \cdot \text{Specific Volume}$$

$\text{Time}^4 \cdot \text{Acceleration}^2 \cdot \text{Pressure,Stress} = \text{Force}$

$\text{Time}^2 \cdot \text{Force} \cdot \text{Acceleration} = \text{Energy,Work,Heat Quantity}$

$\text{Time} \cdot \text{Force} \cdot \text{Acceleration} = \text{Power, Radiance Flux}$

$\text{Time}^3 \cdot \text{Acceleration}^2 \cdot \text{Dynamic Viscosity} = \text{Force}$

$\text{Time}^6 \cdot \text{Acceleration}^4 \cdot \text{Mass Density} = \text{Force}$

$\text{Time} \cdot \text{Force} = \text{Mass} \cdot \text{Velocity}$

$\text{Time}^4 \cdot \text{Force}^2 = \text{Mass}^2 \cdot \text{Area}$

$\text{Time}^6 \cdot \text{Force}^3 = \text{Mass}^3 \cdot \text{Volume}$

$\text{Time}^2 \cdot \text{Force} \cdot \text{Wavenumber} = \text{Mass}$

$\text{Time}^6 \cdot \text{Force}^3 = \text{Mass}^4 \cdot \text{Specific Volume}$

$\text{Time}^4 \cdot \text{Force} \cdot \text{Pressure,Stress} = \text{Mass}^2$

$\text{Time}^2 \cdot \text{Force}^2 = \text{Mass} \cdot \text{Energy,Work,Heat Quantity}$

$\text{Time} \cdot \text{Force}^2 = \text{Mass} \cdot \text{Power, Radiance Flux}$

$\text{Time}^2 \cdot \text{Force}^2 = \text{Mass}^2 \cdot \text{Absorbed Dose}$

$\text{Time}^3 \cdot \text{Force} \cdot \text{Dynamic Viscosity} = \text{Mass}^2$

$\text{Time}^3 \cdot \text{Force}^2 = \text{Mass}^2 \cdot \text{Kinematic Viscosity}$

$\text{Time}^6 \cdot \text{Force}^3 \cdot \text{Mass Density} = \text{Mass}^4$

$\text{Time}^2 \cdot \text{Velocity}^4 = \text{Force} \cdot \text{Specific Volume}$

$\text{Time}^2 \cdot \text{Velocity}^2 \cdot \text{Pressure,Stress} = \text{Force}$

$\text{Time} \cdot \text{Force} \cdot \text{Velocity} = \text{Energy,Work,Heat Quantity}$

$\text{Time} \cdot \text{Velocity}^2 \cdot \text{Dynamic Viscosity} = \text{Force}$

$\text{Time}^2 \cdot \text{Velocity}^4 \cdot \text{Mass Density} = \text{Force}$

$$\text{Time}^2 \cdot \text{Force} \cdot \text{Specific Volume} = \text{Area}^2$$

$$\text{Time}^2 \cdot \text{Power, Radiance Flux}^2 = \text{Force}^2 \cdot \text{Area}$$

$$\text{Time} \cdot \text{Force} = \text{Area} \cdot \text{Dynamic Viscosity}$$

$$\text{Time}^2 \cdot \text{Force} = \text{Area}^2 \cdot \text{Mass Density}$$

$$\text{Time}^6 \cdot \text{Force}^3 \cdot \text{Specific Volume}^3 = \text{Volume}^4$$

$$\text{Time}^3 \cdot \text{Power, Radiance Flux}^3 = \text{Force}^3 \cdot \text{Volume}$$

$$\text{Time}^3 \cdot \text{Force}^3 = \text{Volume}^2 \cdot \text{Dynamic Viscosity}^3$$

$$\text{Time}^6 \cdot \text{Force}^3 = \text{Volume}^4 \cdot \text{Mass Density}^3$$

$$\text{Time}^2 \cdot \text{Force} \cdot \text{Wavenumber}^4 \cdot \text{Specific Volume} = \text{A Constant}$$

$$\text{Time} \cdot \text{Wavenumber} \cdot \text{Power, Radiance Flux} = \text{Force}$$

$$\text{Time} \cdot \text{Force} \cdot \text{Wavenumber}^2 = \text{Dynamic Viscosity}$$

$$\text{Time}^2 \cdot \text{Force} \cdot \text{Wavenumber}^4 = \text{Mass Density}$$

$$\text{Time}^2 \cdot \text{Specific Volume} \cdot \text{Pressure,Stress}^2 = \text{Force}$$

$$\text{Time}^2 \cdot \text{Force}^5 \cdot \text{Specific Volume} = \text{Energy,Work,Heat Quantity}^4$$

$$\text{Time}^2 \cdot \text{Power, Radiance Flux}^4 = \text{Force}^5 \cdot \text{Specific Volume}$$

$$\text{Time}^2 \cdot \text{Absorbed Dose}^2 = \text{Force} \cdot \text{Specific Volume}$$

$$\text{Time} \cdot \text{Magnetic Field Strength} \cdot \text{Electric Potential Difference} = \text{Force}$$

$$\text{Time}^2 \cdot \text{Pressure,Stress} \cdot \text{Power, Radiance Flux}^2 = \text{Force}^3$$

$$\text{Time}^2 \cdot \text{Pressure,Stress} \cdot \text{Absorbed Dose} = \text{Force}$$

$$\text{Time} \cdot \text{Pressure,Stress} \cdot \text{Kinematic Viscosity} = \text{Force}$$

$$\text{Time}^2 \cdot \text{Pressure,Stress}^2 = \text{Force} \cdot \text{Mass Density}$$

$$\text{Time}^2 \cdot \text{Force}^2 \cdot \text{Absorbed Dose} = \text{Energy,Work,Heat Quantity}^2$$

$$\text{Time} \cdot \text{Force}^3 = \text{Energy,Work,Heat Quantity}^2 \cdot \text{Dynamic Viscosity}$$

$\text{Time} \cdot \text{Force}^2 \cdot \text{Kinematic Viscosity} = \text{Energy,Work,Heat Quantity}^2$

$\text{Time}^2 \cdot \text{Force}^5 = \text{Energy,Work,Heat Quantity}^4 \cdot \text{Mass Density}$

$\text{Time} \cdot \text{Power, Radiance Flux}^2 \cdot \text{Dynamic Viscosity} = \text{Force}^3$

$\text{Time} \cdot \text{Power, Radiance Flux}^2 = \text{Force}^2 \cdot \text{Kinematic Viscosity}$

$\text{Time}^2 \cdot \text{Power, Radiance Flux}^4 \cdot \text{Mass Density} = \text{Force}^5$

$\text{Time} \cdot \text{Absorbed Dose} \cdot \text{Dynamic Viscosity} = \text{Force}$

$\text{Time}^2 \cdot \text{Absorbed Dose}^2 \cdot \text{Mass Density} = \text{Force}$

$\text{Time}^2 \cdot \text{Acceleration}^2 \cdot \text{Mass} = \text{Energy}$

$\text{Time}^8 \cdot \text{Acceleration}^5 = \text{Energy} \cdot \text{Specific Volume}$

$\text{Time}^6 \cdot \text{Acceleration}^3 \cdot \text{Pressure,Stress} = \text{Energy}$

$\text{Time}^5 \cdot \text{Acceleration}^3 \cdot \text{Dynamic Viscosity} = \text{Energy}$

$\text{Time}^8 \cdot \text{Acceleration}^5 \cdot \text{Mass Density} = \text{Energy}$

$\text{Time}^2 \cdot \text{Energy} = \text{Mass} \cdot \text{Area}$

$\text{Time}^6 \cdot \text{Energy}^3 = \text{Mass}^3 \cdot \text{Volume}^2$

$\text{Time}^2 \cdot \text{Energy} \cdot \text{Wavenumber}^2 = \text{Mass}$

$\text{Time}^6 \cdot \text{Energy}^3 = \text{Mass}^5 \cdot \text{Specific Volume}^2$

$\text{Time}^6 \cdot \text{Energy} \cdot \text{Pressure,Stress}^2 = \text{Mass}^3$

$\text{Time}^4 \cdot \text{Energy} \cdot \text{Dynamic Viscosity}^2 = \text{Mass}^3$

$\text{Time} \cdot \text{Energy} = \text{Mass} \cdot \text{Kinematic Viscosity}$

$\text{Time}^6 \cdot \text{Energy}^3 \cdot \text{Mass Density}^2 = \text{Mass}^5$

$\text{Time}^3 \cdot \text{Velocity}^5 = \text{Energy} \cdot \text{Specific Volume}$

$\text{Time}^3 \cdot \text{Velocity}^3 \cdot \text{Pressure,Stress} = \text{Energy}$

Time^2 · Velocity^3 · Dynamic Viscosity = Energy

Time^3 · Velocity^5 · Mass Density = Energy

Time^4 · Energy^2 · Specific Volume^2 = Area^5

Time^2 · Energy^2 = Area^3 · Dynamic Viscosity^2

Time^4 · Energy^2 = Area^5 · Mass Density^2

Time^6 · Energy^3 · Specific Volume^3 = Volume^5

Time · Energy = Volume · Dynamic Viscosity

Time^6 · Energy^3 = Volume^5 · Mass Density^3

Time^2 · Energy · Wavenumber^5 · Specific Volume = A Constant

Time · Energy · Wavenumber^3 = Dynamic Viscosity

Time^2 · Energy · Wavenumber^5 = Mass Density

Time^6 · Specific Volume^3 · Pressure,Stress^5 = Energy^2

Time^6 · Absorbed Dose^5 = Energy^2 · Specific Volume^2

Time · Specific Volume^3 · Dynamic Viscosity^5 = Energy^2

Time · Kinematic Viscosity^5 = Energy^2 · Specific Volume^2

Time^6 · Pressure,Stress^2 · Absorbed Dose^3 = Energy^2

Time^3 · Pressure,Stress^2 · Kinematic Viscosity^3 = Energy^2

Time^6 · Pressure,Stress^5 = Energy^2 · Mass Density^3

Time · Energy = Electric Charge^2 · Electric Resistance

Time^2 · Energy = Electric Charge^2 · Inductance

Time · Energy = Electric Charge · Magnetic Flux

Time · Electric Potential Difference^2 = Energy · Electric Resistance

Time^2 · Electric Potential Difference^2 = Energy · Inductance

$Time^2 \cdot Energy = Capacitance \cdot Magnetic\ Flux^2$

$Time \cdot Energy \cdot Electric\ Resistance = Magnetic\ Flux^2$

$Time^4 \cdot Absorbed\ Dose^3 \cdot Dynamic\ Viscosity^2 = Energy^2$

$Time^6 \cdot Absorbed\ Dose^5 \cdot Mass\ Density^2 = Energy^2$

$Time \cdot Dynamic\ Viscosity^2 \cdot Kinematic\ Viscosity^3 = Energy^2$

$Time \cdot Dynamic\ Viscosity^5 = Energy^2 \cdot Mass\ Density^3$

$Time \cdot Kinematic\ Viscosity^5 \cdot Mass\ Density^2 = Energy^2$

$Time^6 \cdot Acceleration^3 = Mass \cdot Specific\ Volume$

$Time^4 \cdot Acceleration \cdot Pressure,Stress = Mass$

$Time^2 \cdot Acceleration^2 \cdot Mass = Energy,Work,Heat\ Quantity$

$Time \cdot Acceleration^2 \cdot Mass = Power,\ Radiance\ Flux$

$Time^3 \cdot Acceleration \cdot Dynamic\ Viscosity = Mass$

$Time^6 \cdot Acceleration^3 \cdot Mass\ Density = Mass$

$Time^2 \cdot Acceleration^2 = Specific\ Volume \cdot Pressure,Stress$

$Time^8 \cdot Acceleration^5 = Specific\ Volume \cdot Energy,Work,Heat\ Quantity$

$Time^7 \cdot Acceleration^5 = Specific\ Volume \cdot Power,\ Radiance\ Flux$

$Time^3 \cdot Acceleration^2 = Specific\ Volume \cdot Dynamic\ Viscosity$

$Time^6 \cdot Acceleration^3 \cdot Magnetic\ Field\ Strength = Current\ Density$

$Time^3 \cdot Acceleration^2 \cdot Electric\ Charge = Current\ Density$

$Time^3 \cdot Acceleration \cdot Magnetic\ Field\ Strength = Electric\ Charge$

$Time^6 \cdot Acceleration^3 \cdot Pressure,Stress = Energy,Work,Heat\ Quantity$

$Time^5 \cdot Acceleration^3 \cdot Pressure,Stress = Power,\ Radiance\ Flux$

Time^2 · Acceleration^2 · Mass Density = Pressure,Stress

Time^5 · Acceleration^3 · Dynamic Viscosity = Energy,Work,Heat Quantity

Time^8 · Acceleration^5 · Mass Density = Energy,Work,Heat Quantity

Time^4 · Acceleration^3 · Dynamic Viscosity = Power, Radiance Flux

Time^7 · Acceleration^5 · Mass Density = Power, Radiance Flux

Time^3 · Acceleration^2 · Magnetic Flux Density = Electric Potential Difference

Time^4 · Acceleration^2 · Magnetic Flux Density = Magnetic Flux

Time^3 · Acceleration^2 · Mass Density = Dynamic Viscosity

Time^3 · Velocity^3 = Mass · Specific Volume

Time^3 · Velocity · Pressure,Stress = Mass

Time · Power, Radiance Flux = Mass · Velocity^2

Time^2 · Velocity · Dynamic Viscosity = Mass

Time^3 · Velocity^3 · Mass Density = Mass

Time^4 · Area · Pressure,Stress^2 = Mass^2

Time^2 · Energy,Work,Heat Quantity = Mass · Area

Time^3 · Power, Radiance Flux = Mass · Area

Time^2 · Area · Dynamic Viscosity^2 = Mass^2

Time^6 · Volume · Pressure,Stress^3 = Mass^3

Time^6 · Energy,Work,Heat Quantity^3 = Mass^3 · Volume^2

Time^9 · Power, Radiance Flux^3 = Mass^3 · Volume^2

Time^3 · Volume · Dynamic Viscosity^3 = Mass^3

Time^2 · Pressure,Stress = Mass · Wavenumber

Time^2 · Wavenumber^2 · Energy,Work,Heat Quantity = Mass

$Time^3 \cdot Wavenumber^2 \cdot Power, Radiance\ Flux = Mass$

$Time \cdot Dynamic\ Viscosity = Mass \cdot Wavenumber$

$Time^6 \cdot Specific\ Volume \cdot Pressure, Stress^3 = Mass^2$

$Time^6 \cdot Energy, Work, Heat\ Quantity^3 = Mass^5 \cdot Specific\ Volume^2$

$Time^9 \cdot Power, Radiance\ Flux^3 = Mass^5 \cdot Specific\ Volume^2$

$Time^6 \cdot Absorbed\ Dose^3 = Mass^2 \cdot Specific\ Volume^2$

$Time^3 \cdot Specific\ Volume \cdot Dynamic\ Viscosity^3 = Mass^2$

$Time^3 \cdot Kinematic\ Viscosity^3 = Mass^2 \cdot Specific\ Volume^2$

$Time^4 \cdot Magnetic\ Field\ Strength^2 = Mass \cdot Capacitance$

$Time^3 \cdot Magnetic\ Field\ Strength^2 \cdot Electric\ Resistance = Mass$

$Time^2 \cdot Magnetic\ Field\ Strength^2 \cdot Inductance = Mass$

$Time^6 \cdot Pressure, Stress^2 \cdot Energy, Work, Heat\ Quantity = Mass^3$

$Time^7 \cdot Pressure, Stress^2 \cdot Power, Radiance\ Flux = Mass^3$

$Time^6 \cdot Pressure, Stress^2 \cdot Absorbed\ Dose = Mass^2$

$Time^5 \cdot Pressure, Stress^2 \cdot Kinematic\ Viscosity = Mass^2$

$Time^6 \cdot Pressure, Stress^3 = Mass^2 \cdot Mass\ Density$

$Time^4 \cdot Energy, Work, Heat\ Quantity \cdot Dynamic\ Viscosity^2 = Mass^3$

$Time \cdot Energy, Work, Heat\ Quantity = Mass \cdot Kinematic\ Viscosity$

$Time^6 \cdot Energy, Work, Heat\ Quantity^3 \cdot Mass\ Density^2 = Mass^5$

$Time \cdot Power, Radiance\ Flux = Mass \cdot Absorbed\ Dose$

$Time^5 \cdot Power, Radiance\ Flux \cdot Dynamic\ Viscosity^2 = Mass^3$

$Time^2 \cdot Power, Radiance\ Flux = Mass \cdot Kinematic\ Viscosity$

$Time^9 \cdot Power, Radiance\ Flux^3 \cdot Mass\ Density^2 = Mass^5$

$Time \cdot Electric\ Charge \cdot Magnetic\ Flux\ Density = Mass$

$Time^4 \cdot Absorbed\ Dose \cdot Dynamic\ Viscosity^2 = Mass^2$

$Time^6 \cdot Absorbed\ Dose^3 \cdot Mass\ Density^2 = Mass^2$

$Time^3 \cdot Dynamic\ Viscosity^2 \cdot Kinematic\ Viscosity = Mass^2$

$Time^3 \cdot Dynamic\ Viscosity^3 = Mass^2 \cdot Mass\ Density$

$Time^3 \cdot Kinematic\ Viscosity^3 \cdot Mass\ Density^2 = Mass^2$

$Time^3 \cdot Velocity^5 = Specific\ Volume \cdot Energy, Work, Heat\ Quantity$

$Time^2 \cdot Velocity^5 = Specific\ Volume \cdot Power, Radiance\ Flux$

$Time \cdot Velocity^2 = Specific\ Volume \cdot Dynamic\ Viscosity$

$Time^3 \cdot Velocity^3 \cdot Magnetic\ Field\ Strength = Current\ Density$

$Time \cdot Velocity^2 \cdot Electric\ Charge = Current\ Density$

$Length^5 \cdot Force = Volumetric\ Velocity^2 \cdot Mass$

$Length^2 \cdot Force \cdot Specific\ Volume = Volumetric\ Velocity^2$

$Length^2 \cdot Power, Radiance\ Flux = Volumetric\ Velocity \cdot Force$

$Length \cdot Force = Volumetric\ Velocity \cdot Dynamic\ Viscosity$

$Length^2 \cdot Force = Volumetric\ Velocity^2 \cdot Mass\ Density$

$Length^4 \cdot Energy = Volumetric\ Velocity^2 \cdot Mass$

$Length \cdot Energy \cdot Specific\ Volume = Volumetric\ Velocity^2$

$Length^3 \cdot Power, Radiance\ Flux = Volumetric\ Velocity \cdot Energy$

$Length \cdot Energy = Volumetric\ Velocity^2 \cdot Mass\ Density$

$Length^7 \cdot Pressure, Stress = Volumetric\ Velocity^2 \cdot Mass$

$Length^4 \cdot Energy, Work, Heat\ Quantity = Volumetric\ Velocity^2 \cdot Mass$

Length^7 · Power, Radiance Flux = Volumetric Velocity^3 · Mass

Length^4 · Dynamic Viscosity = Volumetric Velocity · Mass

Length^4 · Specific Volume · Pressure,Stress = Volumetric Velocity^2

Length · Specific Volume · Energy,Work,Heat Quantity = Volumetric Velocity^2

Length^4 · Specific Volume · Power, Radiance Flux = Volumetric Velocity^3

Length · Specific Volume · Dynamic Viscosity = Volumetric Velocity

Length · Current Density = Volumetric Velocity · Electric Charge

Length^4 · Magnetic Field Strength = Volumetric Velocity · Electric Charge

Length^3 · Pressure,Stress = Volumetric Velocity · Dynamic Viscosity

Length^4 · Pressure,Stress = Volumetric Velocity^2 · Mass Density

Length^3 · Power, Radiance Flux = Volumetric Velocity · Energy,Work,Heat Quantity

Length · Energy,Work,Heat Quantity = Volumetric Velocity^2 · Mass Density

Length^3 · Power, Radiance Flux = Volumetric Velocity^2 · Dynamic Viscosity

Length^4 · Power, Radiance Flux = Volumetric Velocity^3 · Mass Density

Length · Electric Potential Difference = Volumetric Velocity · Magnetic Flux Density

Length^3 · Electric Potential Difference = Volumetric Velocity · Magnetic Flux

Length^3 = Volumetric Velocity · Capacitance · Electric Resistance

Length^6 = Volumetric Velocity^2 · Capacitance · Inductance

Length^3 · Electric Resistance = Volumetric Velocity · Inductance

Length · Dynamic Viscosity = Volumetric Velocity · Mass Density

Length^3 · Acceleration = Force · Specific Volume

Length · Force^2 · Acceleration = Power, Radiance Flux^2

Length^3 · Acceleration · Dynamic Viscosity^2 = Force^2

Length^3 · Acceleration · Mass Density = Force

Length · Force = Mass · Velocity^2

Length · Mass · Frequency^2 = Force

Length · Force^3 = Mass · Power, Radiance Flux^2

Length · Force = Mass · Absorbed Dose

Length^3 · Dynamic Viscosity^2 = Force · Mass

Length^3 · Force = Mass · Kinematic Viscosity^2

Length^2 · Velocity^2 = Force · Specific Volume

Length · Velocity · Dynamic Viscosity = Force

Length^2 · Velocity^2 · Mass Density = Force

Length^4 · Frequency^2 = Force · Specific Volume

Length^2 · Power, Radiance Flux^2 = Force^3 · Specific Volume

Length^2 · Absorbed Dose = Force · Specific Volume

Length · Force = Current Density · Magnetic Flux Density

Length^5 · Force = Current Density^2 · Inductance

Length^3 · Force = Current Density · Magnetic Flux

Length^2 · Magnetic Field Strength · Magnetic Flux Density = Force

Length · Magnetic Field Strength^2 · Inductance = Force

Length · Force · Frequency = Power, Radiance Flux

Length^2 · Frequency · Dynamic Viscosity = Force

Length^4 · Frequency^2 · Mass Density = Force

Length · Power, Radiance Flux · Dynamic Viscosity = Force^2

Length $\cdot$ Power, Radiance Flux = Force $\cdot$ Kinematic Viscosity

Length^2 $\cdot$ Power, Radiance Flux^2 $\cdot$ Mass Density = Force^3

Length $\cdot$ Force = Electric Charge $\cdot$ Electric Potential Difference

Length $\cdot$ Force $\cdot$ Capacitance = Electric Charge^2

Length $\cdot$ Force = Electric Potential Difference^2 $\cdot$ Capacitance

Length^3 $\cdot$ Magnetic Flux Density^2 = Force $\cdot$ Inductance

Length $\cdot$ Force $\cdot$ Inductance = Magnetic Flux^2

Length^2 $\cdot$ Absorbed Dose $\cdot$ Dynamic Viscosity^2 = Force^2

Length^2 $\cdot$ Absorbed Dose $\cdot$ Mass Density = Force

Length $\cdot$ Acceleration $\cdot$ Mass = Energy

Length^4 $\cdot$ Acceleration = Energy $\cdot$ Specific Volume

Length $\cdot$ Power, Radiance Flux^2 = Energy^2 $\cdot$ Acceleration

Length^5 $\cdot$ Acceleration $\cdot$ Dynamic Viscosity^2 = Energy^2

Length^4 $\cdot$ Acceleration $\cdot$ Mass Density = Energy

Length^2 $\cdot$ Mass $\cdot$ Frequency^2 = Energy

Length^2 $\cdot$ Mass $\cdot$ Power, Radiance Flux^2 = Energy^3

Length^4 $\cdot$ Dynamic Viscosity^2 = Energy $\cdot$ Mass

Length^2 $\cdot$ Energy = Mass $\cdot$ Kinematic Viscosity^2

Length^3 $\cdot$ Velocity^2 = Energy $\cdot$ Specific Volume

Length $\cdot$ Power, Radiance Flux = Energy $\cdot$ Velocity

Length^2 $\cdot$ Velocity $\cdot$ Dynamic Viscosity = Energy

Length^3 $\cdot$ Velocity^2 $\cdot$ Mass Density = Energy

Length^5 · Frequency^2 = Energy · Specific Volume

Length^5 · Power, Radiance Flux^2 = Energy^3 · Specific Volume

Length^3 · Absorbed Dose = Energy · Specific Volume

Length · Specific Volume · Dynamic Viscosity^2 = Energy

Length · Kinematic Viscosity^2 = Energy · Specific Volume

Length^4 · Energy = Current Density^2 · Inductance

Length^2 · Energy = Current Density · Magnetic Flux

Length^3 · Magnetic Field Strength · Magnetic Flux Density = Energy

Length^2 · Magnetic Field Strength^2 · Inductance = Energy

Length · Magnetic Field Strength · Magnetic Flux = Energy

Length^3 · Frequency · Dynamic Viscosity = Energy

Length^5 · Frequency^2 · Mass Density = Energy

Length^2 · Power, Radiance Flux^2 = Energy^2 · Absorbed Dose

Length^3 · Power, Radiance Flux · Dynamic Viscosity = Energy^2

Length^2 · Power, Radiance Flux = Energy · Kinematic Viscosity

Length^5 · Power, Radiance Flux^2 · Mass Density = Energy^3

Length^4 · Magnetic Flux Density^2 = Energy · Inductance

Length^4 · Absorbed Dose · Dynamic Viscosity^2 = Energy^2

Length^3 · Absorbed Dose · Mass Density = Energy

Length · Dynamic Viscosity · Kinematic Viscosity = Energy

Length · Dynamic Viscosity^2 = Energy · Mass Density

Length · Kinematic Viscosity^2 · Mass Density = Energy

Length^2 · Pressure,Stress = Acceleration · Mass

Length $\cdot$ Acceleration $\cdot$ Mass = Energy,Work,Heat Quantity

Length $\cdot$ Acceleration^3 $\cdot$ Mass^2 = Power, Radiance Flux^2

Length^3 $\cdot$ Dynamic Viscosity^2 = Acceleration $\cdot$ Mass^2

Length $\cdot$ Acceleration = Specific Volume $\cdot$ Pressure,Stress

Length^4 $\cdot$ Acceleration = Specific Volume $\cdot$ Energy,Work,Heat Quantity

Length^7 $\cdot$ Acceleration^3 = Specific Volume^2 $\cdot$ Power, Radiance Flux^2

Length^3 $\cdot$ Acceleration = Specific Volume^2 $\cdot$ Dynamic Viscosity^2

Length^3 $\cdot$ Acceleration $\cdot$ Electric Charge^2 = Current Density^2

Length^3 $\cdot$ Magnetic Field Strength^2 = Acceleration $\cdot$ Electric Charge^2

Length^5 $\cdot$ Acceleration $\cdot$ Pressure,Stress^2 = Power, Radiance Flux^2

Length $\cdot$ Pressure,Stress^2 = Acceleration $\cdot$ Dynamic Viscosity^2

Length $\cdot$ Acceleration $\cdot$ Mass Density = Pressure,Stress

Length $\cdot$ Power, Radiance Flux^2 = Acceleration $\cdot$ Energy,Work,Heat Quantity^2

Length^5 $\cdot$ Acceleration $\cdot$ Dynamic Viscosity^2 = Energy,Work,Heat Quantity^2

Length^4 $\cdot$ Acceleration $\cdot$ Mass Density = Energy,Work,Heat Quantity

Length^2 $\cdot$ Acceleration $\cdot$ Dynamic Viscosity = Power, Radiance Flux

Length^7 $\cdot$ Acceleration^3 $\cdot$ Mass Density^2 = Power, Radiance Flux^2

Length^3 $\cdot$ Acceleration $\cdot$ Magnetic Flux Density^2 = Electric Potential Difference^2

Length $\cdot$ Electric Potential Difference^2 = Acceleration $\cdot$ Magnetic Flux^2

Length = Acceleration $\cdot$ Capacitance^2 $\cdot$ Electric Resistance^2

Length = Acceleration $\cdot$ Capacitance $\cdot$ Inductance

Length $\cdot$ Electric Resistance^2 = Acceleration $\cdot$ Inductance^2

Length^3 · Acceleration · Mass Density^2 = Dynamic Viscosity^2

Length^3 · Pressure,Stress = Mass · Velocity^2

Length · Power, Radiance Flux = Mass · Velocity^3

Length^2 · Dynamic Viscosity = Mass · Velocity

Length · Pressure,Stress = Mass · Frequency^2

Length^2 · Mass · Frequency^2 = Energy,Work,Heat Quantity

Length^2 · Mass · Frequency^3 = Power, Radiance Flux

Length · Dynamic Viscosity = Mass · Frequency

Length^7 · Pressure,Stress^3 = Mass · Power, Radiance Flux^2

Amount of a Substance · Absorbed Dose = Catalytic Activity · Kinematic Viscosity

Volumetric Velocity^2 · Force^5 = Energy^5 · Acceleration

Volumetric Velocity^2 · Force^4 · Mass = Energy^5

Volumetric Velocity · Force^2 = Energy^2 · Velocity

Volumetric Velocity^2 · Force = Energy^2 · Specific Volume

Volumetric Velocity · Force^3 = Energy^3 · Frequency

Volumetric Velocity · Force^3 = Energy^2 · Power, Radiance Flux

Volumetric Velocity^2 · Force^4 = Energy^4 · Absorbed Dose

Volumetric Velocity · Force = Energy · Kinematic Viscosity

Volumetric Velocity^2 · Force · Mass Density = Energy^2

Volumetric Velocity^6 · Acceleration^2 = Force^5 · Specific Volume^5

Volumetric Velocity^4 · Pressure,Stress^5 = Force^5 · Acceleration^2

Volumetric Velocity^2 · Force^5 = Acceleration · Energy,Work,Heat Quantity^5

Volumetric Velocity · Force^5 · Acceleration^2 = Power, Radiance Flux^5

Volumetric Velocity^3 · Acceleration · Dynamic Viscosity^5 = Force^5

Volumetric Velocity^6 · Acceleration^2 · Mass Density^5 = Force^5

Volumetric Velocity · Force^2 = Mass^2 · Velocity^5

Volumetric Velocity^4 · Mass^2 = Force^2 · Area^5

Volumetric Velocity^6 · Mass^3 = Force^3 · Volume^5

Volumetric Velocity^2 · Mass · Wavenumber^5 = Force

Volumetric Velocity^6 = Force^3 · Mass^2 · Specific Volume^5

Volumetric Velocity · Mass^3 · Frequency^5 = Force^3

Volumetric Velocity^4 · Mass^2 · Pressure,Stress^5 = Force^7

Volumetric Velocity^2 · Force^4 · Mass = Energy,Work,Heat Quantity^5

Volumetric Velocity · Force^7 = Mass^2 · Power, Radiance Flux^5

Volumetric Velocity^2 · Force^4 = Mass^4 · Absorbed Dose^5

Volumetric Velocity^3 · Dynamic Viscosity^5 = Force^4 · Mass

Volumetric Velocity^3 · Force = Mass · Kinematic Viscosity^5

Volumetric Velocity^6 · Mass Density^5 = Force^3 · Mass^2

Volumetric Velocity · Velocity = Force · Specific Volume

Volumetric Velocity · Pressure,Stress = Force · Velocity

Volumetric Velocity · Force^2 = Velocity · Energy,Work,Heat Quantity^2

Volumetric Velocity · Velocity · Dynamic Viscosity^2 = Force^2

Volumetric Velocity · Velocity · Mass Density = Force

Volumetric Velocity^2 = Force · Area · Specific Volume

Volumetric Velocity · Force = Area · Power, Radiance Flux

Volumetric Velocity^2 · Dynamic Viscosity^2 = Force^2 · Area

Volumetric Velocity^2 · Mass Density = Force · Area

Volumetric Velocity^6 = Force^3 · Volume^2 · Specific Volume^3

Volumetric Velocity^3 · Force^3 = Volume^2 · Power, Radiance Flux^3

Volumetric Velocity^3 · Dynamic Viscosity^3 = Force^3 · Volume

Volumetric Velocity^6 · Mass Density^3 = Force^3 · Volume^2

Volumetric Velocity^2 · Wavenumber^2 = Force · Specific Volume

Volumetric Velocity · Force · Wavenumber^2 = Power, Radiance Flux

Volumetric Velocity · Wavenumber · Dynamic Viscosity = Force

Volumetric Velocity^2 · Wavenumber^2 · Mass Density = Force

Volumetric Velocity^4 · Frequency^2 = Force^3 · Specific Volume^3

Volumetric Velocity^2 · Pressure,Stress = Force^2 · Specific Volume

Volumetric Velocity^2 · Force = Specific Volume · Energy,Work,Heat Quantity^2

Volumetric Velocity · Power, Radiance Flux = Force^2 · Specific Volume

Volumetric Velocity^2 · Absorbed Dose = Force^2 · Specific Volume^2

Volumetric Velocity · Force = Current Density · Electric Potential Difference

Volumetric Velocity^2 · Pressure,Stress^3 = Force^3 · Frequency^2

Volumetric Velocity · Force^3 = Frequency · Energy,Work,Heat Quantity^3

Volumetric Velocity · Force^3 · Frequency^2 = Power, Radiance Flux^3

Volumetric Velocity^2 · Frequency · Dynamic Viscosity^3 = Force^3

Volumetric Velocity^4 · Frequency^2 · Mass Density^3 = Force^3

Volumetric Velocity^2 · Pressure,Stress^2 = Force^2 · Absorbed Dose

Volumetric Velocity^2 · Pressure,Stress · Dynamic Viscosity^2 = Force^3

Volumetric Velocity$^2 \cdot$ Pressure,Stress = Force $\cdot$ Kinematic Viscosity^2

Volumetric Velocity$^2 \cdot$ Pressure,Stress $\cdot$ Mass Density = Force^2

Volumetric Velocity $\cdot$ Force^3 = Energy,Work,Heat Quantity$^2 \cdot$ Power, Radiance Flux

Volumetric Velocity$^2 \cdot$ Force^4 = Energy,Work,Heat Quantity$^4 \cdot$ Absorbed Dose

Volumetric Velocity $\cdot$ Force = Energy,Work,Heat Quantity $\cdot$ Kinematic Viscosity

Volumetric Velocity$^2 \cdot$ Force $\cdot$ Mass Density = Energy,Work,Heat Quantity^2

Volumetric Velocity $\cdot$ Power, Radiance Flux $\cdot$ Dynamic Viscosity^2 = Force^3

Volumetric Velocity $\cdot$ Power, Radiance Flux = Force $\cdot$ Kinematic Viscosity^2

Volumetric Velocity $\cdot$ Power, Radiance Flux $\cdot$ Mass Density = Force^2

Volumetric Velocity $\cdot$ Magnetic Flux Density^2 = Force $\cdot$ Electric Resistance

Volumetric Velocity$^2 \cdot$ Absorbed Dose $\cdot$ Dynamic Viscosity^4 = Force^4

Volumetric Velocity$^2 \cdot$ Absorbed Dose $\cdot$ Mass Density^2 = Force^2

Volumetric Velocity$^2 \cdot$ Acceleration$^4 \cdot$ Mass^5 = Energy^5

Volumetric Velocity$^8 \cdot$ Acceleration = Energy$^5 \cdot$ Specific Volume^5

Volumetric Velocity$^6 \cdot$ Pressure,Stress^5 = Energy$^5 \cdot$ Acceleration^3

Volumetric Velocity $\cdot$ Power, Radiance Flux^5 = Energy$^5 \cdot$ Acceleration^3

Volumetric Velocity$^8 \cdot$ Acceleration $\cdot$ Mass Density^5 = Energy^5

Volumetric Velocity$^2 \cdot$ Mass = Energy $\cdot$ Area^2

Volumetric Velocity$^6 \cdot$ Mass^3 = Energy$^3 \cdot$ Volume^4

Volumetric Velocity$^2 \cdot$ Mass $\cdot$ Wavenumber^4 = Energy

Volumetric Velocity^6 = Energy$^3 \cdot$ Mass $\cdot$ Specific Volume^4

Volumetric Velocity$^2 \cdot$ Mass$^3 \cdot$ Frequency^4 = Energy^3

Volumetric Velocity$^6 \cdot$ Mass$^3 \cdot$ Pressure,Stress^4 = Energy^7

Volumetric Velocity$^2 \cdot$ Mass$^3 \cdot$ Power, Radiance Flux^4 = Energy^7

Volumetric Velocity$^2 \cdot$ Energy = Mass $\cdot$ Kinematic Viscosity^4

Volumetric Velocity$^6 \cdot$ Mass Density^4 = Energy$^3 \cdot$ Mass

Volumetric Velocity$^3 \cdot$ Velocity = Energy$^2 \cdot$ Specific Volume^2

Volumetric Velocity$^3 \cdot$ Pressure,Stress^2 = Energy$^2 \cdot$ Velocity^3

Volumetric Velocity $\cdot$ Power, Radiance Flux^2 = Energy$^2 \cdot$ Velocity^3

Volumetric Velocity$^3 \cdot$ Velocity $\cdot$ Mass Density^2 = Energy^2

Volumetric Velocity^4 = Energy$^2 \cdot$ Area $\cdot$ Specific Volume^2

Volumetric Velocity$^2 \cdot$ Energy^2 = Area$^3 \cdot$ Power, Radiance Flux^2

Volumetric Velocity$^4 \cdot$ Mass Density^2 = Energy$^2 \cdot$ Area

Volumetric Velocity^6 = Energy$^3 \cdot$ Volume $\cdot$ Specific Volume^3

Volumetric Velocity $\cdot$ Energy = Volume $\cdot$ Power, Radiance Flux

Volumetric Velocity$^6 \cdot$ Mass Density^3 = Energy$^3 \cdot$ Volume

Volumetric Velocity$^2 \cdot$ Wavenumber = Energy $\cdot$ Specific Volume

Volumetric Velocity $\cdot$ Energy $\cdot$ Wavenumber^3 = Power, Radiance Flux

Volumetric Velocity$^2 \cdot$ Wavenumber $\cdot$ Mass Density = Energy

Volumetric Velocity$^5 \cdot$ Frequency = Energy$^3 \cdot$ Specific Volume^3

Volumetric Velocity$^6 \cdot$ Pressure,Stress = Energy$^4 \cdot$ Specific Volume^3

Volumetric Velocity$^5 \cdot$ Power, Radiance Flux = Energy$^4 \cdot$ Specific Volume^3

Volumetric Velocity$^6 \cdot$ Absorbed Dose = Energy$^4 \cdot$ Specific Volume^4

Volumetric Velocity $\cdot$ Kinematic Viscosity = Energy $\cdot$ Specific Volume

Volumetric Velocity $\cdot$ Pressure,Stress = Energy $\cdot$ Frequency

Volumetric Velocity^5 · Frequency · Mass Density^3 = Energy^3

Volumetric Velocity^6 · Pressure,Stress^4 = Energy^4 · Absorbed Dose^3

Volumetric Velocity^3 · Pressure,Stress = Energy · Kinematic Viscosity^3

Volumetric Velocity^6 · Pressure,Stress · Mass Density^3 = Energy^4

Volumetric Velocity^2 · Power, Radiance Flux^4 = Energy^4 · Absorbed Dose^3

Volumetric Velocity^2 · Power, Radiance Flux = Energy · Kinematic Viscosity^3

Volumetric Velocity^5 · Power, Radiance Flux · Mass Density^3 = Energy^4

Volumetric Velocity^6 · Absorbed Dose · Mass Density^4 = Energy^4

Volumetric Velocity · Kinematic Viscosity · Mass Density = Energy

Volumetric Velocity^6 = Acceleration^3 · Mass^5 · Specific Volume^5

Volumetric Velocity^4 · Pressure,Stress^5 = Acceleration^7 · Mass^5

Volumetric Velocity^2 · Acceleration^4 · Mass^5 = Energy,Work,Heat Quantity^5

Volumetric Velocity · Acceleration^7 · Mass^5 = Power, Radiance Flux^5

Volumetric Velocity^3 · Dynamic Viscosity^5 = Acceleration^4 · Mass^5

Volumetric Velocity^6 · Mass Density^5 = Acceleration^3 · Mass^5

Volumetric Velocity^2 · Acceleration^4 = Specific Volume^5 · Pressure,Stress^5

Volumetric Velocity^8 · Acceleration = Specific Volume^5 · Energy,Work,Heat Quantity^5

Volumetric Velocity^7 · Acceleration^4 = Specific Volume^5 · Power, Radiance Flux^5

Volumetric Velocity^3 · Acceleration = Specific Volume^5 · Dynamic Viscosity^5

Volumetric Velocity^6 · Magnetic Field Strength^5 = Acceleration^3 · Current Density^5

Volumetric Velocity^3 · Acceleration · Electric Charge^5 = Current Density^5

Volumetric Velocity^3 · Magnetic Field Strength^5 = Acceleration^4 · Electric Charge^5

Volumetric Velocity^6 · Pressure,Stress^5 = Acceleration^3 · Energy,Work,Heat Quantity^5

Volumetric Velocity · Pressure,Stress^5 = Acceleration^3 · Dynamic Viscosity^5

Volumetric Velocity^2 · Acceleration^4 · Mass Density^5 = Pressure,Stress^5

Volumetric Velocity · Power, Radiance Flux^5 = Acceleration^3 · Energy,Work,Heat Quantity^5

Volumetric Velocity^8 · Acceleration · Mass Density^5 = Energy,Work,Heat Quantity^5

Volumetric Velocity^4 · Acceleration^3 · Dynamic Viscosity^5 = Power, Radiance Flux^5

Volumetric Velocity^7 · Acceleration^4 · Mass Density^5 = Power, Radiance Flux^5

Volumetric Velocity^3 · Acceleration · Magnetic Flux Density^5 = Electric Potential Difference^5

Volumetric Velocity · Electric Potential Difference^5 = Acceleration^3 · Magnetic Flux^5

Volumetric Velocity = Acceleration^3 · Capacitance^5 · Electric Resistance^5

Volumetric Velocity^2 = Acceleration^6 · Capacitance^5 · Inductance^5

Volumetric Velocity · Electric Resistance^5 = Acceleration^3 · Inductance^5

Volumetric Velocity^4 · Magnetic Flux Density^5 = Acceleration^2 · Magnetic Flux^5

Volumetric Velocity^3 · Acceleration · Mass Density^5 = Dynamic Viscosity^5

Volumetric Velocity^3 = Mass^2 · Velocity^3 · Specific Volume^2

Volumetric Velocity^3 · Pressure,Stress^2 = Mass^2 · Velocity^7

Volumetric Velocity · Power, Radiance Flux^2 = Mass^2 · Velocity^7

Volumetric Velocity · Dynamic Viscosity = Mass · Velocity^2

Volumetric Velocity^3 · Mass Density^2 = Mass^2 · Velocity^3

Volumetric Velocity^4 · Mass^2 = Area^7 · Pressure,Stress^2

Volumetric Velocity^2 · Mass = Area^2 · Energy,Work,Heat Quantity

Volumetric Velocity^6 · Mass^2 = Area^7 · Power, Radiance Flux^2

Volumetric Velocity · Mass = Area^2 · Dynamic Viscosity

Volumetric Velocity^6 · Mass^3 = Volume^7 · Pressure,Stress^3

Volumetric Velocity^6 · Mass^3 = Volume^4 · Energy,Work,Heat Quantity^3

Volumetric Velocity^9 · Mass^3 = Volume^7 · Power, Radiance Flux^3

Volumetric Velocity^3 · Mass^3 = Volume^4 · Dynamic Viscosity^3

Volumetric Velocity^2 · Mass · Wavenumber^7 = Pressure,Stress

Volumetric Velocity^2 · Mass · Wavenumber^4 = Energy,Work,Heat Quantity

Volumetric Velocity^3 · Mass · Wavenumber^7 = Power, Radiance Flux

Volumetric Velocity · Mass · Wavenumber^4 = Dynamic Viscosity

Volumetric Velocity = Mass · Specific Volume · Frequency

Volumetric Velocity^6 = Mass^4 · Specific Volume^7 · Pressure,Stress^3

Volumetric Velocity^6 = Mass · Specific Volume^4 · Energy,Work,Heat Quantity^3

Volumetric Velocity^9 = Mass^4 · Specific Volume^7 · Power, Radiance Flux^3

Volumetric Velocity^6 = Mass^4 · Specific Volume^4 · Absorbed Dose^3

Volumetric Velocity^3 = Mass · Specific Volume^4 · Dynamic Viscosity^3

Volumetric Velocity^3 = Mass · Specific Volume · Kinematic Viscosity^3

Volumetric Velocity^2 · Mass = Current Density^2 · Inductance

Volumetric Velocity · Pressure,Stress^3 = Mass^3 · Frequency^7

Volumetric Velocity^2 · Mass^3 · Frequency^4 = Energy,Work,Heat Quantity^3

Volumetric Velocity^2 · Mass^3 · Frequency^7 = Power, Radiance Flux^3

Volumetric Velocity · Dynamic Viscosity^3 = Mass^3 · Frequency^4

Volumetric Velocity · Mass Density = Mass · Frequency

Volumetric Velocity^6 · Mass^3 · Pressure,Stress^4 = Energy,Work,Heat Quantity^7

Volumetric Velocity^6 · Pressure,Stress^4 = Mass^4 · Absorbed Dose^7

Volumetric Velocity · Dynamic Viscosity^7 = Mass^3 · Pressure,Stress^4

Volumetric Velocity^5 · Pressure,Stress = Mass · Kinematic Viscosity^7

Volumetric Velocity^6 · Mass Density^7 = Mass^4 · Pressure,Stress^3

Volumetric Velocity^2 · Mass^3 · Power, Radiance Flux^4 = Energy,Work,Heat Quantity^7

Volumetric Velocity^2 · Energy,Work,Heat Quantity = Mass · Kinematic Viscosity^4

Volumetric Velocity^6 · Mass Density^4 = Mass · Energy,Work,Heat Quantity^3

Volumetric Velocity^2 · Power, Radiance Flux^4 = Mass^4 · Absorbed Dose^7

Volumetric Velocity^5 · Dynamic Viscosity^7 = Mass^3 · Power, Radiance Flux^4

Volumetric Velocity^4 · Power, Radiance Flux = Mass · Kinematic Viscosity^7

Volumetric Velocity^9 · Mass Density^7 = Mass^4 · Power, Radiance Flux^3

Volumetric Velocity · Dynamic Viscosity = Mass · Absorbed Dose

Volumetric Velocity^6 · Mass Density^4 = Mass^4 · Absorbed Dose^3

Volumetric Velocity^3 · Dynamic Viscosity = Mass · Kinematic Viscosity^4

Volumetric Velocity^3 · Mass Density^4 = Mass · Dynamic Viscosity^3

Volumetric Velocity^3 · Mass Density = Mass · Kinematic Viscosity^3

Volumetric Velocity^3 · Velocity = Specific Volume^2 · Energy,Work,Heat Quantity^2

Volumetric Velocity · Velocity^2 = Specific Volume · Power, Radiance Flux

Volumetric Velocity · Velocity = Specific Volume^2 · Dynamic Viscosity^2

Volumetric Velocity^3 · Magnetic Field Strength^2 = Velocity^3 · Current Density^2

Volumetric Velocity · Velocity · Electric Charge^2 = Current Density^2

Volumetric Velocity · Magnetic Field Strength = Velocity^2 · Electric Charge

Volumetric Velocity^3 · Pressure,Stress^2 = Velocity^3 · Energy,Work,Heat Quantity^2

Volumetric Velocity $\cdot$ Pressure,Stress^2 = Velocity^3 $\cdot$ Dynamic Viscosity^2

Volumetric Velocity $\cdot$ Power, Radiance Flux^2 = Velocity^3 $\cdot$ Energy,Work,Heat Quantity^2

Volumetric Velocity^3 $\cdot$ Velocity $\cdot$ Mass Density^2 = Energy,Work,Heat Quantity^2

Volumetric Velocity $\cdot$ Velocity^3 $\cdot$ Dynamic Viscosity^2 = Power, Radiance Flux^2

Volumetric Velocity $\cdot$ Velocity^2 $\cdot$ Mass Density = Power, Radiance Flux

Volumetric Velocity $\cdot$ Velocity $\cdot$ Magnetic Flux Density^2 = Electric Potential Difference^2

Volumetric Velocity $\cdot$ Electric Potential Difference^2 = Velocity^3 $\cdot$ Magnetic Flux^2

Volumetric Velocity = Velocity^3 $\cdot$ Capacitance^2 $\cdot$ Electric Resistance^2

Volumetric Velocity = Velocity^3 $\cdot$ Capacitance $\cdot$ Inductance

Volumetric Velocity $\cdot$ Electric Resistance^2 = Velocity^3 $\cdot$ Inductance^2

Volumetric Velocity $\cdot$ Magnetic Flux Density = Velocity $\cdot$ Magnetic Flux

Volumetric Velocity $\cdot$ Velocity $\cdot$ Mass Density^2 = Dynamic Viscosity^2

Volumetric Velocity^2 = Area^2 $\cdot$ Specific Volume $\cdot$ Pressure,Stress

Volumetric Velocity^4 = Area $\cdot$ Specific Volume^2 $\cdot$ Energy,Work,Heat Quantity^2

Volumetric Velocity^3 = Area^2 $\cdot$ Specific Volume $\cdot$ Power, Radiance Flux

Volumetric Velocity^2 = Area $\cdot$ Specific Volume^2 $\cdot$ Dynamic Viscosity^2

Volumetric Velocity^2 $\cdot$ Electric Charge^2 = Area $\cdot$ Current Density^2

Volumetric Velocity $\cdot$ Electric Charge = Area^2 $\cdot$ Magnetic Field Strength

$h^3 \cdot G_c{}^3 \cdot$ Time^4 = Volumetric Velocity^5

$h^4 = G_c \cdot$ Time^8 $\cdot$ Force^5

$h \cdot G_c =$ Time^7 $\cdot$ Acceleration^5

$h^3 = G_c{}^2 \cdot$ Time $\cdot$ Mass^5

147

$h \cdot G_c = Time^2 \cdot Velocity^5$

$h^2 \cdot G_c{}^2 \cdot Time^6 = Area^5$

$h^3 \cdot G_c{}^3 \cdot Time^9 = Volume^5$

$h \cdot G_c \cdot Time^3 \cdot Wavenumber^5 = A\ Constant$

$h^2 \cdot G_c{}^2 = Time^4 \cdot Absorbed\ Dose^5$

$h^2 = G_c{}^3 \cdot Time^9 \cdot Dynamic\ Viscosity^5$

$h^2 \cdot G_c{}^2 \cdot Time = Kinematic\ Viscosity^5$

$h \cdot G_c \cdot Length^4 = Volumetric\ Velocity^3$

$h^4 \cdot G_c = Length^8 \cdot Force^3$

$h^4 \cdot G_c = Length^5 \cdot Energy^3$

$h^2 \cdot G_c{}^2 = Length^7 \cdot Acceleration^3$

$h^2 = G_c \cdot Length \cdot Mass^3$

$h \cdot G_c = Length^2 \cdot Velocity^3$

$h \cdot G_c = Length^5 \cdot Frequency^3$

$h^4 \cdot G_c = Length^5 \cdot Energy, Work, Heat\ Quantity^3$

$h^2 \cdot G_c{}^2 = Length^4 \cdot Absorbed\ Dose^3$

$h \cdot G_c \cdot Length = Kinematic\ Viscosity^3$

$h^2 \cdot G_c = Volumetric\ Velocity^2 \cdot Force$

$h^7 \cdot G_c{}^3 = Volumetric\ Velocity^5 \cdot Energy^4$

$h^5 \cdot G_c{}^5 = Volumetric\ Velocity^7 \cdot Acceleration^4$

$h^3 = G_c \cdot Volumetric\ Velocity \cdot Mass^4$

$h \cdot G_c = Volumetric\ Velocity \cdot Velocity^2$

$h \cdot G_c \cdot Area^2 = Volumetric\ Velocity^3$

148

$h^3 \cdot G_c^3 \cdot \text{Volume}^4 = \text{Volumetric Velocity}^9$

$h \cdot G_c = \text{Volumetric Velocity}^3 \cdot \text{Wavenumber}^4$

$h^3 \cdot G_c \cdot \text{Specific Volume}^2 = \text{Volumetric Velocity}^5$

$h^3 \cdot G_c^3 = \text{Volumetric Velocity}^5 \cdot \text{Frequency}^4$

$h^5 \cdot G_c^3 = \text{Volumetric Velocity}^7 \cdot \text{Pressure,Stress}^2$

$h^7 \cdot G_c^3 = \text{Volumetric Velocity}^5 \cdot \text{Energy,Work,Heat Quantity}^4$

$h^5 \cdot G_c^3 = \text{Volumetric Velocity}^5 \cdot \text{Power, Radiance Flux}^2$

$h \cdot G_c = \text{Volumetric Velocity} \cdot \text{Absorbed Dose}$

$h^7 \cdot G_c^3 = \text{Volumetric Velocity}^9 \cdot \text{Dynamic Viscosity}^4$

$h \cdot G_c \cdot \text{Volumetric Velocity} = \text{Kinematic Viscosity}^4$

$h^3 \cdot G_c = \text{Volumetric Velocity}^5 \cdot \text{Mass Density}^2$

$h^4 \cdot G_c \cdot \text{Force}^5 = \text{Energy}^8$

$h^4 \cdot \text{Acceleration}^8 = G_c^3 \cdot \text{Force}^7$

$h^4 \cdot \text{Force} = G_c^3 \cdot \text{Mass}^8$

$h^4 \cdot G_c = \text{Force}^3 \cdot \text{Area}^4$

$h^4 \cdot G_c \cdot \text{Wavenumber}^8 = \text{Force}^3$

$h^4 \cdot G_c^3 = \text{Force}^5 \cdot \text{Specific Volume}^4$

$h^4 \cdot \text{Frequency}^8 = G_c \cdot \text{Force}^5$

$h^4 \cdot G_c \cdot \text{Pressure,Stress}^4 = \text{Force}^7$

$h^4 \cdot G_c \cdot \text{Force}^5 = \text{Energy,Work,Heat Quantity}^8$

$h^4 \cdot G_c^3 \cdot \text{Dynamic Viscosity}^8 = \text{Force}^9$

$h^4 \cdot G_c^3 = \text{Force} \cdot \text{Kinematic Viscosity}^8$

$h^4 \cdot G_c{}^3 \cdot \text{Mass Density}^4 = \text{Force}^5$

$h^6 \cdot \text{Acceleration}^5 = G_c \cdot \text{Energy}^7$

$h^2 \cdot \text{Energy} = G_c{}^2 \cdot \text{Mass}^5$

$h \cdot \text{Velocity}^5 = G_c \cdot \text{Energy}^2$

$h^8 \cdot G_c{}^2 = \text{Energy}^6 \cdot \text{Area}^5$

$h^4 \cdot G_c \cdot \text{Wavenumber}^5 = \text{Energy}^3$

$h^2 \cdot G_c = \text{Energy}^2 \cdot \text{Specific Volume}$

$h^2 \cdot \text{Absorbed Dose}^5 = G_c{}^2 \cdot \text{Energy}^4$

$h^7 \cdot G_c{}^3 \cdot \text{Dynamic Viscosity}^5 = \text{Energy}^9$

$h^3 \cdot G_c{}^2 = \text{Energy} \cdot \text{Kinematic Viscosity}^5$

$h^2 \cdot G_c \cdot \text{Mass Density} = \text{Energy}^2$

$h^4 \cdot \text{Acceleration} = G_c{}^3 \cdot \text{Mass}^7$

$h \cdot G_c \cdot \text{Acceleration}^2 = \text{Velocity}^7$

$h^4 \cdot G_c{}^4 = \text{Acceleration}^6 \cdot \text{Area}^7$

$h^6 \cdot G_c{}^6 = \text{Acceleration}^9 \cdot \text{Volume}^7$

$h^2 \cdot G_c{}^2 \cdot \text{Wavenumber}^7 = \text{Acceleration}^3$

$h \cdot G_c \cdot \text{Frequency}^7 = \text{Acceleration}^5$

$h^6 \cdot \text{Acceleration}^5 = G_c \cdot \text{Energy,Work,Heat Quantity}^7$

$h^2 \cdot G_c{}^2 \cdot \text{Acceleration}^4 = \text{Absorbed Dose}^7$

$h \cdot \text{Acceleration}^9 = G_c{}^6 \cdot \text{Dynamic Viscosity}^7$

$h^3 \cdot G_c{}^3 = \text{Acceleration} \cdot \text{Kinematic Viscosity}^7$

$h \cdot \text{Velocity} = G_c \cdot \text{Mass}^2$

$h^4 = G_c{}^2 \cdot \text{Mass}^6 \cdot \text{Area}$

$h^6 = G_c^3 \cdot Mass^9 \cdot Volume$

$h^2 \cdot Wavenumber = G_c \cdot Mass^3$

$h^3 \cdot Frequency = G_c^2 \cdot Mass^5$

$h^2 \cdot Energy, Work, Heat\ Quantity = G_c^2 \cdot Mass^5$

$h^2 \cdot Absorbed\ Dose = G_c^2 \cdot Mass^4$

$h^5 \cdot Dynamic\ Viscosity = G_c^3 \cdot Mass^9$

$h \cdot G_c = Velocity^3 \cdot Area$

$h^3 \cdot G_c^3 = Velocity^9 \cdot Volume^2$

$h \cdot G_c \cdot Wavenumber^2 = Velocity^3$

$h \cdot G_c^2 = Velocity^5 \cdot Specific\ Volume$

$h \cdot G_c \cdot Frequency^2 = Velocity^5$

$h \cdot G_c^2 \cdot Pressure, Stress = Velocity^7$

$h \cdot Velocity^5 = G_c \cdot Energy, Work, Heat\ Quantity^2$

$h \cdot G_c^3 \cdot Dynamic\ Viscosity^2 = Velocity^9$

$h \cdot G_c = Velocity \cdot Kinematic\ Viscosity^2$

$h \cdot G_c^2 \cdot Mass\ Density = Velocity^5$

$h^2 \cdot Specific\ Volume^3 = G_c \cdot Area^5$

$h^2 \cdot G_c^2 = Area^5 \cdot Frequency^6$

$h^4 \cdot G_c = Area^7 \cdot Pressure, Stress^3$

$h^8 \cdot G_c^2 = Area^5 \cdot Energy, Work, Heat\ Quantity^6$

$h^5 \cdot G_c^2 = Area^5 \cdot Power,\ Radiance\ Flux^3$

$h^2 \cdot G_c^2 = Area^2 \cdot Absorbed\ Dose^3$

$h^2 \cdot G_c^2 \cdot \text{Area} = \text{Kinematic Viscosity}^6$

$h^2 = G_c \cdot \text{Area}^5 \cdot \text{Mass Density}^3$

$h^3 \cdot G_c^3 = \text{Volume}^5 \cdot \text{Frequency}^9$

$h^6 \cdot G_c^6 = \text{Volume}^4 \cdot \text{Absorbed Dose}^9$

$h^3 \cdot G_c^3 \cdot \text{Volume} = \text{Kinematic Viscosity}^9$

$h \cdot G_c \cdot \text{Wavenumber}^5 = \text{Frequency}^3$

$h^4 \cdot G_c \cdot \text{Wavenumber}^5 = \text{Energy,Work,Heat Quantity}^3$

$h^2 \cdot G_c^2 \cdot \text{Wavenumber}^4 = \text{Absorbed Dose}^3$

$h \cdot G_c = \text{Wavenumber} \cdot \text{Kinematic Viscosity}^3$

$h^2 \cdot G_c^4 = \text{Specific Volume}^7 \cdot \text{Pressure,Stress}^5$

$h^2 \cdot G_c = \text{Specific Volume} \cdot \text{Energy,Work,Heat Quantity}^2$

$h \cdot G_c = \text{Specific Volume} \cdot \text{Power, Radiance Flux}$

$h^2 \cdot G_c^4 = \text{Specific Volume}^2 \cdot \text{Absorbed Dose}^5$

$h^2 \cdot G_c^2 \cdot \text{Frequency}^4 = \text{Absorbed Dose}^5$

$h^2 \cdot \text{Frequency}^9 = G_c^3 \cdot \text{Dynamic Viscosity}^5$

$h^2 \cdot G_c^2 = \text{Frequency} \cdot \text{Kinematic Viscosity}^5$

$h^5 \cdot G_c^3 \cdot \text{Pressure,Stress}^5 = \text{Power, Radiance Flux}^7$

$h^2 \cdot G_c^4 \cdot \text{Pressure,Stress}^2 = \text{Absorbed Dose}^7$

$h^2 \cdot G_c^4 \cdot \text{Mass Density}^7 = \text{Pressure,Stress}^5$

$h^2 \cdot \text{Absorbed Dose}^5 = G_c^2 \cdot \text{Energy,Work,Heat Quantity}^4$

$h^7 \cdot G_c^3 \cdot \text{Dynamic Viscosity}^5 = \text{Energy,Work,Heat Quantity}^9$

$h^3 \cdot G_c^2 = \text{Energy,Work,Heat Quantity} \cdot \text{Kinematic Viscosity}^5$

$h^2 \cdot G_c \cdot \text{Mass Density} = \text{Energy,Work,Heat Quantity}^2$

$h \cdot G_c \cdot$ Mass Density = Power, Radiance Flux

$h^2 \cdot G_c{}^6 \cdot$ Dynamic Viscosity^4 = Absorbed Dose^9

$h^2 \cdot G_c{}^2$ = Absorbed Dose $\cdot$ Kinematic Viscosity^4

$h^2 \cdot G_c{}^4 \cdot$ Mass Density^2 = Absorbed Dose^5

$h^4 \cdot G_c{}^3$ = Dynamic Viscosity $\cdot$ Kinematic Viscosity^9

$h^2 \cdot$ Illuminance^3 = Luminous Intensity$^3 \cdot$ Dynamic Viscosity^2

$h \cdot$ Catalytic Activity = Amount of a Substance $\cdot$ Energy

$h \cdot$ Catalytic Activity = Amount of a Substance $\cdot$ Energy,Work,Heat Quantity

$h \cdot$ Catalytic Activity^2 = Amount of a Substance$^2 \cdot$ Power, Radiance Flux

h = Electric Current $\cdot$ Time$^2 \cdot$ Electric Potential Difference

$h \cdot$ Capacitance = Electric Current$^2 \cdot$ Time^3

h = Electric Current$^2 \cdot$ Time$^2 \cdot$ Electric Resistance

h = Electric Current$^2 \cdot$ Time $\cdot$ Inductance

h = Electric Current $\cdot$ Time $\cdot$ Magnetic Flux

$h \cdot$ Electric Current = Energy $\cdot$ Electric Charge

$h \cdot$ Electric Current $\cdot$ Electric Potential Difference = Energy^2

$h^2 \cdot$ Electric Current^2 = Energy$^3 \cdot$ Capacitance

$h \cdot$ Electric Current$^2 \cdot$ Electric Resistance = Energy^2

$h^2 \cdot$ Electric Current^3 = Current Density$^3 \cdot$ Dynamic Viscosity^2

$h \cdot$ Magnetic Field Strength^3 = Electric Current$^3 \cdot$ Dynamic Viscosity

$h \cdot$ Frequency^2 = Electric Current $\cdot$ Electric Potential Difference

$h \cdot$ Frequency$^3 \cdot$ Capacitance = Electric Current^2

$h \cdot \text{Frequency}^2 = \text{Electric Current}^2 \cdot \text{Electric Resistance}$

$h \cdot \text{Frequency} = \text{Electric Current}^2 \cdot \text{Inductance}$

$h \cdot \text{Frequency} = \text{Electric Current} \cdot \text{Magnetic Flux}$

$h \cdot \text{Electric Current} = \text{Energy,Work,Heat Quantity} \cdot \text{Electric Charge}$

$h \cdot \text{Electric Current} \cdot \text{Electric Potential Difference} = \text{Energy,Work,Heat Quantity}^2$

$h^2 \cdot \text{Electric Current}^2 = \text{Energy,Work,Heat Quantity}^3 \cdot \text{Capacitance}$

$h \cdot \text{Electric Current}^2 \cdot \text{Electric Resistance} = \text{Energy,Work,Heat Quantity}^2$

$h \cdot \text{Electric Current}^2 = \text{Power, Radiance Flux} \cdot \text{Electric Charge}^2$

$h \cdot \text{Electric Current}^4 = \text{Power, Radiance Flux}^3 \cdot \text{Capacitance}^2$

$h \cdot \text{Power, Radiance Flux} = \text{Electric Current}^4 \cdot \text{Inductance}^2$

$h \cdot \text{Power, Radiance Flux} = \text{Electric Current}^2 \cdot \text{Magnetic Flux}^2$

$h \cdot \text{Electric Current} = \text{Electric Charge}^2 \cdot \text{Electric Potential Difference}$

$h \cdot \text{Electric Current} \cdot \text{Capacitance} = \text{Electric Charge}^3$

$h = \text{Electric Current} \cdot \text{Electric Charge} \cdot \text{Inductance}$

$h \cdot \text{Electric Current} = \text{Electric Potential Difference}^3 \cdot \text{Capacitance}^2$

$h \cdot \text{Electric Potential Difference} = \text{Electric Current}^3 \cdot \text{Inductance}^2$

$h \cdot \text{Electric Potential Difference} = \text{Electric Current} \cdot \text{Magnetic Flux}^2$

$h = \text{Electric Current}^2 \cdot \text{Capacitance}^2 \cdot \text{Electric Resistance}^3$

$h^2 = \text{Electric Current}^4 \cdot \text{Capacitance} \cdot \text{Inductance}^3$

$h^2 = \text{Electric Current} \cdot \text{Capacitance} \cdot \text{Magnetic Flux}^3$

$h \cdot \text{Electric Resistance} = \text{Electric Current}^2 \cdot \text{Inductance}^2$

$h = \text{Time} \cdot \text{Length} \cdot \text{Force}$

$h \cdot \text{Time} = \text{Length}^2 \cdot \text{Mass}$

$h \cdot$ Time $\cdot$ Specific Volume $=$ Length^5

$h =$ Time $\cdot$ Length$^3 \cdot$ Pressure,Stress

$h \cdot$ Time $=$ Length$^5 \cdot$ Mass Density

$h^3 =$ Time$^4 \cdot$ Volumetric Velocity $\cdot$ Force^3

$h^3 \cdot$ Time $=$ Volumetric Velocity$^2 \cdot$ Mass^3

$h^3 \cdot$ Specific Volume$^3 =$ Time$^2 \cdot$ Volumetric Velocity^5

$h =$ Time$^2 \cdot$ Volumetric Velocity $\cdot$ Pressure,Stress

$h =$ Time $\cdot$ Volumetric Velocity $\cdot$ Dynamic Viscosity

$h^3 =$ Time$^2 \cdot$ Volumetric Velocity$^5 \cdot$ Mass Density^3

$h =$ Time$^3 \cdot$ Force $\cdot$ Acceleration

$h \cdot$ Mass $=$ Time$^3 \cdot$ Force^2

$h =$ Time$^2 \cdot$ Force $\cdot$ Velocity

$h^2 =$ Time$^2 \cdot$ Force$^2 \cdot$ Area

$h^3 =$ Time$^3 \cdot$ Force$^3 \cdot$ Volume

$h \cdot$ Wavenumber $=$ Time $\cdot$ Force

$h^4 =$ Time$^6 \cdot$ Force$^5 \cdot$ Specific Volume

$h^2 \cdot$ Pressure,Stress $=$ Time$^2 \cdot$ Force^3

$h^2 =$ Time$^4 \cdot$ Force$^2 \cdot$ Absorbed Dose

$h^2 \cdot$ Dynamic Viscosity $=$ Time$^3 \cdot$ Force^3

$h^2 =$ Time$^3 \cdot$ Force$^2 \cdot$ Kinematic Viscosity

$h^4 \cdot$ Mass Density $=$ Time$^6 \cdot$ Force^5

$h =$ Time$^3 \cdot$ Acceleration$^2 \cdot$ Mass

$h \cdot \text{Specific Volume} = \text{Time}^9 \cdot \text{Acceleration}^5$

$h = \text{Time}^7 \cdot \text{Acceleration}^3 \cdot \text{Pressure,Stress}$

$h = \text{Time}^6 \cdot \text{Acceleration}^3 \cdot \text{Dynamic Viscosity}$

$h = \text{Time}^9 \cdot \text{Acceleration}^5 \cdot \text{Mass Density}$

$h = \text{Time} \cdot \text{Mass} \cdot \text{Velocity}^2$

$h \cdot \text{Time} = \text{Mass} \cdot \text{Area}$

$h^3 \cdot \text{Time}^3 = \text{Mass}^3 \cdot \text{Volume}^2$

$h \cdot \text{Time} \cdot \text{Wavenumber}^2 = \text{Mass}$

$h^3 \cdot \text{Time}^3 = \text{Mass}^5 \cdot \text{Specific Volume}^2$

$h \cdot \text{Time}^5 \cdot \text{Pressure,Stress}^2 = \text{Mass}^3$

$h = \text{Time} \cdot \text{Mass} \cdot \text{Absorbed Dose}$

$h \cdot \text{Time}^3 \cdot \text{Dynamic Viscosity}^2 = \text{Mass}^3$

$h^3 \cdot \text{Time}^3 \cdot \text{Mass Density}^2 = \text{Mass}^5$

$h \cdot \text{Specific Volume} = \text{Time}^4 \cdot \text{Velocity}^5$

$h = \text{Time}^4 \cdot \text{Velocity}^3 \cdot \text{Pressure,Stress}$

$h = \text{Time}^3 \cdot \text{Velocity}^3 \cdot \text{Dynamic Viscosity}$

$h = \text{Time}^4 \cdot \text{Velocity}^5 \cdot \text{Mass Density}$

$h^2 \cdot \text{Time}^2 \cdot \text{Specific Volume}^2 = \text{Area}^5$

$h^2 = \text{Time}^2 \cdot \text{Area}^3 \cdot \text{Pressure,Stress}^2$

$h^2 \cdot \text{Time}^2 = \text{Area}^5 \cdot \text{Mass Density}^2$

$h^3 \cdot \text{Time}^3 \cdot \text{Specific Volume}^3 = \text{Volume}^5$

$h = \text{Time} \cdot \text{Volume} \cdot \text{Pressure,Stress}$

$h^3 \cdot \text{Time}^3 = \text{Volume}^5 \cdot \text{Mass Density}^3$

$h \cdot Time \cdot Wavenumber^5 \cdot Specific\ Volume = A\ Constant$

$h \cdot Wavenumber^3 = Time \cdot Pressure,Stress$

$h \cdot Time \cdot Wavenumber^5 = Mass\ Density$

$h^2 = Time^8 \cdot Specific\ Volume^3 \cdot Pressure,Stress^5$

$h^2 \cdot Specific\ Volume^2 = Time^8 \cdot Absorbed\ Dose^5$

$h^2 = Time^3 \cdot Specific\ Volume^3 \cdot Dynamic\ Viscosity^5$

$h^2 \cdot Specific\ Volume^2 = Time^3 \cdot Kinematic\ Viscosity^5$

$h = Time \cdot Current\ Density \cdot Magnetic\ Flux\ Density$

$h^2 = Time^8 \cdot Pressure,Stress^2 \cdot Absorbed\ Dose^3$

$h^2 = Time^5 \cdot Pressure,Stress^2 \cdot Kinematic\ Viscosity^3$

$h^2 \cdot Mass\ Density^3 = Time^8 \cdot Pressure,Stress^5$

$h = Time \cdot Electric\ Charge \cdot Electric\ Potential\ Difference$

$h \cdot Capacitance = Time \cdot Electric\ Charge^2$

$h \cdot Time = Electric\ Charge^2 \cdot Inductance$

$h = Time \cdot Electric\ Potential\ Difference^2 \cdot Capacitance$

$h \cdot Electric\ Resistance = Time^2 \cdot Electric\ Potential\ Difference^2$

$h \cdot Inductance = Time^3 \cdot Electric\ Potential\ Difference^2$

$h \cdot Time = Capacitance \cdot Magnetic\ Flux^2$

$h \cdot Inductance = Time \cdot Magnetic\ Flux^2$

$h^2 = Time^6 \cdot Absorbed\ Dose^3 \cdot Dynamic\ Viscosity^2$

$h^2 = Time^8 \cdot Absorbed\ Dose^5 \cdot Mass\ Density^2$

$h^2 = Time^3 \cdot Dynamic\ Viscosity^2 \cdot Kinematic\ Viscosity^3$

$h^2 \cdot \text{Mass Density}^3 = \text{Time}^3 \cdot \text{Dynamic Viscosity}^5$

$h^2 = \text{Time}^3 \cdot \text{Kinematic Viscosity}^5 \cdot \text{Mass Density}^2$

$h \cdot \text{Volumetric Velocity} = \text{Length}^4 \cdot \text{Force}$

$h \cdot \text{Volumetric Velocity} = \text{Length}^3 \cdot \text{Energy}$

$h \cdot \text{Length} = \text{Volumetric Velocity} \cdot \text{Mass}$

$h \cdot \text{Specific Volume} = \text{Length}^2 \cdot \text{Volumetric Velocity}$

$h \cdot \text{Volumetric Velocity} = \text{Length}^6 \cdot \text{Pressure,Stress}$

$h \cdot \text{Volumetric Velocity} = \text{Length}^3 \cdot \text{Energy,Work,Heat Quantity}$

$h \cdot \text{Volumetric Velocity}^2 = \text{Length}^6 \cdot \text{Power, Radiance Flux}$

$h = \text{Length}^2 \cdot \text{Volumetric Velocity} \cdot \text{Mass Density}$

$h^2 \cdot \text{Acceleration} = \text{Length}^3 \cdot \text{Force}^2$

$h^2 = \text{Length}^3 \cdot \text{Force} \cdot \text{Mass}$

$h \cdot \text{Velocity} = \text{Length}^2 \cdot \text{Force}$

$h^2 \cdot \text{Specific Volume} = \text{Length}^6 \cdot \text{Force}$

$h \cdot \text{Frequency} = \text{Length} \cdot \text{Force}$

$h \cdot \text{Power, Radiance Flux} = \text{Length}^2 \cdot \text{Force}^2$

$h^2 \cdot \text{Absorbed Dose} = \text{Length}^4 \cdot \text{Force}^2$

$h \cdot \text{Kinematic Viscosity} = \text{Length}^3 \cdot \text{Force}$

$h^2 = \text{Length}^6 \cdot \text{Force} \cdot \text{Mass Density}$

$h^2 \cdot \text{Acceleration} = \text{Length} \cdot \text{Energy}^2$

$h^2 = \text{Length}^2 \cdot \text{Energy} \cdot \text{Mass}$

$h \cdot \text{Velocity} = \text{Length} \cdot \text{Energy}$

$h^2 \cdot \text{Specific Volume} = \text{Length}^5 \cdot \text{Energy}$

$h^2 \cdot$ Absorbed Dose $=$ Length$^2 \cdot$ Energy^2

$h \cdot$ Kinematic Viscosity $=$ Length$^2 \cdot$ Energy

$h^2 =$ Length$^5 \cdot$ Energy $\cdot$ Mass Density

$h^2 =$ Length$^3 \cdot$ Acceleration $\cdot$ Mass^2

$h^2 \cdot$ Specific Volume$^2 =$ Length$^9 \cdot$ Acceleration

$h^2 \cdot$ Acceleration $=$ Length$^7 \cdot$ Pressure,Stress^2

$h^2 \cdot$ Acceleration $=$ Length $\cdot$ Energy,Work,Heat Quantity^2

$h \cdot$ Acceleration $=$ Length $\cdot$ Power, Radiance Flux

$h^2 =$ Length$^9 \cdot$ Acceleration $\cdot$ Mass Density^2

$h =$ Length $\cdot$ Mass $\cdot$ Velocity

$h =$ Length$^2 \cdot$ Mass $\cdot$ Frequency

$h^2 =$ Length$^5 \cdot$ Mass $\cdot$ Pressure,Stress

$h^2 =$ Length$^2 \cdot$ Mass $\cdot$ Energy,Work,Heat Quantity

$h^3 =$ Length$^4 \cdot$ Mass$^2 \cdot$ Power, Radiance Flux

$h^2 =$ Length$^2 \cdot$ Mass$^2 \cdot$ Absorbed Dose

$h \cdot$ Specific Volume $=$ Length$^4 \cdot$ Velocity

$h \cdot$ Velocity $=$ Length$^4 \cdot$ Pressure,Stress

$h \cdot$ Velocity $=$ Length $\cdot$ Energy,Work,Heat Quantity

$h \cdot$ Velocity$^2 =$ Length$^2 \cdot$ Power, Radiance Flux

$h =$ Length$^4 \cdot$ Velocity $\cdot$ Mass Density

$h \cdot$ Specific Volume $=$ Length$^5 \cdot$ Frequency

$h^2 \cdot$ Specific Volume $=$ Length$^8 \cdot$ Pressure,Stress

$h^2 \cdot$ Specific Volume = Length$^5 \cdot$ Energy,Work,Heat Quantity

$h^2 \cdot$ Specific Volume^2 = Length$^8 \cdot$ Absorbed Dose

$h \cdot$ Specific Volume = Length$^3 \cdot$ Kinematic Viscosity

$h \cdot$ Frequency = Length$^3 \cdot$ Pressure,Stress

h = Length$^5 \cdot$ Frequency $\cdot$ Mass Density

$h \cdot$ Power, Radiance Flux = Length$^6 \cdot$ Pressure,Stress^2

$h^2 \cdot$ Absorbed Dose = Length$^8 \cdot$ Pressure,Stress^2

$h \cdot$ Kinematic Viscosity = Length$^5 \cdot$ Pressure,Stress

h^2 = Length$^8 \cdot$ Pressure,Stress $\cdot$ Mass Density

$h^2 \cdot$ Absorbed Dose = Length$^2 \cdot$ Energy,Work,Heat Quantity^2

$h \cdot$ Kinematic Viscosity = Length$^2 \cdot$ Energy,Work,Heat Quantity

h^2 = Length$^5 \cdot$ Energy,Work,Heat Quantity $\cdot$ Mass Density

$h \cdot$ Absorbed Dose = Length$^2 \cdot$ Power, Radiance Flux

$h \cdot$ Kinematic Viscosity^2 = Length$^4 \cdot$ Power, Radiance Flux

h = Length$^2 \cdot$ Electric Charge $\cdot$ Magnetic Flux Density

$h \cdot$ Electric Resistance = Length$^4 \cdot$ Magnetic Flux Density^2

h^2 = Length$^8 \cdot$ Absorbed Dose $\cdot$ Mass Density^2

h = Length$^3 \cdot$ Kinematic Viscosity $\cdot$ Mass Density

$h \cdot$ Volumetric Velocity $\cdot$ Force^3 = Energy^4

$h^5 \cdot$ Acceleration^4 = Volumetric Velocity$^3 \cdot$ Force^5

h^5 = Volumetric Velocity$^3 \cdot$ Force $\cdot$ Mass^4

$h \cdot$ Velocity^2 = Volumetric Velocity $\cdot$ Force

$h \cdot$ Volumetric Velocity = Force $\cdot$ Area^2

$h^3 \cdot \text{Volumetric Velocity}^3 = \text{Force}^3 \cdot \text{Volume}^4$

$h \cdot \text{Volumetric Velocity} \cdot \text{Wavenumber}^4 = \text{Force}$

$h \cdot \text{Force} \cdot \text{Specific Volume}^2 = \text{Volumetric Velocity}^3$

$h^3 \cdot \text{Frequency}^4 = \text{Volumetric Velocity} \cdot \text{Force}^3$

$h \cdot \text{Volumetric Velocity} \cdot \text{Pressure,Stress}^2 = \text{Force}^3$

$h \cdot \text{Volumetric Velocity} \cdot \text{Force}^3 = \text{Energy,Work,Heat Quantity}^4$

$h \cdot \text{Power, Radiance Flux}^2 = \text{Volumetric Velocity} \cdot \text{Force}^3$

$h \cdot \text{Absorbed Dose} = \text{Volumetric Velocity} \cdot \text{Force}$

$h \cdot \text{Force}^3 = \text{Volumetric Velocity}^3 \cdot \text{Dynamic Viscosity}^4$

$h \cdot \text{Kinematic Viscosity}^4 = \text{Volumetric Velocity}^3 \cdot \text{Force}$

$h \cdot \text{Force} = \text{Volumetric Velocity}^3 \cdot \text{Mass Density}^2$

$h^5 \cdot \text{Acceleration}^3 = \text{Volumetric Velocity} \cdot \text{Energy}^5$

$h^4 = \text{Volumetric Velocity}^2 \cdot \text{Energy} \cdot \text{Mass}^3$

$h^2 \cdot \text{Velocity}^3 = \text{Volumetric Velocity} \cdot \text{Energy}^2$

$h^2 \cdot \text{Volumetric Velocity}^2 = \text{Energy}^2 \cdot \text{Area}^3$

$h \cdot \text{Volumetric Velocity} = \text{Energy} \cdot \text{Volume}$

$h \cdot \text{Volumetric Velocity} \cdot \text{Wavenumber}^3 = \text{Energy}$

$h \cdot \text{Energy}^2 \cdot \text{Specific Volume}^3 = \text{Volumetric Velocity}^5$

$h \cdot \text{Volumetric Velocity} \cdot \text{Pressure,Stress} = \text{Energy}^2$

$h^4 \cdot \text{Absorbed Dose}^3 = \text{Volumetric Velocity}^2 \cdot \text{Energy}^4$

$h \cdot \text{Kinematic Viscosity}^3 = \text{Volumetric Velocity}^2 \cdot \text{Energy}$

$h \cdot \text{Energy}^2 = \text{Volumetric Velocity}^5 \cdot \text{Mass Density}^3$

$h^5 = \text{Volumetric Velocity}^3 \cdot \text{Acceleration} \cdot \text{Mass}^5$

$h^5 \cdot \text{Acceleration}^2 \cdot \text{Specific Volume}^5 = \text{Volumetric Velocity}^9$

$h^5 \cdot \text{Acceleration}^6 = \text{Volumetric Velocity}^7 \cdot \text{Pressure,Stress}^5$

$h^5 \cdot \text{Acceleration}^3 = \text{Volumetric Velocity} \cdot \text{Energy,Work,Heat Quantity}^5$

$h^5 \cdot \text{Acceleration}^6 = \text{Volumetric Velocity}^2 \cdot \text{Power, Radiance Flux}^5$

$h^5 \cdot \text{Acceleration}^3 = \text{Volumetric Velocity}^6 \cdot \text{Dynamic Viscosity}^5$

$h^5 \cdot \text{Acceleration}^2 = \text{Volumetric Velocity}^9 \cdot \text{Mass Density}^5$

$h^2 = \text{Volumetric Velocity} \cdot \text{Mass}^2 \cdot \text{Velocity}$

$h^2 \cdot \text{Area} = \text{Volumetric Velocity}^2 \cdot \text{Mass}^2$

$h^3 \cdot \text{Volume} = \text{Volumetric Velocity}^3 \cdot \text{Mass}^3$

$h = \text{Volumetric Velocity} \cdot \text{Mass} \cdot \text{Wavenumber}$

$h^3 \cdot \text{Specific Volume} = \text{Volumetric Velocity}^3 \cdot \text{Mass}^2$

$h^3 = \text{Volumetric Velocity}^2 \cdot \text{Mass}^3 \cdot \text{Frequency}$

$h^7 = \text{Volumetric Velocity}^5 \cdot \text{Mass}^6 \cdot \text{Pressure,Stress}$

$h^4 = \text{Volumetric Velocity}^2 \cdot \text{Mass}^3 \cdot \text{Energy,Work,Heat Quantity}$

$h^7 = \text{Volumetric Velocity}^4 \cdot \text{Mass}^6 \cdot \text{Power, Radiance Flux}$

$h^4 = \text{Volumetric Velocity}^2 \cdot \text{Mass}^4 \cdot \text{Absorbed Dose}$

$h^4 = \text{Volumetric Velocity}^3 \cdot \text{Mass}^3 \cdot \text{Dynamic Viscosity}$

$h^3 = \text{Volumetric Velocity}^3 \cdot \text{Mass}^2 \cdot \text{Mass Density}$

$h \cdot \text{Velocity} \cdot \text{Specific Volume} = \text{Volumetric Velocity}^2$

$h \cdot \text{Velocity}^3 = \text{Volumetric Velocity}^2 \cdot \text{Pressure,Stress}$

$h^2 \cdot \text{Velocity}^3 = \text{Volumetric Velocity} \cdot \text{Energy,Work,Heat Quantity}^2$

$h \cdot \text{Velocity}^3 = \text{Volumetric Velocity} \cdot \text{Power, Radiance Flux}$

$h^2 \cdot \text{Velocity}^3 = \text{Volumetric Velocity}^3 \cdot \text{Dynamic Viscosity}^2$

$h \cdot \text{Velocity} = \text{Volumetric Velocity}^2 \cdot \text{Mass Density}$

$h \cdot \text{Specific Volume} = \text{Volumetric Velocity} \cdot \text{Area}$

$h \cdot \text{Volumetric Velocity} = \text{Area}^3 \cdot \text{Pressure,Stress}$

$h^2 \cdot \text{Volumetric Velocity}^2 = \text{Area}^3 \cdot \text{Energy,Work,Heat Quantity}^2$

$h \cdot \text{Volumetric Velocity}^2 = \text{Area}^3 \cdot \text{Power, Radiance Flux}$

$h = \text{Volumetric Velocity} \cdot \text{Area} \cdot \text{Mass Density}$

$h^3 \cdot \text{Specific Volume}^3 = \text{Volumetric Velocity}^3 \cdot \text{Volume}^2$

$h \cdot \text{Volumetric Velocity} = \text{Volume}^2 \cdot \text{Pressure,Stress}$

$h \cdot \text{Volumetric Velocity} = \text{Volume} \cdot \text{Energy,Work,Heat Quantity}$

$h \cdot \text{Volumetric Velocity}^2 = \text{Volume}^2 \cdot \text{Power, Radiance Flux}$

$h^3 = \text{Volumetric Velocity}^3 \cdot \text{Volume}^2 \cdot \text{Mass Density}^3$

$h \cdot \text{Wavenumber}^2 \cdot \text{Specific Volume} = \text{Volumetric Velocity}$

$h \cdot \text{Volumetric Velocity} \cdot \text{Wavenumber}^6 = \text{Pressure,Stress}$

$h \cdot \text{Volumetric Velocity} \cdot \text{Wavenumber}^3 = \text{Energy,Work,Heat Quantity}$

$h \cdot \text{Volumetric Velocity}^2 \cdot \text{Wavenumber}^6 = \text{Power, Radiance Flux}$

$h \cdot \text{Wavenumber}^2 = \text{Volumetric Velocity} \cdot \text{Mass Density}$

$h^3 \cdot \text{Specific Volume}^3 \cdot \text{Frequency}^2 = \text{Volumetric Velocity}^5$

$h^2 \cdot \text{Specific Volume}^3 \cdot \text{Pressure,Stress} = \text{Volumetric Velocity}^4$

$h \cdot \text{Specific Volume}^3 \cdot \text{Energy,Work,Heat Quantity}^2 = \text{Volumetric Velocity}^5$

$h^2 \cdot \text{Specific Volume}^3 \cdot \text{Power, Radiance Flux} = \text{Volumetric Velocity}^5$

$h^2 \cdot \text{Specific Volume}^2 \cdot \text{Absorbed Dose} = \text{Volumetric Velocity}^4$

$h \cdot \text{Specific Volume}^3 \cdot \text{Dynamic Viscosity}^2 = \text{Volumetric Velocity}^3$

$h \cdot \text{Specific Volume} \cdot \text{Kinematic Viscosity}^2 = \text{Volumetric Velocity}^3$

$h \cdot \text{Frequency}^2 = \text{Volumetric Velocity} \cdot \text{Pressure,Stress}$

$h \cdot \text{Frequency} = \text{Volumetric Velocity} \cdot \text{Dynamic Viscosity}$

$h^3 \cdot \text{Frequency}^2 = \text{Volumetric Velocity}^5 \cdot \text{Mass Density}^3$

$h \cdot \text{Volumetric Velocity} \cdot \text{Pressure,Stress} = \text{Energy,Work,Heat Quantity}^2$

$h^2 \cdot \text{Absorbed Dose}^3 = \text{Volumetric Velocity}^4 \cdot \text{Pressure,Stress}^2$

$h \cdot \text{Pressure,Stress} = \text{Volumetric Velocity} \cdot \text{Dynamic Viscosity}^2$

$h \cdot \text{Kinematic Viscosity}^6 = \text{Volumetric Velocity}^5 \cdot \text{Pressure,Stress}$

$h^2 \cdot \text{Pressure,Stress} = \text{Volumetric Velocity}^4 \cdot \text{Mass Density}^3$

$h^4 \cdot \text{Absorbed Dose}^3 = \text{Volumetric Velocity}^2 \cdot \text{Energy,Work,Heat Quantity}^4$

$h \cdot \text{Kinematic Viscosity}^3 = \text{Volumetric Velocity}^2 \cdot \text{Energy,Work,Heat Quantity}$

$h \cdot \text{Energy,Work,Heat Quantity}^2 = \text{Volumetric Velocity}^5 \cdot \text{Mass Density}^3$

$h^2 \cdot \text{Absorbed Dose}^3 = \text{Volumetric Velocity}^2 \cdot \text{Power, Radiance Flux}^2$

$h \cdot \text{Power, Radiance Flux} = \text{Volumetric Velocity}^2 \cdot \text{Dynamic Viscosity}^2$

$h \cdot \text{Kinematic Viscosity}^6 = \text{Volumetric Velocity}^4 \cdot \text{Power, Radiance Flux}$

$h^2 \cdot \text{Power, Radiance Flux} = \text{Volumetric Velocity}^5 \cdot \text{Mass Density}^3$

$h^4 \cdot \text{Absorbed Dose}^3 = \text{Volumetric Velocity}^6 \cdot \text{Dynamic Viscosity}^4$

$h^2 \cdot \text{Absorbed Dose} = \text{Volumetric Velocity}^4 \cdot \text{Mass Density}^2$

$h \cdot \text{Kinematic Viscosity}^3 = \text{Volumetric Velocity}^3 \cdot \text{Dynamic Viscosity}$

$h \cdot \text{Dynamic Viscosity}^2 = \text{Volumetric Velocity}^3 \cdot \text{Mass Density}^3$

$h \cdot \text{Kinematic Viscosity}^2 = \text{Volumetric Velocity}^3 \cdot \text{Mass Density}$

$h^2 \cdot \text{Force} \cdot \text{Acceleration} = \text{Energy}^3$

$$h^2 \cdot \text{Force}^2 = \text{Energy}^3 \cdot \text{Mass}$$

$$h \cdot \text{Force} \cdot \text{Velocity} = \text{Energy}^2$$

$$h^2 \cdot \text{Force}^5 \cdot \text{Specific Volume} = \text{Energy}^6$$

$$h^2 \cdot \text{Force}^2 \cdot \text{Absorbed Dose} = \text{Energy}^4$$

$$h \cdot \text{Force}^3 = \text{Energy}^3 \cdot \text{Dynamic Viscosity}$$

$$h \cdot \text{Force}^2 \cdot \text{Kinematic Viscosity} = \text{Energy}^3$$

$$h^2 \cdot \text{Force}^5 = \text{Energy}^6 \cdot \text{Mass Density}$$

$$h \cdot \text{Acceleration}^2 = \text{Force} \cdot \text{Velocity}^3$$

$$h^4 \cdot \text{Acceleration}^2 = \text{Force}^4 \cdot \text{Area}^3$$

$$h^2 \cdot \text{Acceleration} = \text{Force}^2 \cdot \text{Volume}$$

$$h^2 \cdot \text{Acceleration} \cdot \text{Wavenumber}^3 = \text{Force}^2$$

$$h^2 \cdot \text{Acceleration}^2 = \text{Force}^3 \cdot \text{Specific Volume}$$

$$h \cdot \text{Frequency}^3 = \text{Force} \cdot \text{Acceleration}$$

$$h^4 \cdot \text{Acceleration}^2 \cdot \text{Pressure,Stress}^3 = \text{Force}^7$$

$$h^2 \cdot \text{Force} \cdot \text{Acceleration} = \text{Energy,Work,Heat Quantity}^3$$

$$h \cdot \text{Force}^2 \cdot \text{Acceleration}^2 = \text{Power, Radiance Flux}^3$$

$$h^2 \cdot \text{Acceleration}^4 = \text{Force}^2 \cdot \text{Absorbed Dose}^3$$

$$h \cdot \text{Acceleration} \cdot \text{Dynamic Viscosity} = \text{Force}^2$$

$$h \cdot \text{Acceleration} = \text{Force} \cdot \text{Kinematic Viscosity}$$

$$h^2 \cdot \text{Acceleration}^2 \cdot \text{Mass Density} = \text{Force}^3$$

$$h \cdot \text{Force} = \text{Mass}^2 \cdot \text{Velocity}^3$$

$$h^4 = \text{Force}^2 \cdot \text{Mass}^2 \cdot \text{Area}^3$$

$h^2 = \text{Force} \cdot \text{Mass} \cdot \text{Volume}$

$h^2 \cdot \text{Wavenumber}^3 = \text{Force} \cdot \text{Mass}$

$h^2 = \text{Force} \cdot \text{Mass}^2 \cdot \text{Specific Volume}$

$h \cdot \text{Mass} \cdot \text{Frequency}^3 = \text{Force}^2$

$h^4 \cdot \text{Pressure,Stress}^3 = \text{Force}^5 \cdot \text{Mass}^2$

$h^2 \cdot \text{Force}^2 = \text{Mass} \cdot \text{Energy,Work,Heat Quantity}^3$

$h \cdot \text{Force}^4 = \text{Mass}^2 \cdot \text{Power, Radiance Flux}^3$

$h^2 \cdot \text{Force}^2 = \text{Mass}^4 \cdot \text{Absorbed Dose}^3$

$h \cdot \text{Dynamic Viscosity} = \text{Force} \cdot \text{Mass}$

$h^2 \cdot \text{Mass Density} = \text{Force} \cdot \text{Mass}^2$

$h \cdot \text{Velocity} = \text{Force} \cdot \text{Area}$

$h^3 \cdot \text{Velocity}^3 = \text{Force}^3 \cdot \text{Volume}^2$

$h \cdot \text{Velocity} \cdot \text{Wavenumber}^2 = \text{Force}$

$h \cdot \text{Velocity}^3 = \text{Force}^2 \cdot \text{Specific Volume}$

$h \cdot \text{Frequency}^2 = \text{Force} \cdot \text{Velocity}$

$h \cdot \text{Velocity} \cdot \text{Pressure,Stress} = \text{Force}^2$

$h \cdot \text{Force} \cdot \text{Velocity} = \text{Energy,Work,Heat Quantity}^2$

$h \cdot \text{Velocity}^3 \cdot \text{Dynamic Viscosity}^2 = \text{Force}^3$

$h \cdot \text{Velocity}^3 = \text{Force} \cdot \text{Kinematic Viscosity}^2$

$h \cdot \text{Velocity}^3 \cdot \text{Mass Density} = \text{Force}^2$

$h^2 \cdot \text{Specific Volume} = \text{Force} \cdot \text{Area}^3$

$h^2 \cdot \text{Frequency}^2 = \text{Force}^2 \cdot \text{Area}$

$h \cdot \text{Power, Radiance Flux} = \text{Force}^2 \cdot \text{Area}$

$h^2 \cdot$ Absorbed Dose $=$ Force$^2 \cdot$ Area^2

$h^2 \cdot$ Kinematic Viscosity$^2 =$ Force$^2 \cdot$ Area^3

$h^2 =$ Force $\cdot$ Area$^3 \cdot$ Mass Density

$h^2 \cdot$ Specific Volume $=$ Force $\cdot$ Volume^2

$h^3 \cdot$ Frequency$^3 =$ Force$^3 \cdot$ Volume

$h^3 \cdot$ Power, Radiance Flux$^3 =$ Force$^6 \cdot$ Volume^2

$h^6 \cdot$ Absorbed Dose$^3 =$ Force$^6 \cdot$ Volume^4

$h \cdot$ Kinematic Viscosity $=$ Force $\cdot$ Volume

$h^2 =$ Force $\cdot$ Volume$^2 \cdot$ Mass Density

$h^2 \cdot$ Wavenumber$^6 \cdot$ Specific Volume $=$ Force

$h \cdot$ Wavenumber $\cdot$ Frequency $=$ Force

$h \cdot$ Wavenumber$^2 \cdot$ Power, Radiance Flux $=$ Force^2

$h^2 \cdot$ Wavenumber$^4 \cdot$ Absorbed Dose $=$ Force^2

$h \cdot$ Wavenumber$^3 \cdot$ Kinematic Viscosity $=$ Force

$h^2 \cdot$ Wavenumber$^6 =$ Force $\cdot$ Mass Density

$h^4 \cdot$ Frequency$^6 =$ Force$^5 \cdot$ Specific Volume

$h^2 \cdot$ Specific Volume $\cdot$ Pressure,Stress$^3 =$ Force^4

$h^2 \cdot$ Force$^5 \cdot$ Specific Volume $=$ Energy,Work,Heat Quantity^6

$h \cdot$ Power, Radiance Flux$^3 =$ Force$^5 \cdot$ Specific Volume

$h^2 \cdot$ Absorbed Dose$^3 =$ Force$^4 \cdot$ Specific Volume^2

$h \cdot$ Magnetic Field Strength $=$ Force $\cdot$ Electric Charge

$h \cdot$ Magnetic Field Strength$^2 \cdot$ Electric Resistance $=$ Force^2

$h^2 \cdot \text{Frequency}^2 \cdot \text{Pressure,Stress} = \text{Force}^3$

$h^2 \cdot \text{Frequency}^4 = \text{Force}^2 \cdot \text{Absorbed Dose}$

$h^2 \cdot \text{Frequency}^3 \cdot \text{Dynamic Viscosity} = \text{Force}^3$

$h^2 \cdot \text{Frequency}^3 = \text{Force}^2 \cdot \text{Kinematic Viscosity}$

$h^4 \cdot \text{Frequency}^6 \cdot \text{Mass Density} = \text{Force}^5$

$h \cdot \text{Pressure,Stress} \cdot \text{Power, Radiance Flux} = \text{Force}^3$

$h^2 \cdot \text{Pressure,Stress}^2 \cdot \text{Absorbed Dose} = \text{Force}^4$

$h^2 \cdot \text{Pressure,Stress}^3 = \text{Force}^3 \cdot \text{Dynamic Viscosity}^2$

$h^2 \cdot \text{Pressure,Stress}^3 \cdot \text{Kinematic Viscosity}^2 = \text{Force}^5$

$h^2 \cdot \text{Pressure,Stress}^3 = \text{Force}^4 \cdot \text{Mass Density}$

$h^2 \cdot \text{Force}^2 \cdot \text{Absorbed Dose} = \text{Energy,Work,Heat Quantity}^4$

$h \cdot \text{Force}^3 = \text{Energy,Work,Heat Quantity}^3 \cdot \text{Dynamic Viscosity}$

$h \cdot \text{Force}^2 \cdot \text{Kinematic Viscosity} = \text{Energy,Work,Heat Quantity}^3$

$h^2 \cdot \text{Force}^5 = \text{Energy,Work,Heat Quantity}^6 \cdot \text{Mass Density}$

$h \cdot \text{Power, Radiance Flux}^3 \cdot \text{Dynamic Viscosity}^2 = \text{Force}^6$

$h \cdot \text{Power, Radiance Flux}^3 = \text{Force}^4 \cdot \text{Kinematic Viscosity}^2$

$h \cdot \text{Power, Radiance Flux}^3 \cdot \text{Mass Density} = \text{Force}^5$

$h^2 \cdot \text{Absorbed Dose}^3 \cdot \text{Dynamic Viscosity}^4 = \text{Force}^6$

$h^2 \cdot \text{Absorbed Dose}^3 = \text{Force}^2 \cdot \text{Kinematic Viscosity}^4$

$h^2 \cdot \text{Absorbed Dose}^3 \cdot \text{Mass Density}^2 = \text{Force}^4$

$h^2 \cdot \text{Acceleration}^2 \cdot \text{Mass} = \text{Energy}^3$

$h \cdot \text{Acceleration} = \text{Energy} \cdot \text{Velocity}$

$h^4 \cdot \text{Acceleration}^2 = \text{Energy}^4 \cdot \text{Area}$

$$h^6 \cdot \text{Acceleration}^3 = \text{Energy}^6 \cdot \text{Volume}$$

$$h^2 \cdot \text{Acceleration} \cdot \text{Wavenumber} = \text{Energy}^2$$

$$h^8 \cdot \text{Acceleration}^5 = \text{Energy}^9 \cdot \text{Specific Volume}$$

$$h^6 \cdot \text{Acceleration}^3 \cdot \text{Pressure,Stress} = \text{Energy}^7$$

$$h^2 \cdot \text{Acceleration}^2 = \text{Energy}^2 \cdot \text{Absorbed Dose}$$

$$h^5 \cdot \text{Acceleration}^3 \cdot \text{Dynamic Viscosity} = \text{Energy}^6$$

$$h^3 \cdot \text{Acceleration}^2 = \text{Energy}^3 \cdot \text{Kinematic Viscosity}$$

$$h^8 \cdot \text{Acceleration}^5 \cdot \text{Mass Density} = \text{Energy}^9$$

$$h^2 = \text{Energy} \cdot \text{Mass} \cdot \text{Area}$$

$$h^6 = \text{Energy}^3 \cdot \text{Mass}^3 \cdot \text{Volume}^2$$

$$h^2 \cdot \text{Wavenumber}^2 = \text{Energy} \cdot \text{Mass}$$

$$h^6 = \text{Energy}^3 \cdot \text{Mass}^5 \cdot \text{Specific Volume}^2$$

$$h^6 \cdot \text{Pressure,Stress}^2 = \text{Energy}^5 \cdot \text{Mass}^3$$

$$h^4 \cdot \text{Dynamic Viscosity}^2 = \text{Energy}^3 \cdot \text{Mass}^3$$

$$h^6 \cdot \text{Mass Density}^2 = \text{Energy}^3 \cdot \text{Mass}^5$$

$$h^2 \cdot \text{Velocity}^2 = \text{Energy}^2 \cdot \text{Area}$$

$$h^3 \cdot \text{Velocity}^3 = \text{Energy}^3 \cdot \text{Volume}$$

$$h \cdot \text{Velocity} \cdot \text{Wavenumber} = \text{Energy}$$

$$h^3 \cdot \text{Velocity}^5 = \text{Energy}^4 \cdot \text{Specific Volume}$$

$$h^3 \cdot \text{Velocity}^3 \cdot \text{Pressure,Stress} = \text{Energy}^4$$

$$h^2 \cdot \text{Velocity}^3 \cdot \text{Dynamic Viscosity} = \text{Energy}^3$$

$$h \cdot \text{Velocity}^2 = \text{Energy} \cdot \text{Kinematic Viscosity}$$

$h^3 \cdot \text{Velocity}^5 \cdot \text{Mass Density} = \text{Energy}^4$

$h^4 \cdot \text{Specific Volume}^2 = \text{Energy}^2 \cdot \text{Area}^5$

$h^2 \cdot \text{Absorbed Dose} = \text{Energy}^2 \cdot \text{Area}$

$h \cdot \text{Kinematic Viscosity} = \text{Energy} \cdot \text{Area}$

$h^4 = \text{Energy}^2 \cdot \text{Area}^5 \cdot \text{Mass Density}^2$

$h^6 \cdot \text{Specific Volume}^3 = \text{Energy}^3 \cdot \text{Volume}^5$

$h^6 \cdot \text{Absorbed Dose}^3 = \text{Energy}^6 \cdot \text{Volume}^2$

$h^3 \cdot \text{Kinematic Viscosity}^3 = \text{Energy}^3 \cdot \text{Volume}^2$

$h^6 = \text{Energy}^3 \cdot \text{Volume}^5 \cdot \text{Mass Density}^3$

$h^2 \cdot \text{Wavenumber}^5 \cdot \text{Specific Volume} = \text{Energy}$

$h^2 \cdot \text{Wavenumber}^2 \cdot \text{Absorbed Dose} = \text{Energy}^2$

$h \cdot \text{Wavenumber}^2 \cdot \text{Kinematic Viscosity} = \text{Energy}$

$h^2 \cdot \text{Wavenumber}^5 = \text{Energy} \cdot \text{Mass Density}$

$h^6 \cdot \text{Specific Volume}^3 \cdot \text{Pressure,Stress}^5 = \text{Energy}^8$

$h^6 \cdot \text{Absorbed Dose}^5 = \text{Energy}^8 \cdot \text{Specific Volume}^2$

$h \cdot \text{Specific Volume}^3 \cdot \text{Dynamic Viscosity}^5 = \text{Energy}^3$

$h \cdot \text{Kinematic Viscosity}^5 = \text{Energy}^3 \cdot \text{Specific Volume}^2$

$h^6 \cdot \text{Pressure,Stress}^2 \cdot \text{Absorbed Dose}^3 = \text{Energy}^8$

$h \cdot \text{Pressure,Stress} = \text{Energy} \cdot \text{Dynamic Viscosity}$

$h^3 \cdot \text{Pressure,Stress}^2 \cdot \text{Kinematic Viscosity}^3 = \text{Energy}^5$

$h^6 \cdot \text{Pressure,Stress}^5 = \text{Energy}^8 \cdot \text{Mass Density}^3$

$h^2 = \text{Energy} \cdot \text{Electric Charge}^2 \cdot \text{Inductance}$

$h \cdot \text{Electric Potential Difference}^2 = \text{Energy}^2 \cdot \text{Electric Resistance}$

$h^2 \cdot \text{Electric Potential Difference}^2 = \text{Energy}^3 \cdot \text{Inductance}$

$h \cdot \text{Electric Potential Difference} = \text{Energy} \cdot \text{Magnetic Flux}$

$h = \text{Energy} \cdot \text{Capacitance} \cdot \text{Electric Resistance}$

$h^2 = \text{Energy}^2 \cdot \text{Capacitance} \cdot \text{Inductance}$

$h^2 = \text{Energy} \cdot \text{Capacitance} \cdot \text{Magnetic Flux}^2$

$h \cdot \text{Electric Resistance} = \text{Energy} \cdot \text{Inductance}$

$h^4 \cdot \text{Absorbed Dose}^3 \cdot \text{Dynamic Viscosity}^2 = \text{Energy}^6$

$h \cdot \text{Absorbed Dose} = \text{Energy} \cdot \text{Kinematic Viscosity}$

$h^6 \cdot \text{Absorbed Dose}^5 \cdot \text{Mass Density}^2 = \text{Energy}^8$

$h \cdot \text{Dynamic Viscosity}^2 \cdot \text{Kinematic Viscosity}^3 = \text{Energy}^3$

$h \cdot \text{Dynamic Viscosity}^5 = \text{Energy}^3 \cdot \text{Mass Density}^3$

$h \cdot \text{Kinematic Viscosity}^5 \cdot \text{Mass Density}^2 = \text{Energy}^3$

$h \cdot \text{Acceleration} = \text{Mass} \cdot \text{Velocity}^3$

$h^4 = \text{Acceleration}^2 \cdot \text{Mass}^4 \cdot \text{Area}^3$

$h^2 = \text{Acceleration} \cdot \text{Mass}^2 \cdot \text{Volume}$

$h^2 \cdot \text{Wavenumber}^3 = \text{Acceleration} \cdot \text{Mass}^2$

$h^2 = \text{Acceleration} \cdot \text{Mass}^3 \cdot \text{Specific Volume}$

$h \cdot \text{Frequency}^3 = \text{Acceleration}^2 \cdot \text{Mass}$

$h^4 \cdot \text{Pressure,Stress}^3 = \text{Acceleration}^5 \cdot \text{Mass}^7$

$h^2 \cdot \text{Acceleration}^2 \cdot \text{Mass} = \text{Energy,Work,Heat Quantity}^3$

$h \cdot \text{Acceleration}^4 \cdot \text{Mass}^2 = \text{Power, Radiance Flux}^3$

$h^2 \cdot \text{Acceleration}^2 = \text{Mass}^2 \cdot \text{Absorbed Dose}^3$

$h \cdot$ Dynamic Viscosity = Acceleration $\cdot$ Mass^2

$h^2 \cdot$ Mass Density = Acceleration $\cdot$ Mass^3

$h \cdot$ Acceleration$^4 \cdot$ Specific Volume = Velocity^9

$h \cdot$ Acceleration^4 = Velocity$^7 \cdot$ Pressure,Stress

$h \cdot$ Acceleration = Velocity $\cdot$ Energy,Work,Heat Quantity

$h \cdot$ Acceleration^2 = Velocity$^2 \cdot$ Power, Radiance Flux

$h \cdot$ Acceleration^3 = Velocity$^6 \cdot$ Dynamic Viscosity

$h \cdot$ Acceleration^4 = Velocity$^9 \cdot$ Mass Density

$h^4 \cdot$ Specific Volume^4 = Acceleration$^2 \cdot$ Area^9

$h^4 \cdot$ Acceleration^2 = Area$^7 \cdot$ Pressure,Stress^4

$h^4 \cdot$ Acceleration^2 = Area $\cdot$ Energy,Work,Heat Quantity^4

$h^2 \cdot$ Acceleration^2 = Area $\cdot$ Power, Radiance Flux^2

h^4 = Acceleration$^2 \cdot$ Area$^9 \cdot$ Mass Density^4

$h^2 \cdot$ Specific Volume^2 = Acceleration $\cdot$ Volume^3

$h^6 \cdot$ Acceleration^3 = Volume$^7 \cdot$ Pressure,Stress^6

$h^6 \cdot$ Acceleration^3 = Volume $\cdot$ Energy,Work,Heat Quantity^6

$h^3 \cdot$ Acceleration^3 = Volume $\cdot$ Power, Radiance Flux^3

h^2 = Acceleration $\cdot$ Volume$^3 \cdot$ Mass Density^2

$h^2 \cdot$ Wavenumber$^9 \cdot$ Specific Volume^2 = Acceleration

$h^2 \cdot$ Acceleration $\cdot$ Wavenumber^7 = Pressure,Stress^2

$h^2 \cdot$ Acceleration $\cdot$ Wavenumber = Energy,Work,Heat Quantity^2

$h \cdot$ Acceleration $\cdot$ Wavenumber = Power, Radiance Flux

$h^2 \cdot$ Wavenumber^9 = Acceleration $\cdot$ Mass Density^2

$h \cdot$ Specific Volume $\cdot$ Frequency^9 = Acceleration^5

$h^2 \cdot$ Acceleration^8 = Specific Volume$^7 \cdot$ Pressure,Stress^9

$h^8 \cdot$ Acceleration^5 = Specific Volume $\cdot$ Energy,Work,Heat Quantity^9

$h^2 \cdot$ Acceleration$^8 \cdot$ Specific Volume^2 = Absorbed Dose^9

$h \cdot$ Acceleration = Specific Volume$^2 \cdot$ Dynamic Viscosity^3

$h \cdot$ Acceleration $\cdot$ Specific Volume = Kinematic Viscosity^3

$h \cdot$ Frequency^7 = Acceleration$^3 \cdot$ Pressure,Stress

$h \cdot$ Frequency^6 = Acceleration$^3 \cdot$ Dynamic Viscosity

$h \cdot$ Frequency^9 = Acceleration$^5 \cdot$ Mass Density

$h^6 \cdot$ Acceleration$^3 \cdot$ Pressure,Stress = Energy,Work,Heat Quantity^7

$h^5 \cdot$ Acceleration$^6 \cdot$ Pressure,Stress^2 = Power, Radiance Flux^7

$h^2 \cdot$ Acceleration^8 = Pressure,Stress$^2 \cdot$ Absorbed Dose^7

$h \cdot$ Pressure,Stress^6 = Acceleration$^3 \cdot$ Dynamic Viscosity^7

$h^3 \cdot$ Acceleration^5 = Pressure,Stress$^3 \cdot$ Kinematic Viscosity^7

$h^2 \cdot$ Acceleration$^8 \cdot$ Mass Density^7 = Pressure,Stress^9

$h^2 \cdot$ Acceleration^2 = Energy,Work,Heat Quantity$^2 \cdot$ Absorbed Dose

$h^5 \cdot$ Acceleration$^3 \cdot$ Dynamic Viscosity = Energy,Work,Heat Quantity^6

$h^3 \cdot$ Acceleration^2 = Energy,Work,Heat Quantity$^3 \cdot$ Kinematic Viscosity

$h^8 \cdot$ Acceleration$^5 \cdot$ Mass Density = Energy,Work,Heat Quantity^9

$h \cdot$ Acceleration^2 = Power, Radiance Flux $\cdot$ Absorbed Dose

$h^2 \cdot$ Acceleration$^3 \cdot$ Dynamic Viscosity = Power, Radiance Flux^3

$h^3 \cdot$ Acceleration^4 = Power, Radiance Flux$^3 \cdot$ Kinematic Viscosity^2

$h \cdot \text{Acceleration}^3 = \text{Absorbed Dose}^3 \cdot \text{Dynamic Viscosity}$

$h^2 \cdot \text{Acceleration}^8 = \text{Absorbed Dose}^9 \cdot \text{Mass Density}^2$

$h \cdot \text{Acceleration} = \text{Dynamic Viscosity} \cdot \text{Kinematic Viscosity}^2$

$h \cdot \text{Acceleration} \cdot \text{Mass Density}^2 = \text{Dynamic Viscosity}^3$

$h \cdot \text{Acceleration} = \text{Kinematic Viscosity}^3 \cdot \text{Mass Density}$

$h^2 = \text{Mass}^2 \cdot \text{Velocity}^2 \cdot \text{Area}$

$h^3 = \text{Mass}^3 \cdot \text{Velocity}^3 \cdot \text{Volume}$

$h \cdot \text{Wavenumber} = \text{Mass} \cdot \text{Velocity}$

$h^3 = \text{Mass}^4 \cdot \text{Velocity}^3 \cdot \text{Specific Volume}$

$h \cdot \text{Frequency} = \text{Mass} \cdot \text{Velocity}^2$

$h^3 \cdot \text{Pressure,Stress} = \text{Mass}^4 \cdot \text{Velocity}^5$

$h \cdot \text{Power, Radiance Flux} = \text{Mass}^2 \cdot \text{Velocity}^4$

$h^2 \cdot \text{Dynamic Viscosity} = \text{Mass}^3 \cdot \text{Velocity}^3$

$h^3 \cdot \text{Mass Density} = \text{Mass}^4 \cdot \text{Velocity}^3$

$h = \text{Mass} \cdot \text{Area} \cdot \text{Frequency}$

$h^4 = \text{Mass}^2 \cdot \text{Area}^5 \cdot \text{Pressure,Stress}^2$

$h^2 = \text{Mass} \cdot \text{Area} \cdot \text{Energy,Work,Heat Quantity}$

$h^3 = \text{Mass}^2 \cdot \text{Area}^2 \cdot \text{Power, Radiance Flux}$

$h^2 = \text{Mass}^2 \cdot \text{Area} \cdot \text{Absorbed Dose}$

$h^3 = \text{Mass}^3 \cdot \text{Volume}^2 \cdot \text{Frequency}^3$

$h^6 = \text{Mass}^3 \cdot \text{Volume}^5 \cdot \text{Pressure,Stress}^3$

$h^6 = \text{Mass}^3 \cdot \text{Volume}^2 \cdot \text{Energy,Work,Heat Quantity}^3$

$h^9 = \text{Mass}^6 \cdot \text{Volume}^4 \cdot \text{Power, Radiance Flux}^3$

$h^6 = \text{Mass}^6 \cdot \text{Volume}^2 \cdot \text{Absorbed Dose}^3$

$h \cdot \text{Wavenumber}^2 = \text{Mass} \cdot \text{Frequency}$

$h^2 \cdot \text{Wavenumber}^5 = \text{Mass} \cdot \text{Pressure,Stress}$

$h^2 \cdot \text{Wavenumber}^2 = \text{Mass} \cdot \text{Energy,Work,Heat Quantity}$

$h^3 \cdot \text{Wavenumber}^4 = \text{Mass}^2 \cdot \text{Power, Radiance Flux}$

$h^2 \cdot \text{Wavenumber}^2 = \text{Mass}^2 \cdot \text{Absorbed Dose}$

$h^3 = \text{Mass}^5 \cdot \text{Specific Volume}^2 \cdot \text{Frequency}^3$

$h^6 = \text{Mass}^8 \cdot \text{Specific Volume}^5 \cdot \text{Pressure,Stress}^3$

$h^6 = \text{Mass}^5 \cdot \text{Specific Volume}^2 \cdot \text{Energy,Work,Heat Quantity}^3$

$h^6 = \text{Mass}^8 \cdot \text{Specific Volume}^2 \cdot \text{Absorbed Dose}^3$

$h = \text{Mass} \cdot \text{Specific Volume} \cdot \text{Dynamic Viscosity}$

$h \cdot \text{Electric Charge} = \text{Mass} \cdot \text{Current Density}$

$h^3 = \text{Mass}^2 \cdot \text{Current Density}^2 \cdot \text{Electric Resistance}$

$h^2 = \text{Mass} \cdot \text{Current Density} \cdot \text{Magnetic Flux}$

$h \cdot \text{Pressure,Stress}^2 = \text{Mass}^3 \cdot \text{Frequency}^5$

$h \cdot \text{Frequency} = \text{Mass} \cdot \text{Absorbed Dose}$

$h \cdot \text{Dynamic Viscosity}^2 = \text{Mass}^3 \cdot \text{Frequency}^3$

$h^3 \cdot \text{Mass Density}^2 = \text{Mass}^5 \cdot \text{Frequency}^3$

$h^6 \cdot \text{Pressure,Stress}^2 = \text{Mass}^3 \cdot \text{Energy,Work,Heat Quantity}^5$

$h^7 \cdot \text{Pressure,Stress}^4 = \text{Mass}^6 \cdot \text{Power, Radiance Flux}^5$

$h^6 \cdot \text{Pressure,Stress}^2 = \text{Mass}^8 \cdot \text{Absorbed Dose}^5$

$h \cdot \text{Dynamic Viscosity}^5 = \text{Mass}^3 \cdot \text{Pressure,Stress}^3$

$h^6 \cdot \text{Mass Density}^5 = \text{Mass}^8 \cdot \text{Pressure,Stress}^3$

$h^4 \cdot \text{Dynamic Viscosity}^2 = \text{Mass}^3 \cdot \text{Energy,Work,Heat Quantity}^3$

$h^6 \cdot \text{Mass Density}^2 = \text{Mass}^5 \cdot \text{Energy,Work,Heat Quantity}^3$

$h \cdot \text{Power, Radiance Flux} = \text{Mass}^2 \cdot \text{Absorbed Dose}^2$

$h^5 \cdot \text{Dynamic Viscosity}^4 = \text{Mass}^6 \cdot \text{Power, Radiance Flux}^3$

$h \cdot \text{Magnetic Flux Density} = \text{Mass} \cdot \text{Electric Potential Difference}$

$h^4 \cdot \text{Dynamic Viscosity}^2 = \text{Mass}^6 \cdot \text{Absorbed Dose}^3$

$h^6 \cdot \text{Mass Density}^2 = \text{Mass}^8 \cdot \text{Absorbed Dose}^3$

$h \cdot \text{Mass Density} = \text{Mass} \cdot \text{Dynamic Viscosity}$

$h \cdot \text{Specific Volume} = \text{Velocity} \cdot \text{Area}^2$

$h \cdot \text{Velocity} = \text{Area}^2 \cdot \text{Pressure,Stress}$

$h^2 \cdot \text{Velocity}^2 = \text{Area} \cdot \text{Energy,Work,Heat Quantity}^2$

$h \cdot \text{Velocity}^2 = \text{Area} \cdot \text{Power, Radiance Flux}$

$h = \text{Velocity} \cdot \text{Area}^2 \cdot \text{Mass Density}$

$h^3 \cdot \text{Specific Volume}^3 = \text{Velocity}^3 \cdot \text{Volume}^4$

$h^3 \cdot \text{Velocity}^3 = \text{Volume}^4 \cdot \text{Pressure,Stress}^3$

$h^3 \cdot \text{Velocity}^3 = \text{Volume} \cdot \text{Energy,Work,Heat Quantity}^3$

$h^3 \cdot \text{Velocity}^6 = \text{Volume}^2 \cdot \text{Power, Radiance Flux}^3$

$h^3 = \text{Velocity}^3 \cdot \text{Volume}^4 \cdot \text{Mass Density}^3$

$h \cdot \text{Wavenumber}^4 \cdot \text{Specific Volume} = \text{Velocity}$

$h \cdot \text{Velocity} \cdot \text{Wavenumber}^4 = \text{Pressure,Stress}$

$h \cdot \text{Velocity} \cdot \text{Wavenumber} = \text{Energy,Work,Heat Quantity}$

$h \cdot \text{Velocity}^2 \cdot \text{Wavenumber}^2 = \text{Power, Radiance Flux}$

$h \cdot \text{Wavenumber}^4 = \text{Velocity} \cdot \text{Mass Density}$

$h \cdot \text{Specific Volume} \cdot \text{Frequency}^4 = \text{Velocity}^5$

$h^3 \cdot \text{Velocity}^5 = \text{Specific Volume} \cdot \text{Energy,Work,Heat Quantity}^4$

$h \cdot \text{Velocity}^5 = \text{Specific Volume} \cdot \text{Power, Radiance Flux}^2$

$h \cdot \text{Velocity}^3 = \text{Specific Volume}^3 \cdot \text{Dynamic Viscosity}^4$

$h \cdot \text{Velocity}^3 \cdot \text{Specific Volume} = \text{Kinematic Viscosity}^4$

$h \cdot \text{Velocity}^2 = \text{Current Density} \cdot \text{Electric Potential Difference}$

$h \cdot \text{Frequency}^4 = \text{Velocity}^3 \cdot \text{Pressure,Stress}$

$h \cdot \text{Frequency}^3 = \text{Velocity}^3 \cdot \text{Dynamic Viscosity}$

$h \cdot \text{Frequency}^4 = \text{Velocity}^5 \cdot \text{Mass Density}$

$h^3 \cdot \text{Velocity}^3 \cdot \text{Pressure,Stress} = \text{Energy,Work,Heat Quantity}^4$

$h \cdot \text{Velocity}^3 \cdot \text{Pressure,Stress} = \text{Power, Radiance Flux}^2$

$h \cdot \text{Pressure,Stress}^3 = \text{Velocity}^3 \cdot \text{Dynamic Viscosity}^4$

$h \cdot \text{Velocity}^5 = \text{Pressure,Stress} \cdot \text{Kinematic Viscosity}^4$

$h^2 \cdot \text{Velocity}^3 \cdot \text{Dynamic Viscosity} = \text{Energy,Work,Heat Quantity}^3$

$h \cdot \text{Velocity}^2 = \text{Energy,Work,Heat Quantity} \cdot \text{Kinematic Viscosity}$

$h^3 \cdot \text{Velocity}^5 \cdot \text{Mass Density} = \text{Energy,Work,Heat Quantity}^4$

$h \cdot \text{Velocity}^6 \cdot \text{Dynamic Viscosity}^2 = \text{Power, Radiance Flux}^3$

$h \cdot \text{Velocity}^4 = \text{Power, Radiance Flux} \cdot \text{Kinematic Viscosity}^2$

$h \cdot \text{Velocity}^5 \cdot \text{Mass Density} = \text{Power, Radiance Flux}^2$

$h \cdot \text{Velocity}^3 = \text{Dynamic Viscosity} \cdot \text{Kinematic Viscosity}^3$

$h \cdot \text{Velocity}^3 \cdot \text{Mass Density}^3 = \text{Dynamic Viscosity}^4$

$h \cdot \text{Velocity}^3 = \text{Kinematic Viscosity}^4 \cdot \text{Mass Density}$

$h^2 \cdot \text{Specific Volume}^2 = \text{Area}^5 \cdot \text{Frequency}^2$

$h^2 \cdot \text{Specific Volume} = \text{Area}^4 \cdot \text{Pressure,Stress}$

$h^4 \cdot \text{Specific Volume}^2 = \text{Area}^5 \cdot \text{Energy,Work,Heat Quantity}^2$

$h^3 \cdot \text{Specific Volume}^2 = \text{Area}^5 \cdot \text{Power, Radiance Flux}$

$h^2 \cdot \text{Specific Volume}^2 = \text{Area}^4 \cdot \text{Absorbed Dose}$

$h^2 \cdot \text{Specific Volume}^2 = \text{Area}^3 \cdot \text{Kinematic Viscosity}^2$

$h^2 \cdot \text{Frequency}^2 = \text{Area}^3 \cdot \text{Pressure,Stress}^2$

$h^2 = \text{Area}^5 \cdot \text{Frequency}^2 \cdot \text{Mass Density}^2$

$h \cdot \text{Power, Radiance Flux} = \text{Area}^3 \cdot \text{Pressure,Stress}^2$

$h^2 \cdot \text{Absorbed Dose} = \text{Area}^4 \cdot \text{Pressure,Stress}^2$

$h^2 \cdot \text{Kinematic Viscosity}^2 = \text{Area}^5 \cdot \text{Pressure,Stress}^2$

$h^2 = \text{Area}^4 \cdot \text{Pressure,Stress} \cdot \text{Mass Density}$

$h^2 \cdot \text{Absorbed Dose} = \text{Area} \cdot \text{Energy,Work,Heat Quantity}^2$

$h \cdot \text{Kinematic Viscosity} = \text{Area} \cdot \text{Energy,Work,Heat Quantity}$

$h^4 = \text{Area}^5 \cdot \text{Energy,Work,Heat Quantity}^2 \cdot \text{Mass Density}^2$

$h \cdot \text{Absorbed Dose} = \text{Area} \cdot \text{Power, Radiance Flux}$

$h \cdot \text{Kinematic Viscosity}^2 = \text{Area}^2 \cdot \text{Power, Radiance Flux}$

$h^3 = \text{Area}^5 \cdot \text{Power, Radiance Flux} \cdot \text{Mass Density}^2$

$h = \text{Area} \cdot \text{Electric Charge} \cdot \text{Magnetic Flux Density}$

$h \cdot \text{Electric Resistance} = \text{Area}^2 \cdot \text{Magnetic Flux Density}^2$

$h^2 = \text{Area}^4 \cdot \text{Absorbed Dose} \cdot \text{Mass Density}^2$

$h^2 = \text{Area}^3 \cdot \text{Kinematic Viscosity}^2 \cdot \text{Mass Density}^2$

$h^3 \cdot \text{Specific Volume}^3 = \text{Volume}^5 \cdot \text{Frequency}^3$

$h^6 \cdot \text{Specific Volume}^3 = \text{Volume}^8 \cdot \text{Pressure,Stress}^3$

$h^6 \cdot \text{Specific Volume}^3 = \text{Volume}^5 \cdot \text{Energy,Work,Heat Quantity}^3$

$h^6 \cdot \text{Specific Volume}^6 = \text{Volume}^8 \cdot \text{Absorbed Dose}^3$

$h \cdot \text{Specific Volume} = \text{Volume} \cdot \text{Kinematic Viscosity}$

$h \cdot \text{Frequency} = \text{Volume} \cdot \text{Pressure,Stress}$

$h^3 = \text{Volume}^5 \cdot \text{Frequency}^3 \cdot \text{Mass Density}^3$

$h \cdot \text{Power, Radiance Flux} = \text{Volume}^2 \cdot \text{Pressure,Stress}^2$

$h^6 \cdot \text{Absorbed Dose}^3 = \text{Volume}^8 \cdot \text{Pressure,Stress}^6$

$h^3 \cdot \text{Kinematic Viscosity}^3 = \text{Volume}^5 \cdot \text{Pressure,Stress}^3$

$h^6 = \text{Volume}^8 \cdot \text{Pressure,Stress}^3 \cdot \text{Mass Density}^3$

$h^6 \cdot \text{Absorbed Dose}^3 = \text{Volume}^2 \cdot \text{Energy,Work,Heat Quantity}^6$

$h^3 \cdot \text{Kinematic Viscosity}^3 = \text{Volume}^2 \cdot \text{Energy,Work,Heat Quantity}^3$

$h^6 = \text{Volume}^5 \cdot \text{Energy,Work,Heat Quantity}^3 \cdot \text{Mass Density}^3$

$h^3 \cdot \text{Absorbed Dose}^3 = \text{Volume}^2 \cdot \text{Power, Radiance Flux}^3$

$h^3 \cdot \text{Kinematic Viscosity}^6 = \text{Volume}^4 \cdot \text{Power, Radiance Flux}^3$

$h^3 = \text{Volume}^2 \cdot \text{Electric Charge}^3 \cdot \text{Magnetic Flux Density}^3$

$h^3 \cdot \text{Electric Resistance}^3 = \text{Volume}^4 \cdot \text{Magnetic Flux Density}^6$

$h^6 = \text{Volume}^8 \cdot \text{Absorbed Dose}^3 \cdot \text{Mass Density}^6$

$h = \text{Volume} \cdot \text{Kinematic Viscosity} \cdot \text{Mass Density}$

$h \cdot \text{Wavenumber}^5 \cdot \text{Specific Volume} = \text{Frequency}$

$h^2 \cdot \text{Wavenumber}^8 \cdot \text{Specific Volume} = \text{Pressure,Stress}$

179

$h^2 \cdot \text{Wavenumber}^5 \cdot \text{Specific Volume} = \text{Energy,Work,Heat Quantity}$

$h^2 \cdot \text{Wavenumber}^8 \cdot \text{Specific Volume}^2 = \text{Absorbed Dose}$

$h \cdot \text{Wavenumber}^3 \cdot \text{Specific Volume} = \text{Kinematic Viscosity}$

$h \cdot \text{Wavenumber}^3 \cdot \text{Frequency} = \text{Pressure,Stress}$

$h \cdot \text{Wavenumber}^5 = \text{Frequency} \cdot \text{Mass Density}$

$h \cdot \text{Wavenumber}^6 \cdot \text{Power, Radiance Flux} = \text{Pressure,Stress}^2$

$h^2 \cdot \text{Wavenumber}^8 \cdot \text{Absorbed Dose} = \text{Pressure,Stress}^2$

$h \cdot \text{Wavenumber}^5 \cdot \text{Kinematic Viscosity} = \text{Pressure,Stress}$

$h^2 \cdot \text{Wavenumber}^8 = \text{Pressure,Stress} \cdot \text{Mass Density}$

$h^2 \cdot \text{Wavenumber}^2 \cdot \text{Absorbed Dose} = \text{Energy,Work,Heat Quantity}^2$

$h \cdot \text{Wavenumber}^2 \cdot \text{Kinematic Viscosity} = \text{Energy,Work,Heat Quantity}$

$h^2 \cdot \text{Wavenumber}^5 = \text{Energy,Work,Heat Quantity} \cdot \text{Mass Density}$

$h \cdot \text{Wavenumber}^2 \cdot \text{Absorbed Dose} = \text{Power, Radiance Flux}$

$h \cdot \text{Wavenumber}^4 \cdot \text{Kinematic Viscosity}^2 = \text{Power, Radiance Flux}$

$h \cdot \text{Wavenumber}^2 = \text{Electric Charge} \cdot \text{Magnetic Flux Density}$

$h \cdot \text{Wavenumber}^4 \cdot \text{Electric Resistance} = \text{Magnetic Flux Density}^2$

$h^2 \cdot \text{Wavenumber}^8 = \text{Absorbed Dose} \cdot \text{Mass Density}^2$

$h \cdot \text{Wavenumber}^3 = \text{Kinematic Viscosity} \cdot \text{Mass Density}$

$h^2 \cdot \text{Frequency}^8 = \text{Specific Volume}^3 \cdot \text{Pressure,Stress}^5$

$h^2 \cdot \text{Specific Volume}^2 \cdot \text{Frequency}^8 = \text{Absorbed Dose}^5$

$h^2 \cdot \text{Frequency}^3 = \text{Specific Volume}^3 \cdot \text{Dynamic Viscosity}^5$

$h^2 \cdot \text{Specific Volume}^2 \cdot \text{Frequency}^3 = \text{Kinematic Viscosity}^5$

$h^6 \cdot \text{Specific Volume}^3 \cdot \text{Pressure,Stress}^5 = \text{Energy,Work,Heat Quantity}^8$

$h^2 \cdot$ Specific Volume$^3 \cdot$ Pressure,Stress$^5 =$ Power, Radiance Flux^4

$h^2 \cdot$ Pressure,Stress$^3 =$ Specific Volume$^3 \cdot$ Dynamic Viscosity^8

$h^2 \cdot$ Specific Volume$^5 \cdot$ Pressure,Stress$^3 =$ Kinematic Viscosity^8

$h^6 \cdot$ Absorbed Dose$^5 =$ Specific Volume$^2 \cdot$ Energy,Work,Heat Quantity^8

$h \cdot$ Specific Volume$^3 \cdot$ Dynamic Viscosity$^5 =$ Energy,Work,Heat Quantity^3

$h \cdot$ Kinematic Viscosity$^5 =$ Specific Volume$^2 \cdot$ Energy,Work,Heat Quantity^3

$h^2 \cdot$ Absorbed Dose$^5 =$ Specific Volume$^2 \cdot$ Power, Radiance Flux^4

$h^2 \cdot$ Absorbed Dose$^3 =$ Specific Volume$^6 \cdot$ Dynamic Viscosity^8

$h^2 \cdot$ Specific Volume$^2 \cdot$ Absorbed Dose$^3 =$ Kinematic Viscosity^8

$h \cdot$ Magnetic Field Strength $=$ Current Density $\cdot$ Dynamic Viscosity

$h \cdot$ Frequency $=$ Current Density $\cdot$ Magnetic Flux Density

$h \cdot$ Power, Radiance Flux $=$ Current Density$^2 \cdot$ Magnetic Flux Density^2

$h \cdot$ Absorbed Dose $=$ Current Density $\cdot$ Electric Potential Difference

$h \cdot$ Kinematic Viscosity$^2 =$ Current Density$^2 \cdot$ Electric Resistance

$h \cdot$ Kinematic Viscosity $=$ Current Density $\cdot$ Magnetic Flux

$h^2 \cdot$ Frequency$^8 =$ Pressure,Stress$^2 \cdot$ Absorbed Dose^3

$h^2 \cdot$ Frequency$^5 =$ Pressure,Stress$^2 \cdot$ Kinematic Viscosity^3

$h^2 \cdot$ Frequency$^8 \cdot$ Mass Density$^3 =$ Pressure,Stress^5

$h \cdot$ Frequency $=$ Electric Charge $\cdot$ Electric Potential Difference

$h \cdot$ Frequency $\cdot$ Capacitance $=$ Electric Charge^2

$h =$ Frequency $\cdot$ Electric Charge$^2 \cdot$ Inductance

$h \cdot$ Frequency $=$ Electric Potential Difference$^2 \cdot$ Capacitance

$h \cdot \text{Frequency}^2 \cdot \text{Electric Resistance} = \text{Electric Potential Difference}^2$

$h \cdot \text{Frequency}^3 \cdot \text{Inductance} = \text{Electric Potential Difference}^2$

$h = \text{Frequency} \cdot \text{Capacitance} \cdot \text{Magnetic Flux}^2$

$h \cdot \text{Frequency} \cdot \text{Inductance} = \text{Magnetic Flux}^2$

$h^2 \cdot \text{Frequency}^6 = \text{Absorbed Dose}^3 \cdot \text{Dynamic Viscosity}^2$

$h^2 \cdot \text{Frequency}^8 = \text{Absorbed Dose}^5 \cdot \text{Mass Density}^2$

$h^2 \cdot \text{Frequency}^3 = \text{Dynamic Viscosity}^2 \cdot \text{Kinematic Viscosity}^3$

$h^2 \cdot \text{Frequency}^3 \cdot \text{Mass Density}^3 = \text{Dynamic Viscosity}^5$

$h^2 \cdot \text{Frequency}^3 = \text{Kinematic Viscosity}^5 \cdot \text{Mass Density}^2$

$h^6 \cdot \text{Pressure,Stress}^2 \cdot \text{Absorbed Dose}^3 = \text{Energy,Work,Heat Quantity}^8$

$h \cdot \text{Pressure,Stress} = \text{Energy,Work,Heat Quantity} \cdot \text{Dynamic Viscosity}$

$h^3 \cdot \text{Pressure,Stress}^2 \cdot \text{Kinematic Viscosity}^3 = \text{Energy,Work,Heat Quantity}^5$

$h^6 \cdot \text{Pressure,Stress}^5 = \text{Energy,Work,Heat Quantity}^8 \cdot \text{Mass Density}^3$

$h^2 \cdot \text{Pressure,Stress}^2 \cdot \text{Absorbed Dose}^3 = \text{Power, Radiance Flux}^4$

$h \cdot \text{Pressure,Stress}^2 = \text{Power, Radiance Flux} \cdot \text{Dynamic Viscosity}^2$

$h \cdot \text{Pressure,Stress}^4 \cdot \text{Kinematic Viscosity}^6 = \text{Power, Radiance Flux}^5$

$h^2 \cdot \text{Pressure,Stress}^5 = \text{Power, Radiance Flux}^4 \cdot \text{Mass Density}^3$

$h^2 \cdot \text{Pressure,Stress}^6 = \text{Absorbed Dose}^3 \cdot \text{Dynamic Viscosity}^8$

$h^2 \cdot \text{Absorbed Dose}^5 = \text{Pressure,Stress}^2 \cdot \text{Kinematic Viscosity}^8$

$h^2 \cdot \text{Pressure,Stress}^3 = \text{Dynamic Viscosity}^5 \cdot \text{Kinematic Viscosity}^3$

$h^2 \cdot \text{Pressure,Stress}^3 \cdot \text{Mass Density}^3 = \text{Dynamic Viscosity}^8$

$h^2 \cdot \text{Pressure,Stress}^3 = \text{Kinematic Viscosity}^8 \cdot \text{Mass Density}^5$

$h^2 = \text{Energy,Work,Heat Quantity} \cdot \text{Electric Charge}^2 \cdot \text{Inductance}$

$h \cdot \text{Electric Potential Difference}^2 = \text{Energy,Work,Heat Quantity}^2 \cdot \text{Electric Resistance}$

$h^2 \cdot \text{Electric Potential Difference}^2 = \text{Energy,Work,Heat Quantity}^3 \cdot \text{Inductance}$

$h \cdot \text{Electric Potential Difference} = \text{Energy,Work,Heat Quantity} \cdot \text{Magnetic Flux}$

$h = \text{Energy,Work,Heat Quantity} \cdot \text{Capacitance} \cdot \text{Electric Resistance}$

$h^2 = \text{Energy,Work,Heat Quantity}^2 \cdot \text{Capacitance} \cdot \text{Inductance}$

$h^2 = \text{Energy,Work,Heat Quantity} \cdot \text{Capacitance} \cdot \text{Magnetic Flux}^2$

$h \cdot \text{Electric Resistance} = \text{Energy,Work,Heat Quantity} \cdot \text{Inductance}$

$h^4 \cdot \text{Absorbed Dose}^3 \cdot \text{Dynamic Viscosity}^2 = \text{Energy,Work,Heat Quantity}^6$

$h \cdot \text{Absorbed Dose} = \text{Energy,Work,Heat Quantity} \cdot \text{Kinematic Viscosity}$

$h^6 \cdot \text{Absorbed Dose}^5 \cdot \text{Mass Density}^2 = \text{Energy,Work,Heat Quantity}^8$

$h \cdot \text{Dynamic Viscosity}^2 \cdot \text{Kinematic Viscosity}^3 = \text{Energy,Work,Heat Quantity}^3$

$h \cdot \text{Dynamic Viscosity}^5 = \text{Energy,Work,Heat Quantity}^3 \cdot \text{Mass Density}^3$

$h \cdot \text{Kinematic Viscosity}^5 \cdot \text{Mass Density}^2 = \text{Energy,Work,Heat Quantity}^3$

$h \cdot \text{Power, Radiance Flux} = \text{Electric Charge}^2 \cdot \text{Electric Potential Difference}^2$

$h \cdot \text{Power, Radiance Flux} \cdot \text{Capacitance}^2 = \text{Electric Charge}^4$

$h^3 = \text{Power, Radiance Flux} \cdot \text{Electric Charge}^4 \cdot \text{Inductance}^2$

$h \cdot \text{Power, Radiance Flux} = \text{Electric Potential Difference}^4 \cdot \text{Capacitance}^2$

$h \cdot \text{Electric Potential Difference}^4 = \text{Power, Radiance Flux}^3 \cdot \text{Inductance}^2$

$h \cdot \text{Electric Potential Difference}^2 = \text{Power, Radiance Flux} \cdot \text{Magnetic Flux}^2$

$h = \text{Power, Radiance Flux} \cdot \text{Capacitance}^2 \cdot \text{Electric Resistance}^2$

$h = \text{Power, Radiance Flux} \cdot \text{Capacitance} \cdot \text{Inductance}$

$h^3 = \text{Power, Radiance Flux} \cdot \text{Capacitance}^2 \cdot \text{Magnetic Flux}^4$

$h \cdot$ Electric Resistance^2 = Power, Radiance Flux $\cdot$ Inductance^2

$h \cdot$ Power, Radiance Flux $\cdot$ Inductance^2 = Magnetic Flux^4

$h \cdot$ Absorbed Dose$^3 \cdot$ Dynamic Viscosity^2 = Power, Radiance Flux^3

$h \cdot$ Absorbed Dose^2 = Power, Radiance Flux $\cdot$ Kinematic Viscosity^2

$h^2 \cdot$ Absorbed Dose$^5 \cdot$ Mass Density^2 = Power, Radiance Flux^4

$h \cdot$ Power, Radiance Flux^3 = Dynamic Viscosity$^4 \cdot$ Kinematic Viscosity^6

h^2 = Electric Charge$^3 \cdot$ Electric Potential Difference $\cdot$ Inductance

$h^2 \cdot$ Capacitance = Electric Charge$^4 \cdot$ Inductance

$h \cdot$ Dynamic Viscosity^2 = Electric Charge$^3 \cdot$ Magnetic Flux Density^3

h = Electric Potential Difference$^2 \cdot$ Capacitance$^2 \cdot$ Electric Resistance

h^2 = Electric Potential Difference$^4 \cdot$ Capacitance$^3 \cdot$ Inductance

h = Electric Potential Difference $\cdot$ Capacitance $\cdot$ Magnetic Flux

$h \cdot$ Electric Resistance^3 = Electric Potential Difference$^2 \cdot$ Inductance^2

$h \cdot$ Electric Potential Difference $\cdot$ Inductance = Magnetic Flux^3

$h^2 \cdot$ Inductance = Capacitance $\cdot$ Magnetic Flux^4

$h \cdot$ Magnetic Flux Density^6 = Electric Resistance$^3 \cdot$ Dynamic Viscosity^4

$h^2 \cdot$ Magnetic Flux Density^3 = Magnetic Flux$^3 \cdot$ Dynamic Viscosity^2

$h^2 \cdot$ Absorbed Dose^3 = Dynamic Viscosity$^2 \cdot$ Kinematic Viscosity^6

$h^2 \cdot$ Absorbed Dose$^3 \cdot$ Mass Density^6 = Dynamic Viscosity^8

$h^2 \cdot$ Absorbed Dose^3 = Kinematic Viscosity$^8 \cdot$ Mass Density^2

$G_c \cdot$ Amount of a Substance^2 = Specific Volume $\cdot$ Catalytic Activity^2

$G_c \cdot$ Amount of a Substance$^2 \cdot$ Mass Density = Catalytic Activity^2

$G_c \cdot$ Force^5 = Electric Current$^4 \cdot$ Electric Potential Difference^4

$G_c \cdot \text{Force}^5 = \text{Electric Current}^8 \cdot \text{Electric Resistance}^4$

$G_c \cdot \text{Electric Current} \cdot \text{Electric Potential Difference} = \text{Velocity}^5$

$G_c \cdot \text{Electric Current}^2 \cdot \text{Electric Resistance} = \text{Velocity}^5$

$G_c \cdot \text{Electric Charge}^2 = \text{Electric Current}^2 \cdot \text{Specific Volume}$

$G_c \cdot \text{Electric Charge}^2 \cdot \text{Mass Density} = \text{Electric Current}^2$

$G_c^2 \cdot \text{Electric Current}^2 \cdot \text{Electric Potential Difference}^2 = \text{Absorbed Dose}^5$

$G_c^2 \cdot \text{Electric Current}^4 \cdot \text{Electric Resistance}^2 = \text{Absorbed Dose}^5$

$G_c \cdot \text{Time}^4 \cdot \text{Force} = \text{Length}^4$

$G_c \cdot \text{Time}^4 \cdot \text{Energy} = \text{Length}^5$

$G_c \cdot \text{Time}^2 \cdot \text{Mass} = \text{Length}^3$

$G_c \cdot \text{Time}^4 \cdot \text{Pressure,Stress} = \text{Length}^2$

$G_c \cdot \text{Time}^4 \cdot \text{Energy,Work,Heat Quantity} = \text{Length}^5$

$G_c \cdot \text{Time}^5 \cdot \text{Power, Radiance Flux} = \text{Length}^5$

$G_c \cdot \text{Time}^3 \cdot \text{Dynamic Viscosity} = \text{Length}^2$

$G_c^3 \cdot \text{Time}^8 \cdot \text{Force}^3 = \text{Volumetric Velocity}^4$

$G_c^3 \cdot \text{Time}^7 \cdot \text{Energy}^3 = \text{Volumetric Velocity}^5$

$G_c \cdot \text{Time} \cdot \text{Mass} = \text{Volumetric Velocity}$

$G_c^3 \cdot \text{Time}^7 \cdot \text{Energy,Work,Heat Quantity}^3 = \text{Volumetric Velocity}^5$

$G_c^3 \cdot \text{Time}^7 \cdot \text{Dynamic Viscosity}^3 = \text{Volumetric Velocity}^2$

$G_c \cdot \text{Time}^4 \cdot \text{Force}^5 = \text{Energy}^4$

$G_c \cdot \text{Force} = \text{Time}^4 \cdot \text{Acceleration}^4$

$G_c \cdot \text{Mass}^4 = \text{Time}^4 \cdot \text{Force}^3$

$$G_c \cdot Time^4 \cdot Force = Area^2$$

$$G_c \cdot Time^4 \cdot Force \cdot Wavenumber^4 = A\ Constant$$

$$G_c \cdot Time^4 \cdot Pressure,Stress^2 = Force$$

$$G_c \cdot Time^4 \cdot Force^5 = Energy,Work,Heat\ Quantity^4$$

$$G_c \cdot Time^2 \cdot Dynamic\ Viscosity^2 = Force$$

$$G_c \cdot Time^2 \cdot Force = Kinematic\ Viscosity^2$$

$$G_c \cdot Energy = Time^6 \cdot Acceleration^5$$

$$G_c^2 \cdot Mass^5 = Time^2 \cdot Energy^3$$

$$G_c \cdot Energy = Time \cdot Velocity^5$$

$$G_c^2 \cdot Time^8 \cdot Energy^2 = Area^5$$

$$G_c \cdot Time^4 \cdot Energy \cdot Wavenumber^5 = A\ Constant$$

$$G_c^2 \cdot Energy^2 = Time^2 \cdot Absorbed\ Dose^5$$

$$G_c^3 \cdot Time^7 \cdot Dynamic\ Viscosity^5 = Energy^2$$

$$G_c^2 \cdot Time^3 \cdot Energy^2 = Kinematic\ Viscosity^5$$

$$G_c \cdot Mass = Time^4 \cdot Acceleration^3$$

$$G_c \cdot Energy,Work,Heat\ Quantity = Time^6 \cdot Acceleration^5$$

$$G_c \cdot Power,\ Radiance\ Flux = Time^5 \cdot Acceleration^5$$

$$G_c \cdot Dynamic\ Viscosity = Time \cdot Acceleration^2$$

$$G_c \cdot Mass = Time \cdot Velocity^3$$

$$G_c^2 \cdot Time^4 \cdot Mass^2 = Area^3$$

$$G_c \cdot Time^2 \cdot Mass = Volume$$

$$G_c \cdot Time^2 \cdot Mass \cdot Wavenumber^3 = A\ Constant$$

$$G_c \cdot Time^8 \cdot Pressure,Stress^3 = Mass^2$$

$G_c^2 \cdot Mass^5 = Time^2 \cdot Energy,Work,Heat\ Quantity^3$

$G_c^2 \cdot Mass^5 = Time^5 \cdot Power,\ Radiance\ Flux^3$

$G_c^2 \cdot Mass^2 = Time^2 \cdot Absorbed\ Dose^3$

$G_c \cdot Time^5 \cdot Dynamic\ Viscosity^3 = Mass^2$

$G_c^2 \cdot Time \cdot Mass^2 = Kinematic\ Viscosity^3$

$G_c \cdot Time^2 \cdot Pressure,Stress = Velocity^2$

$G_c \cdot Energy,Work,Heat\ Quantity = Time \cdot Velocity^5$

$G_c \cdot Time \cdot Dynamic\ Viscosity = Velocity^2$

$G_c \cdot Time^4 \cdot Pressure,Stress = Area$

$G_c^2 \cdot Time^8 \cdot Energy,Work,Heat\ Quantity^2 = Area^5$

$G_c \cdot Time^3 \cdot Dynamic\ Viscosity = Area$

$G_c^3 \cdot Time^9 \cdot Dynamic\ Viscosity^3 = Volume^2$

$G_c \cdot Time^4 \cdot Wavenumber^2 \cdot Pressure,Stress = A\ Constant$

$G_c \cdot Time^4 \cdot Wavenumber^5 \cdot Energy,Work,Heat\ Quantity = A\ Constant$

$G_c \cdot Time^5 \cdot Wavenumber^5 \cdot Power,\ Radiance\ Flux = A\ Constant$

$G_c \cdot Time^3 \cdot Wavenumber^2 \cdot Dynamic\ Viscosity = A\ Constant$

$G_c \cdot Time^2 \cdot Pressure,Stress = Absorbed\ Dose$

$G_c \cdot Time^3 \cdot Pressure,Stress = Kinematic\ Viscosity$

$G_c^2 \cdot Energy,Work,Heat\ Quantity^2 = Time^2 \cdot Absorbed\ Dose^5$

$G_c^3 \cdot Time^7 \cdot Dynamic\ Viscosity^5 = Energy,Work,Heat\ Quantity^2$

$G_c^2 \cdot Time^3 \cdot Energy,Work,Heat\ Quantity^2 = Kinematic\ Viscosity^5$

$G_c^3 \cdot Time^5 \cdot Dynamic\ Viscosity^5 = Power,\ Radiance\ Flux^2$

$G_c{}^2 \cdot \text{Time}^5 \cdot \text{Power, Radiance Flux}^2 = \text{Kinematic Viscosity}^5$

$G_c \cdot \text{Time} \cdot \text{Dynamic Viscosity} = \text{Absorbed Dose}$

$G_c \cdot \text{Time}^2 \cdot \text{Dynamic Viscosity} = \text{Kinematic Viscosity}$

$G_c \cdot \text{Length}^8 \cdot \text{Force} = \text{Volumetric Velocity}^4$

$G_c \cdot \text{Length}^7 \cdot \text{Energy} = \text{Volumetric Velocity}^4$

$G_c \cdot \text{Length}^3 \cdot \text{Mass} = \text{Volumetric Velocity}^2$

$G_c \cdot \text{Length}^6 = \text{Volumetric Velocity}^2 \cdot \text{Specific Volume}$

$G_c \cdot \text{Length}^7 \cdot \text{Energy,Work,Heat Quantity} = \text{Volumetric Velocity}^4$

$G_c \cdot \text{Length}^7 \cdot \text{Dynamic Viscosity} = \text{Volumetric Velocity}^3$

$G_c \cdot \text{Length}^6 \cdot \text{Mass Density} = \text{Volumetric Velocity}^2$

$G_c \cdot \text{Force} = \text{Length}^2 \cdot \text{Acceleration}^2$

$G_c \cdot \text{Mass}^2 = \text{Length}^2 \cdot \text{Force}$

$G_c \cdot \text{Length}^4 = \text{Force} \cdot \text{Specific Volume}^2$

$G_c \cdot \text{Force} = \text{Length}^4 \cdot \text{Frequency}^4$

$G_c \cdot \text{Length}^4 \cdot \text{Dynamic Viscosity}^4 = \text{Force}^3$

$G_c \cdot \text{Length}^4 \cdot \text{Force} = \text{Kinematic Viscosity}^4$

$G_c \cdot \text{Length}^4 \cdot \text{Mass Density}^2 = \text{Force}$

$G_c \cdot \text{Energy} = \text{Length}^3 \cdot \text{Acceleration}^2$

$G_c \cdot \text{Mass}^2 = \text{Length} \cdot \text{Energy}$

$G_c \cdot \text{Energy} = \text{Length} \cdot \text{Velocity}^4$

$G_c \cdot \text{Length}^5 = \text{Energy} \cdot \text{Specific Volume}^2$

$G_c \cdot \text{Energy} = \text{Length}^5 \cdot \text{Frequency}^4$

$G_c \cdot \text{Energy}^5 = \text{Length}^5 \cdot \text{Power, Radiance Flux}^4$

$G_c \cdot$ Energy = Length $\cdot$ Absorbed Dose^2

$G_c \cdot$ Length$^7 \cdot$ Dynamic Viscosity^4 = Energy^3

$G_c \cdot$ Length$^3 \cdot$ Energy = Kinematic Viscosity^4

$G_c \cdot$ Length$^5 \cdot$ Mass Density^2 = Energy

$G_c \cdot$ Mass = Length$^2 \cdot$ Acceleration

$G_c \cdot$ Length = Acceleration $\cdot$ Specific Volume

$G_c \cdot$ Energy,Work,Heat Quantity = Length$^3 \cdot$ Acceleration^2

$G_c^2 \cdot$ Power, Radiance Flux^2 = Length$^5 \cdot$ Acceleration^5

$G_c^2 \cdot$ Dynamic Viscosity^2 = Length $\cdot$ Acceleration^3

$G_c \cdot$ Length $\cdot$ Mass Density = Acceleration

$G_c \cdot$ Mass = Length $\cdot$ Velocity^2

$G_c \cdot$ Mass = Length$^3 \cdot$ Frequency^2

$G_c \cdot$ Mass^2 = Length$^4 \cdot$ Pressure,Stress

$G_c \cdot$ Mass^2 = Length $\cdot$ Energy,Work,Heat Quantity

$G_c^3 \cdot$ Mass^5 = Length$^5 \cdot$ Power, Radiance Flux^2

$G_c \cdot$ Mass = Length $\cdot$ Absorbed Dose

$G_c \cdot$ Mass^3 = Length$^5 \cdot$ Dynamic Viscosity^2

$G_c \cdot$ Length $\cdot$ Mass = Kinematic Viscosity^2

$G_c \cdot$ Length^2 = Velocity$^2 \cdot$ Specific Volume

$G_c \cdot$ Length$^2 \cdot$ Pressure,Stress = Velocity^4

$G_c \cdot$ Energy,Work,Heat Quantity = Length $\cdot$ Velocity^4

$G_c \cdot$ Length $\cdot$ Dynamic Viscosity = Velocity^3

$G_c \cdot \text{Length}^2 \cdot \text{Mass Density} = \text{Velocity}^2$

$G_c \cdot \text{Length}^2 = \text{Specific Volume}^2 \cdot \text{Pressure,Stress}$

$G_c \cdot \text{Length}^5 = \text{Specific Volume}^2 \cdot \text{Energy,Work,Heat Quantity}$

$G_c \cdot \text{Length}^2 = \text{Specific Volume} \cdot \text{Absorbed Dose}$

$G_c \cdot \text{Length}^4 = \text{Specific Volume}^3 \cdot \text{Dynamic Viscosity}^2$

$G_c \cdot \text{Length}^4 = \text{Specific Volume} \cdot \text{Kinematic Viscosity}^2$

$G_c \cdot \text{Pressure,Stress} = \text{Length}^2 \cdot \text{Frequency}^4$

$G_c \cdot \text{Energy,Work,Heat Quantity} = \text{Length}^5 \cdot \text{Frequency}^4$

$G_c \cdot \text{Power, Radiance Flux} = \text{Length}^5 \cdot \text{Frequency}^5$

$G_c \cdot \text{Dynamic Viscosity} = \text{Length}^2 \cdot \text{Frequency}^3$

$G_c \cdot \text{Length}^2 \cdot \text{Pressure,Stress} = \text{Absorbed Dose}^2$

$G_c \cdot \text{Dynamic Viscosity}^4 = \text{Length}^2 \cdot \text{Pressure,Stress}^3$

$G_c \cdot \text{Length}^6 \cdot \text{Pressure,Stress} = \text{Kinematic Viscosity}^4$

$G_c \cdot \text{Length}^2 \cdot \text{Mass Density}^2 = \text{Pressure,Stress}$

$G_c \cdot \text{Energy,Work,Heat Quantity}^5 = \text{Length}^5 \cdot \text{Power, Radiance Flux}^4$

$G_c \cdot \text{Energy,Work,Heat Quantity} = \text{Length} \cdot \text{Absorbed Dose}^2$

$G_c \cdot \text{Length}^7 \cdot \text{Dynamic Viscosity}^4 = \text{Energy,Work,Heat Quantity}^3$

$G_c \cdot \text{Length}^3 \cdot \text{Energy,Work,Heat Quantity} = \text{Kinematic Viscosity}^4$

$G_c \cdot \text{Length}^5 \cdot \text{Mass Density}^2 = \text{Energy,Work,Heat Quantity}$

$G_c^2 \cdot \text{Length}^5 \cdot \text{Dynamic Viscosity}^5 = \text{Power, Radiance Flux}^3$

$G_c \cdot \text{Length}^5 \cdot \text{Power, Radiance Flux} = \text{Kinematic Viscosity}^5$

$G_c^2 \cdot \text{Length}^2 \cdot \text{Dynamic Viscosity}^2 = \text{Absorbed Dose}^3$

$G_c \cdot \text{Length}^2 \cdot \text{Mass Density} = \text{Absorbed Dose}$

$G_c \cdot Length^4 \cdot Dynamic\ Viscosity = Kinematic\ Viscosity^3$

$G_c \cdot Length^4 \cdot Mass\ Density^3 = Dynamic\ Viscosity^2$

$G_c \cdot Length^4 \cdot Mass\ Density = Kinematic\ Viscosity^2$

$G_c \cdot Energy^8 = Volumetric\ Velocity^4 \cdot Force^7$

$G_c^5 \cdot Force^5 = Volumetric\ Velocity^4 \cdot Acceleration^8$

$G_c^5 \cdot Mass^8 = Volumetric\ Velocity^4 \cdot Force^3$

$G_c \cdot Force \cdot Area^4 = Volumetric\ Velocity^4$

$G_c \cdot Force = Volumetric\ Velocity^4 \cdot Wavenumber^8$

$G_c \cdot Volumetric\ Velocity^4 = Force^3 \cdot Specific\ Volume^4$

$G_c^3 \cdot Force^3 = Volumetric\ Velocity^4 \cdot Frequency^8$

$G_c \cdot Force^5 = Volumetric\ Velocity^4 \cdot Pressure,Stress^4$

$G_c \cdot Energy,Work,Heat\ Quantity^8 = Volumetric\ Velocity^4 \cdot Force^7$

$G_c \cdot Volumetric\ Velocity^4 \cdot Dynamic\ Viscosity^8 = Force^7$

$G_c \cdot Volumetric\ Velocity^4 \cdot Force = Kinematic\ Viscosity^8$

$G_c \cdot Volumetric\ Velocity^4 \cdot Mass\ Density^4 = Force^3$

$G_c^5 \cdot Energy^5 = Volumetric\ Velocity^6 \cdot Acceleration^7$

$G_c^4 \cdot Mass^7 = Volumetric\ Velocity^2 \cdot Energy^3$

$G_c^2 \cdot Energy^2 = Volumetric\ Velocity \cdot Velocity^7$

$G_c^2 \cdot Energy^2 \cdot Area^7 = Volumetric\ Velocity^8$

$G_c \cdot Energy = Volumetric\ Velocity^4 \cdot Wavenumber^7$

$G_c^3 \cdot Energy^3 = Volumetric\ Velocity^5 \cdot Frequency^7$

$G_c^4 \cdot Energy^4 = Volumetric\ Velocity^2 \cdot Absorbed\ Dose^7$

$G_c \cdot$ Volumetric Velocity$^3 \cdot$ Energy = Kinematic Viscosity^7

$G_c^5 \cdot$ Mass^5 = Volumetric Velocity$^4 \cdot$ Acceleration^3

$G_c^5 \cdot$ Volumetric Velocity^2 = Acceleration$^6 \cdot$ Specific Volume^5

$G_c^5 \cdot$ Energy,Work,Heat Quantity^5 = Volumetric Velocity$^6 \cdot$ Acceleration^7

$G_c \cdot$ Power, Radiance Flux = Volumetric Velocity $\cdot$ Acceleration^2

$G_c^5 \cdot$ Dynamic Viscosity^5 = Volumetric Velocity $\cdot$ Acceleration^7

$G_c^5 \cdot$ Volumetric Velocity$^2 \cdot$ Mass Density^5 = Acceleration^6

$G_c^2 \cdot$ Mass^2 = Volumetric Velocity $\cdot$ Velocity^3

$G_c^2 \cdot$ Mass$^2 \cdot$ Area^3 = Volumetric Velocity^4

$G_c \cdot$ Mass $\cdot$ Volume = Volumetric Velocity^2

$G_c \cdot$ Mass = Volumetric Velocity$^2 \cdot$ Wavenumber^3

$G_c \cdot$ Mass$^2 \cdot$ Specific Volume = Volumetric Velocity^2

$G_c \cdot$ Mass = Volumetric Velocity $\cdot$ Frequency

$G_c^4 \cdot$ Mass^7 = Volumetric Velocity$^2 \cdot$ Energy,Work,Heat Quantity^3

$G_c^4 \cdot$ Mass^4 = Volumetric Velocity$^2 \cdot$ Absorbed Dose^3

$G_c^4 \cdot$ Mass^7 = Volumetric Velocity$^5 \cdot$ Dynamic Viscosity^3

$G_c \cdot$ Volumetric Velocity $\cdot$ Mass = Kinematic Viscosity^3

$G_c \cdot$ Mass^2 = Volumetric Velocity$^2 \cdot$ Mass Density

$G_c \cdot$ Volumetric Velocity = Velocity$^3 \cdot$ Specific Volume

$G_c \cdot$ Volumetric Velocity $\cdot$ Pressure,Stress = Velocity^5

$G_c^2 \cdot$ Energy,Work,Heat Quantity^2 = Volumetric Velocity $\cdot$ Velocity^7

$G_c^2 \cdot$ Volumetric Velocity $\cdot$ Dynamic Viscosity^2 = Velocity^7

$G_c \cdot$ Volumetric Velocity $\cdot$ Mass Density = Velocity^3

$G_c \cdot Area^3 = Volumetric\ Velocity^2 \cdot Specific\ Volume$

$G_c \cdot Area^5 \cdot Pressure, Stress = Volumetric\ Velocity^4$

$G_c^2 \cdot Area^7 \cdot Energy, Work, Heat\ Quantity^2 = Volumetric\ Velocity^8$

$G_c \cdot Area^5 \cdot Power,\ Radiance\ Flux = Volumetric\ Velocity^5$

$G_c^2 \cdot Area^7 \cdot Dynamic\ Viscosity^2 = Volumetric\ Velocity^6$

$G_c \cdot Area^3 \cdot Mass\ Density = Volumetric\ Velocity^2$

$G_c \cdot Volume^2 = Volumetric\ Velocity^2 \cdot Specific\ Volume$

$G_c^3 \cdot Volume^7 \cdot Dynamic\ Viscosity^3 = Volumetric\ Velocity^9$

$G_c \cdot Volume^2 \cdot Mass\ Density = Volumetric\ Velocity^2$

$G_c = Volumetric\ Velocity^2 \cdot Wavenumber^6 \cdot Specific\ Volume$

$G_c \cdot Energy, Work, Heat\ Quantity = Volumetric\ Velocity^4 \cdot Wavenumber^7$

$G_c \cdot Dynamic\ Viscosity = Volumetric\ Velocity^3 \cdot Wavenumber^7$

$G_c \cdot Mass\ Density = Volumetric\ Velocity^2 \cdot Wavenumber^6$

$G_c^2 \cdot Volumetric\ Velocity^2 = Specific\ Volume^5 \cdot Pressure, Stress^3$

$G_c^2 \cdot Volumetric\ Velocity^5 = Specific\ Volume^5 \cdot Power,\ Radiance\ Flux^3$

$G_c^2 \cdot Volumetric\ Velocity^2 = Specific\ Volume^2 \cdot Absorbed\ Dose^3$

$G_c \cdot Volumetric\ Velocity^4 = Specific\ Volume^7 \cdot Dynamic\ Viscosity^6$

$G_c \cdot Volumetric\ Velocity^4 = Specific\ Volume \cdot Kinematic\ Viscosity^6$

$G_c^3 \cdot Energy, Work, Heat\ Quantity^3 = Volumetric\ Velocity^5 \cdot Frequency^7$

$G_c^3 \cdot Dynamic\ Viscosity^3 = Volumetric\ Velocity^2 \cdot Frequency^7$

$G_c^2 \cdot Volumetric\ Velocity^2 \cdot Pressure, Stress^2 = Absorbed\ Dose^5$

$G_c^2 \cdot Volumetric\ Velocity^2 \cdot Mass\ Density^5 = Pressure, Stress^3$

$G_c{}^4 \cdot \text{Energy,Work,Heat Quantity}^4 = \text{Volumetric Velocity}^2 \cdot \text{Absorbed Dose}^7$

$G_c \cdot \text{Volumetric Velocity}^3 \cdot \text{Energy,Work,Heat Quantity} = \text{Kinematic Viscosity}^7$

$G_c{}^2 \cdot \text{Volumetric Velocity}^5 \cdot \text{Mass Density}^5 = \text{Power, Radiance Flux}^3$

$G_c{}^4 \cdot \text{Volumetric Velocity}^2 \cdot \text{Dynamic Viscosity}^4 = \text{Absorbed Dose}^7$

$G_c{}^2 \cdot \text{Volumetric Velocity}^2 \cdot \text{Mass Density}^2 = \text{Absorbed Dose}^3$

$G_c \cdot \text{Volumetric Velocity}^4 \cdot \text{Dynamic Viscosity} = \text{Kinematic Viscosity}^7$

$G_c \cdot \text{Volumetric Velocity}^4 \cdot \text{Mass Density}^7 = \text{Dynamic Viscosity}^6$

$G_c \cdot \text{Volumetric Velocity}^4 \cdot \text{Mass Density} = \text{Kinematic Viscosity}^6$

$G_c \cdot \text{Force}^3 = \text{Energy}^2 \cdot \text{Acceleration}^2$

$G_c \cdot \text{Force} \cdot \text{Mass}^2 = \text{Energy}^2$

$G_c \cdot \text{Energy}^4 = \text{Force}^5 \cdot \text{Specific Volume}^2$

$G_c \cdot \text{Force}^5 = \text{Energy}^4 \cdot \text{Frequency}^4$

$G_c \cdot \text{Energy}^4 \cdot \text{Dynamic Viscosity}^4 = \text{Force}^7$

$G_c \cdot \text{Energy}^4 = \text{Force}^3 \cdot \text{Kinematic Viscosity}^4$

$G_c \cdot \text{Energy}^4 \cdot \text{Mass Density}^2 = \text{Force}^5$

$G_c \cdot \text{Force} = \text{Acceleration}^2 \cdot \text{Area}$

$G_c{}^3 \cdot \text{Force}^3 = \text{Acceleration}^6 \cdot \text{Volume}^2$

$G_c \cdot \text{Force} \cdot \text{Wavenumber}^2 = \text{Acceleration}^2$

$G_c{}^3 \cdot \text{Force} = \text{Acceleration}^4 \cdot \text{Specific Volume}^2$

$G_c \cdot \text{Force} \cdot \text{Frequency}^4 = \text{Acceleration}^4$

$G_c \cdot \text{Force}^3 = \text{Acceleration}^2 \cdot \text{Energy,Work,Heat Quantity}^2$

$G_c{}^3 \cdot \text{Dynamic Viscosity}^4 = \text{Force} \cdot \text{Acceleration}^4$

$G_c{}^3 \cdot \text{Force}^3 = \text{Acceleration}^4 \cdot \text{Kinematic Viscosity}^4$

$$G_c{}^3 \cdot \text{Force} \cdot \text{Mass Density}^2 = \text{Acceleration}^4$$

$$G_c \cdot \text{Mass}^2 = \text{Force} \cdot \text{Area}$$

$$G_c{}^3 \cdot \text{Mass}^6 = \text{Force}^3 \cdot \text{Volume}^2$$

$$G_c \cdot \text{Mass}^2 \cdot \text{Wavenumber}^2 = \text{Force}$$

$$G_c{}^3 \cdot \text{Mass}^4 = \text{Force}^3 \cdot \text{Specific Volume}^2$$

$$G_c \cdot \text{Mass}^4 \cdot \text{Frequency}^4 = \text{Force}^3$$

$$G_c \cdot \text{Mass}^2 \cdot \text{Pressure,Stress} = \text{Force}^2$$

$$G_c \cdot \text{Force} \cdot \text{Mass}^2 = \text{Energy,Work,Heat Quantity}^2$$

$$G_c{}^3 \cdot \text{Mass}^4 \cdot \text{Dynamic Viscosity}^4 = \text{Force}^5$$

$$G_c{}^3 \cdot \text{Mass}^4 = \text{Force} \cdot \text{Kinematic Viscosity}^4$$

$$G_c{}^3 \cdot \text{Mass}^4 \cdot \text{Mass Density}^2 = \text{Force}^3$$

$$G_c \cdot \text{Area}^2 = \text{Force} \cdot \text{Specific Volume}^2$$

$$G_c \cdot \text{Force} = \text{Area}^2 \cdot \text{Frequency}^4$$

$$G_c \cdot \text{Area}^2 \cdot \text{Dynamic Viscosity}^4 = \text{Force}^3$$

$$G_c \cdot \text{Force} \cdot \text{Area}^2 = \text{Kinematic Viscosity}^4$$

$$G_c \cdot \text{Area}^2 \cdot \text{Mass Density}^2 = \text{Force}$$

$$G_c{}^3 \cdot \text{Volume}^4 = \text{Force}^3 \cdot \text{Specific Volume}^6$$

$$G_c{}^3 \cdot \text{Volume}^4 \cdot \text{Mass Density}^6 = \text{Force}^3$$

$$G_c = \text{Force} \cdot \text{Wavenumber}^4 \cdot \text{Specific Volume}^2$$

$$G_c \cdot \text{Force} \cdot \text{Wavenumber}^4 = \text{Frequency}^4$$

$$G_c \cdot \text{Dynamic Viscosity}^4 = \text{Force}^3 \cdot \text{Wavenumber}^4$$

$$G_c \cdot \text{Force} = \text{Wavenumber}^4 \cdot \text{Kinematic Viscosity}^4$$

$G_c \cdot$ Mass Density^2 = Force $\cdot$ Wavenumber^4

$G_c \cdot$ Force = Specific Volume$^2 \cdot$ Pressure,Stress^2

$G_c \cdot$ Energy,Work,Heat Quantity^4 = Force$^5 \cdot$ Specific Volume^2

$G_c \cdot$ Pressure,Stress^2 = Force $\cdot$ Frequency^4

$G_c \cdot$ Force^5 = Frequency$^4 \cdot$ Energy,Work,Heat Quantity^4

$G_c \cdot$ Dynamic Viscosity^2 = Force $\cdot$ Frequency^2

$G_c \cdot$ Force = Frequency$^2 \cdot$ Kinematic Viscosity^2

$G_c \cdot$ Dynamic Viscosity^4 = Force $\cdot$ Pressure,Stress^2

$G_c \cdot$ Force^3 = Pressure,Stress$^2 \cdot$ Kinematic Viscosity^4

$G_c \cdot$ Force $\cdot$ Mass Density^2 = Pressure,Stress^2

$G_c \cdot$ Energy,Work,Heat Quantity$^4 \cdot$ Dynamic Viscosity^4 = Force^7

$G_c \cdot$ Energy,Work,Heat Quantity^4 = Force$^3 \cdot$ Kinematic Viscosity^4

$G_c \cdot$ Energy,Work,Heat Quantity$^4 \cdot$ Mass Density^2 = Force^5

$G_c \cdot$ Electric Charge$^4 \cdot$ Magnetic Flux Density^4 = Force^3

$G_c \cdot$ Force$^5 \cdot$ Electric Resistance^4 = Electric Potential Difference^8

$G_c \cdot$ Acceleration $\cdot$ Mass^3 = Energy^2

$G_c \cdot$ Energy $\cdot$ Acceleration = Velocity^6

$G_c^2 \cdot$ Energy^2 = Acceleration$^4 \cdot$ Area^3

$G_c \cdot$ Energy = Acceleration$^2 \cdot$ Volume

$G_c \cdot$ Energy $\cdot$ Wavenumber^3 = Acceleration^2

$G_c^4 \cdot$ Energy = Acceleration$^5 \cdot$ Specific Volume^3

$G_c \cdot$ Energy $\cdot$ Frequency^6 = Acceleration^5

$G_c \cdot$ Power, Radiance Flux^6 = Energy$^5 \cdot$ Acceleration^5

$G_c \cdot$ Energy $\cdot$ Acceleration = Absorbed Dose^3

$G_c{}^5 \cdot$ Dynamic Viscosity^6 = Energy $\cdot$ Acceleration^7

$G_c \cdot$ Energy = Acceleration $\cdot$ Kinematic Viscosity^2

$G_c{}^4 \cdot$ Energy $\cdot$ Mass Density^3 = Acceleration^5

$G_c{}^2 \cdot$ Mass^4 = Energy$^2 \cdot$ Area

$G_c{}^3 \cdot$ Mass^6 = Energy$^3 \cdot$ Volume

$G_c \cdot$ Mass$^2 \cdot$ Wavenumber = Energy

$G_c{}^3 \cdot$ Mass^5 = Energy$^3 \cdot$ Specific Volume

$G_c{}^2 \cdot$ Mass$^5 \cdot$ Frequency^2 = Energy^3

$G_c{}^3 \cdot$ Mass$^6 \cdot$ Pressure,Stress = Energy^4

$G_c{}^2 \cdot$ Mass$^5 \cdot$ Power, Radiance Flux^2 = Energy^5

$G_c{}^4 \cdot$ Mass$^7 \cdot$ Dynamic Viscosity^2 = Energy^5

$G_c{}^2 \cdot$ Mass^3 = Energy $\cdot$ Kinematic Viscosity^2

$G_c{}^3 \cdot$ Mass$^5 \cdot$ Mass Density = Energy^3

$G_c{}^2 \cdot$ Energy^2 = Velocity$^8 \cdot$ Area

$G_c \cdot$ Energy $\cdot$ Wavenumber = Velocity^4

$G_c \cdot$ Energy $\cdot$ Frequency = Velocity^5

$G_c{}^2 \cdot$ Energy $\cdot$ Dynamic Viscosity = Velocity^7

$G_c \cdot$ Energy = Velocity$^3 \cdot$ Kinematic Viscosity

$G_c{}^2 \cdot$ Area^5 = Energy$^2 \cdot$ Specific Volume^4

$G_c{}^2 \cdot$ Energy^2 = Area$^5 \cdot$ Frequency^8

$G_c{}^2 \cdot$ Energy^2 = Area $\cdot$ Absorbed Dose^4

$$G_c{}^2 \cdot \text{Area}^7 \cdot \text{Dynamic Viscosity}^8 = \text{Energy}^6$$

$$G_c{}^2 \cdot \text{Energy}^2 \cdot \text{Area}^3 = \text{Kinematic Viscosity}^8$$

$$G_c{}^2 \cdot \text{Area}^5 \cdot \text{Mass Density}^4 = \text{Energy}^2$$

$$G_c{}^3 \cdot \text{Volume}^5 = \text{Energy}^3 \cdot \text{Specific Volume}^6$$

$$G_c{}^3 \cdot \text{Energy}^3 = \text{Volume} \cdot \text{Absorbed Dose}^6$$

$$G_c \cdot \text{Energy} \cdot \text{Volume} = \text{Kinematic Viscosity}^4$$

$$G_c{}^3 \cdot \text{Volume}^5 \cdot \text{Mass Density}^6 = \text{Energy}^3$$

$$G_c = \text{Energy} \cdot \text{Wavenumber}^5 \cdot \text{Specific Volume}^2$$

$$G_c \cdot \text{Energy} \cdot \text{Wavenumber}^5 = \text{Frequency}^4$$

$$G_c \cdot \text{Energy}^5 \cdot \text{Wavenumber}^5 = \text{Power, Radiance Flux}^4$$

$$G_c \cdot \text{Energy} \cdot \text{Wavenumber} = \text{Absorbed Dose}^2$$

$$G_c \cdot \text{Dynamic Viscosity}^4 = \text{Energy}^3 \cdot \text{Wavenumber}^7$$

$$G_c \cdot \text{Energy} = \text{Wavenumber}^3 \cdot \text{Kinematic Viscosity}^4$$

$$G_c \cdot \text{Mass Density}^2 = \text{Energy} \cdot \text{Wavenumber}^5$$

$$G_c{}^3 \cdot \text{Energy}^2 = \text{Specific Volume}^6 \cdot \text{Pressure,Stress}^5$$

$$G_c \cdot \text{Energy}^2 = \text{Specific Volume} \cdot \text{Power, Radiance Flux}^2$$

$$G_c{}^3 \cdot \text{Energy}^2 = \text{Specific Volume} \cdot \text{Absorbed Dose}^5$$

$$G_c{}^2 \cdot \text{Energy}^2 \cdot \text{Frequency}^2 = \text{Absorbed Dose}^5$$

$$G_c{}^3 \cdot \text{Dynamic Viscosity}^5 = \text{Energy}^2 \cdot \text{Frequency}^7$$

$$G_c{}^2 \cdot \text{Energy}^2 = \text{Frequency}^3 \cdot \text{Kinematic Viscosity}^5$$

$$G_c{}^3 \cdot \text{Energy}^2 \cdot \text{Pressure,Stress} = \text{Absorbed Dose}^6$$

$$G_c \cdot \text{Energy}^2 = \text{Pressure,Stress} \cdot \text{Kinematic Viscosity}^4$$

$$G_c{}^3 \cdot \text{Energy}^2 \cdot \text{Mass Density}^6 = \text{Pressure,Stress}^5$$

$G_c{}^3 \cdot$ Energy$^5 \cdot$ Dynamic Viscosity^5 = Power, Radiance Flux^7

$G_c{}^2 \cdot$ Energy^5 = Power, Radiance Flux$^3 \cdot$ Kinematic Viscosity^5

$G_c \cdot$ Energy$^2 \cdot$ Mass Density = Power, Radiance Flux^2

$G_c{}^4 \cdot$ Energy$^2 \cdot$ Dynamic Viscosity^2 = Absorbed Dose^7

Electric Current$^4 \cdot$ Inductance^2 = Area$^3 \cdot$ Pressure,Stress^2

Electric Current$^2 \cdot$ Inductance = Volume $\cdot$ Pressure,Stress

Electric Current$^2 \cdot$ Inductance $\cdot$ Wavenumber^3 = Pressure,Stress

Electric Current$^2 \cdot$ Inductance $\cdot$ Frequency = Power, Radiance Flux

Electric Current $\cdot$ Magnetic Flux = Time $\cdot$ Power, Radiance Flux

Electric Current $\cdot$ Magnetic Flux = Length $\cdot$ Force

Electric Current $\cdot$ Magnetic Flux = Length$^3 \cdot$ Pressure,Stress

Electric Current $\cdot$ Magnetic Flux = Volumetric Velocity $\cdot$ Dynamic Viscosity

Electric Current$^2 \cdot$ Magnetic Flux^2 = Force$^2 \cdot$ Area

Electric Current$^3 \cdot$ Magnetic Flux^3 = Force$^3 \cdot$ Volume

Electric Current $\cdot$ Magnetic Flux $\cdot$ Wavenumber = Force

Electric Current$^2 \cdot$ Magnetic Flux$^2 \cdot$ Pressure,Stress = Force^3

Electric Current $\cdot$ Magnetic Flux = Mass $\cdot$ Velocity^2

Electric Current $\cdot$ Magnetic Flux = Mass $\cdot$ Absorbed Dose

Electric Current$^2 \cdot$ Magnetic Flux^2 = Area$^3 \cdot$ Pressure,Stress^2

Electric Current $\cdot$ Magnetic Flux = Volume $\cdot$ Pressure,Stress

Electric Current $\cdot$ Magnetic Flux $\cdot$ Wavenumber^3 = Pressure,Stress

Electric Current $\cdot$ Magnetic Flux $\cdot$ Frequency = Power, Radiance Flux

$\text{Time}^2 \cdot \text{Velocity} \cdot \text{Magnetic Field Strength} = \text{Electric Charge}$

$\text{Time}^3 \cdot \text{Velocity}^3 \cdot \text{Pressure,Stress} = \text{Energy,Work,Heat Quantity}$

$\text{Time}^2 \cdot \text{Velocity}^3 \cdot \text{Pressure,Stress} = \text{Power, Radiance Flux}$

$\text{Time}^2 \cdot \text{Velocity}^3 \cdot \text{Dynamic Viscosity} = \text{Energy,Work,Heat Quantity}$

$\text{Time}^3 \cdot \text{Velocity}^5 \cdot \text{Mass Density} = \text{Energy,Work,Heat Quantity}$

$\text{Time} \cdot \text{Velocity}^3 \cdot \text{Dynamic Viscosity} = \text{Power, Radiance Flux}$

$\text{Time}^2 \cdot \text{Velocity}^5 \cdot \text{Mass Density} = \text{Power, Radiance Flux}$

$\text{Time} \cdot \text{Velocity}^2 \cdot \text{Magnetic Flux Density} = \text{Electric Potential Difference}$

$\text{Time}^2 \cdot \text{Velocity}^2 \cdot \text{Magnetic Flux Density} = \text{Magnetic Flux}$

$\text{Time} \cdot \text{Velocity}^2 \cdot \text{Mass Density} = \text{Dynamic Viscosity}$

$\text{Time}^2 \cdot \text{Specific Volume} \cdot \text{Pressure,Stress} = \text{Area}$

$\text{Time}^4 \cdot \text{Specific Volume}^2 \cdot \text{Energy,Work,Heat Quantity}^2 = \text{Area}^5$

$\text{Time}^6 \cdot \text{Specific Volume}^2 \cdot \text{Power, Radiance Flux}^2 = \text{Area}^5$

$\text{Time} \cdot \text{Specific Volume} \cdot \text{Dynamic Viscosity} = \text{Area}$

$\text{Time} \cdot \text{Current Density} = \text{Area} \cdot \text{Electric Charge}$

$\text{Time}^2 \cdot \text{Area} \cdot \text{Magnetic Field Strength}^2 = \text{Electric Charge}^2$

$\text{Time}^2 \cdot \text{Power, Radiance Flux}^2 = \text{Area}^3 \cdot \text{Pressure,Stress}^2$

$\text{Time}^2 \cdot \text{Pressure,Stress} = \text{Area} \cdot \text{Mass Density}$

$\text{Time}^2 \cdot \text{Energy,Work,Heat Quantity}^2 = \text{Area}^3 \cdot \text{Dynamic Viscosity}^2$

$\text{Time}^4 \cdot \text{Energy,Work,Heat Quantity}^2 = \text{Area}^5 \cdot \text{Mass Density}^2$

$\text{Time}^4 \cdot \text{Power, Radiance Flux}^2 = \text{Area}^3 \cdot \text{Dynamic Viscosity}^2$

$\text{Time}^6 \cdot \text{Power, Radiance Flux}^2 = \text{Area}^5 \cdot \text{Mass Density}^2$

$\text{Time} \cdot \text{Electric Potential Difference} = \text{Area} \cdot \text{Magnetic Flux Density}$

Time · Dynamic Viscosity = Area · Mass Density

Time^6 · Specific Volume^3 · Pressure,Stress^3 = Volume^2

Time^6 · Specific Volume^3 · Energy,Work,Heat Quantity^3 = Volume^5

Time^9 · Specific Volume^3 · Power, Radiance Flux^3 = Volume^5

Time^3 · Specific Volume^3 · Dynamic Viscosity^3 = Volume^2

Time^3 · Current Density^3 = Volume^2 · Electric Charge^3

Time^3 · Volume · Magnetic Field Strength^3 = Electric Charge^3

Time · Power, Radiance Flux = Volume · Pressure,Stress

Time^6 · Pressure,Stress^3 = Volume^2 · Mass Density^3

Time · Energy,Work,Heat Quantity = Volume · Dynamic Viscosity

Time^6 · Energy,Work,Heat Quantity^3 = Volume^5 · Mass Density^3

Time^2 · Power, Radiance Flux = Volume · Dynamic Viscosity

Time^9 · Power, Radiance Flux^3 = Volume^5 · Mass Density^3

Time^3 · Electric Potential Difference^3 = Volume^2 · Magnetic Flux Density^3

Time^3 · Dynamic Viscosity^3 = Volume^2 · Mass Density^3

Time^2 · Wavenumber^2 · Specific Volume · Pressure,Stress = A Constant

Time^2 · Wavenumber^5 · Specific Volume · Energy,Work,Heat Quantity = A Constant

Time^3 · Wavenumber^5 · Specific Volume · Power, Radiance Flux = A Constant

Time · Wavenumber^2 · Specific Volume · Dynamic Viscosity = A Constant

Time · Wavenumber^2 · Current Density = Electric Charge

Time · Magnetic Field Strength = Wavenumber · Electric Charge

Time · Wavenumber^3 · Power, Radiance Flux = Pressure,Stress

201

$Time^2 \cdot Wavenumber^2 \cdot Pressure,Stress = Mass\ Density$

$Time \cdot Wavenumber^3 \cdot Energy,Work,Heat\ Quantity = Dynamic\ Viscosity$

$Time^2 \cdot Wavenumber^5 \cdot Energy,Work,Heat\ Quantity = Mass\ Density$

$Time^2 \cdot Wavenumber^3 \cdot Power,\ Radiance\ Flux = Dynamic\ Viscosity$

$Time^3 \cdot Wavenumber^5 \cdot Power,\ Radiance\ Flux = Mass\ Density$

$Time \cdot Wavenumber^2 \cdot Electric\ Potential\ Difference = Magnetic\ Flux\ Density$

$Time \cdot Wavenumber^2 \cdot Dynamic\ Viscosity = Mass\ Density$

$Time^6 \cdot Specific\ Volume^3 \cdot Pressure,Stress^5 = Energy,Work,Heat\ Quantity^2$

$Time^4 \cdot Specific\ Volume^3 \cdot Pressure,Stress^5 = Power,\ Radiance\ Flux^2$

$Time \cdot Specific\ Volume \cdot Pressure,Stress = Kinematic\ Viscosity$

$Time^6 \cdot Absorbed\ Dose^5 = Specific\ Volume^2 \cdot Energy,Work,Heat\ Quantity^2$

$Time \cdot Specific\ Volume^3 \cdot Dynamic\ Viscosity^5 = Energy,Work,Heat\ Quantity^2$

$Time \cdot Kinematic\ Viscosity^5 = Specific\ Volume^2 \cdot Energy,Work,Heat\ Quantity^2$

$Time^4 \cdot Absorbed\ Dose^5 = Specific\ Volume^2 \cdot Power,\ Radiance\ Flux^2$

$Time \cdot Power,\ Radiance\ Flux^2 = Specific\ Volume^3 \cdot Dynamic\ Viscosity^5$

$Time \cdot Specific\ Volume^2 \cdot Power,\ Radiance\ Flux^2 = Kinematic\ Viscosity^5$

$Time \cdot Absorbed\ Dose = Specific\ Volume \cdot Dynamic\ Viscosity$

$Time^3 \cdot Current\ Density \cdot Magnetic\ Field\ Strength^2 = Electric\ Charge^3$

$Time^6 \cdot Magnetic\ Field\ Strength^2 \cdot Absorbed\ Dose^3 = Current\ Density^2$

$Time^3 \cdot Magnetic\ Field\ Strength^2 \cdot Kinematic\ Viscosity^3 = Current\ Density^2$

$Time \cdot Power,\ Radiance\ Flux = Current\ Density \cdot Magnetic\ Flux\ Density$

$Time \cdot Electric\ Charge \cdot Absorbed\ Dose = Current\ Density$

$Time^4 \cdot Magnetic\ Field\ Strength^2 \cdot Absorbed\ Dose = Electric\ Charge^2$

$Time^3 \cdot Magnetic\ Field\ Strength^2 \cdot Kinematic\ Viscosity = Electric\ Charge^2$

$Time \cdot Magnetic\ Field\ Strength \cdot Magnetic\ Flux\ Density = Dynamic\ Viscosity$

$Time^6 \cdot Pressure,Stress^2 \cdot Absorbed\ Dose^3 = Energy,Work,Heat\ Quantity^2$

$Time^3 \cdot Pressure,Stress^2 \cdot Kinematic\ Viscosity^3 = Energy,Work,Heat\ Quantity^2$

$Time^6 \cdot Pressure,Stress^5 = Energy,Work,Heat\ Quantity^2 \cdot Mass\ Density^3$

$Time^4 \cdot Pressure,Stress^2 \cdot Absorbed\ Dose^3 = Power,\ Radiance\ Flux^2$

$Time \cdot Pressure,Stress^2 \cdot Kinematic\ Viscosity^3 = Power,\ Radiance\ Flux^2$

$Time^4 \cdot Pressure,Stress^5 = Power,\ Radiance\ Flux^2 \cdot Mass\ Density^3$

$Time \cdot Pressure,Stress = Kinematic\ Viscosity \cdot Mass\ Density$

$Time \cdot Energy,Work,Heat\ Quantity = Electric\ Charge^2 \cdot Electric\ Resistance$

$Time^2 \cdot Energy,Work,Heat\ Quantity = Electric\ Charge^2 \cdot Inductance$

$Time \cdot Energy,Work,Heat\ Quantity = Electric\ Charge \cdot Magnetic\ Flux$

$Time \cdot Electric\ Potential\ Difference^2 = Energy,Work,Heat\ Quantity \cdot Electric\ Resistance$

$Time^2 \cdot Electric\ Potential\ Difference^2 = Energy,Work,Heat\ Quantity \cdot Inductance$

$Time^2 \cdot Energy,Work,Heat\ Quantity = Capacitance \cdot Magnetic\ Flux^2$

$Time \cdot Energy,Work,Heat\ Quantity \cdot Electric\ Resistance = Magnetic\ Flux^2$

$Time^4 \cdot Absorbed\ Dose^3 \cdot Dynamic\ Viscosity^2 = Energy,Work,Heat\ Quantity^2$

$Time^6 \cdot Absorbed\ Dose^5 \cdot Mass\ Density^2 = Energy,Work,Heat\ Quantity^2$

$Time \cdot Dynamic\ Viscosity^2 \cdot Kinematic\ Viscosity^3 = Energy,Work,Heat\ Quantity^2$

$Time \cdot Dynamic\ Viscosity^5 = Energy,Work,Heat\ Quantity^2 \cdot Mass\ Density^3$

$Time \cdot Kinematic\ Viscosity^5 \cdot Mass\ Density^2 = Energy,Work,Heat\ Quantity^2$

$Time \cdot Power,\ Radiance\ Flux = Electric\ Charge \cdot Electric\ Potential\ Difference$

Time $\cdot$ Power, Radiance Flux $\cdot$ Capacitance = Electric Charge^2

Time^2 $\cdot$ Power, Radiance Flux = Electric Charge^2 $\cdot$ Electric Resistance

Time^3 $\cdot$ Power, Radiance Flux = Electric Charge^2 $\cdot$ Inductance

Time^2 $\cdot$ Power, Radiance Flux = Electric Charge $\cdot$ Magnetic Flux

Time $\cdot$ Power, Radiance Flux = Electric Potential Difference^2 $\cdot$ Capacitance

Time $\cdot$ Electric Potential Difference^2 = Power, Radiance Flux $\cdot$ Inductance

Time^3 $\cdot$ Power, Radiance Flux = Capacitance $\cdot$ Magnetic Flux^2

Time^2 $\cdot$ Power, Radiance Flux $\cdot$ Electric Resistance = Magnetic Flux^2

Time $\cdot$ Power, Radiance Flux $\cdot$ Inductance = Magnetic Flux^2

Time^2 $\cdot$ Absorbed Dose^3 $\cdot$ Dynamic Viscosity^2 = Power, Radiance Flux^2

Time^4 $\cdot$ Absorbed Dose^5 $\cdot$ Mass Density^2 = Power, Radiance Flux^2

Time $\cdot$ Power, Radiance Flux^2 = Dynamic Viscosity^2 $\cdot$ Kinematic Viscosity^3

Time $\cdot$ Power, Radiance Flux^2 $\cdot$ Mass Density^3 = Dynamic Viscosity^5

Time $\cdot$ Power, Radiance Flux^2 = Kinematic Viscosity^5 $\cdot$ Mass Density^2

Time $\cdot$ Electric Potential Difference = Electric Charge $\cdot$ Electric Resistance

Time^2 $\cdot$ Electric Potential Difference = Electric Charge $\cdot$ Inductance

Time $\cdot$ Electric Charge = Capacitance $\cdot$ Magnetic Flux

Time $\cdot$ Magnetic Flux = Electric Charge $\cdot$ Inductance

Time $\cdot$ Magnetic Flux Density $\cdot$ Absorbed Dose = Electric Potential Difference

Time^2 $\cdot$ Magnetic Flux Density $\cdot$ Absorbed Dose = Magnetic Flux

Time $\cdot$ Magnetic Flux Density $\cdot$ Kinematic Viscosity = Magnetic Flux

Time $\cdot$ Absorbed Dose $\cdot$ Mass Density = Dynamic Viscosity

Length^3 $\cdot$ Pressure,Stress = Mass $\cdot$ Absorbed Dose

Length $\cdot$ Dynamic Viscosity^2 = Mass $\cdot$ Pressure,Stress

Length^5 $\cdot$ Pressure,Stress = Mass $\cdot$ Kinematic Viscosity^2

Length^2 $\cdot$ Mass $\cdot$ Power, Radiance Flux^2 = Energy,Work,Heat Quantity^3

Length^4 $\cdot$ Dynamic Viscosity^2 = Mass $\cdot$ Energy,Work,Heat Quantity

Length^2 $\cdot$ Energy,Work,Heat Quantity = Mass $\cdot$ Kinematic Viscosity^2

Length^2 $\cdot$ Power, Radiance Flux^2 = Mass^2 $\cdot$ Absorbed Dose^3

Length^5 $\cdot$ Dynamic Viscosity^3 = Mass^2 $\cdot$ Power, Radiance Flux

Length^4 $\cdot$ Power, Radiance Flux = Mass $\cdot$ Kinematic Viscosity^3

Length^2 $\cdot$ Mass = Electric Charge^2 $\cdot$ Inductance

Length^2 $\cdot$ Capacitance $\cdot$ Magnetic Flux Density^2 = Mass

Length^2 $\cdot$ Mass = Capacitance $\cdot$ Magnetic Flux^2

Length^4 $\cdot$ Dynamic Viscosity^2 = Mass^2 $\cdot$ Absorbed Dose

Length^3 $\cdot$ Dynamic Viscosity = Mass $\cdot$ Kinematic Viscosity

Length^3 $\cdot$ Velocity^2 = Specific Volume $\cdot$ Energy,Work,Heat Quantity

Length^2 $\cdot$ Velocity^3 = Specific Volume $\cdot$ Power, Radiance Flux

Length $\cdot$ Velocity = Specific Volume $\cdot$ Dynamic Viscosity

Length $\cdot$ Velocity $\cdot$ Electric Charge = Current Density

Length^2 $\cdot$ Magnetic Field Strength = Velocity $\cdot$ Electric Charge

Length^2 $\cdot$ Velocity $\cdot$ Pressure,Stress = Power, Radiance Flux

Length $\cdot$ Pressure,Stress = Velocity $\cdot$ Dynamic Viscosity

Length $\cdot$ Power, Radiance Flux = Velocity $\cdot$ Energy,Work,Heat Quantity

Length^2 $\cdot$ Velocity $\cdot$ Dynamic Viscosity = Energy,Work,Heat Quantity

$Length^3 \cdot Velocity^2 \cdot Mass\ Density = Energy, Work, Heat\ Quantity$

$Length \cdot Velocity^2 \cdot Dynamic\ Viscosity = Power, Radiance\ Flux$

$Length^2 \cdot Velocity^3 \cdot Mass\ Density = Power, Radiance\ Flux$

$Length \cdot Velocity \cdot Magnetic\ Flux\ Density = Electric\ Potential\ Difference$

$Length \cdot Electric\ Potential\ Difference = Velocity \cdot Magnetic\ Flux$

$Length = Velocity \cdot Capacitance \cdot Electric\ Resistance$

$Length^2 = Velocity^2 \cdot Capacitance \cdot Inductance$

$Length \cdot Electric\ Resistance = Velocity \cdot Inductance$

$Length \cdot Velocity \cdot Mass\ Density = Dynamic\ Viscosity$

$Length^2 \cdot Frequency^2 = Specific\ Volume \cdot Pressure, Stress$

$Length^5 \cdot Frequency^2 = Specific\ Volume \cdot Energy, Work, Heat\ Quantity$

$Length^5 \cdot Frequency^3 = Specific\ Volume \cdot Power, Radiance\ Flux$

$Length^2 \cdot Frequency = Specific\ Volume \cdot Dynamic\ Viscosity$

$Length^4 \cdot Specific\ Volume \cdot Pressure, Stress^3 = Power, Radiance\ Flux^2$

$Length^2 \cdot Pressure, Stress = Specific\ Volume \cdot Dynamic\ Viscosity^2$

$Length^2 \cdot Specific\ Volume \cdot Pressure, Stress = Kinematic\ Viscosity^2$

$Length^5 \cdot Power, Radiance\ Flux^2 = Specific\ Volume \cdot Energy, Work, Heat\ Quantity^3$

$Length^3 \cdot Absorbed\ Dose = Specific\ Volume \cdot Energy, Work, Heat\ Quantity$

$Length \cdot Specific\ Volume \cdot Dynamic\ Viscosity^2 = Energy, Work, Heat\ Quantity$

$Length \cdot Kinematic\ Viscosity^2 = Specific\ Volume \cdot Energy, Work, Heat\ Quantity$

$Length^4 \cdot Absorbed\ Dose^3 = Specific\ Volume^2 \cdot Power, Radiance\ Flux^2$

$Length \cdot Power, Radiance\ Flux = Specific\ Volume^2 \cdot Dynamic\ Viscosity^3$

$Length \cdot Specific\ Volume \cdot Power, Radiance\ Flux = Kinematic\ Viscosity^3$

$Length^5 = Specific\ Volume \cdot Electric\ Charge^2 \cdot Inductance$

$Length = Specific\ Volume \cdot Capacitance \cdot Magnetic\ Flux\ Density^2$

$Length^5 = Specific\ Volume \cdot Capacitance \cdot Magnetic\ Flux^2$

$Length^2 \cdot Absorbed\ Dose = Specific\ Volume^2 \cdot Dynamic\ Viscosity^2$

$Length^2 \cdot Frequency \cdot Electric\ Charge = Current\ Density$

$Length^3 \cdot Pressure,Stress = Current\ Density \cdot Magnetic\ Flux\ Density$

$Length^7 \cdot Pressure,Stress = Current\ Density^2 \cdot Inductance$

$Length^5 \cdot Pressure,Stress = Current\ Density \cdot Magnetic\ Flux$

$Length^4 \cdot Energy,Work,Heat\ Quantity = Current\ Density^2 \cdot Inductance$

$Length^2 \cdot Energy,Work,Heat\ Quantity = Current\ Density \cdot Magnetic\ Flux$

$Length^2 \cdot Power,\ Radiance\ Flux = Current\ Density \cdot Electric\ Potential\ Difference$

$Length^4 \cdot Power,\ Radiance\ Flux = Current\ Density^2 \cdot Electric\ Resistance$

$Length^2 \cdot Electric\ Charge^2 \cdot Absorbed\ Dose = Current\ Density^2$

$Length^2 \cdot Electric\ Potential\ Difference = Current\ Density \cdot Electric\ Resistance$

$Length^4 \cdot Magnetic\ Flux\ Density = Current\ Density \cdot Inductance$

$Length^2 \cdot Magnetic\ Flux = Current\ Density \cdot Inductance$

$Length \cdot Magnetic\ Field\ Strength = Frequency \cdot Electric\ Charge$

$Length \cdot Pressure,Stress = Magnetic\ Field\ Strength^2 \cdot Inductance$

$Length^2 \cdot Pressure,Stress = Magnetic\ Field\ Strength \cdot Magnetic\ Flux$

$Length^3 \cdot Magnetic\ Field\ Strength \cdot Magnetic\ Flux\ Density = Energy,Work,Heat\ Quantity$

$Length^2 \cdot Magnetic\ Field\ Strength^2 \cdot Inductance = Energy,Work,Heat\ Quantity$

$Length \cdot Magnetic\ Field\ Strength \cdot Magnetic\ Flux = Energy,Work,Heat\ Quantity$

Length · Magnetic Field Strength · Electric Potential Difference = Power, Radiance Flux

Length^2 · Magnetic Field Strength^2 · Electric Resistance = Power, Radiance Flux

Length^4 · Magnetic Field Strength^2 = Electric Charge^2 · Absorbed Dose

Length^3 · Magnetic Field Strength = Electric Charge · Kinematic Viscosity

Length · Magnetic Field Strength · Electric Resistance = Electric Potential Difference

Length · Magnetic Flux Density = Magnetic Field Strength · Inductance

Length · Magnetic Field Strength · Inductance = Magnetic Flux

Length^3 · Frequency · Pressure,Stress = Power, Radiance Flux

Length^2 · Frequency^2 · Mass Density = Pressure,Stress

Length^3 · Frequency · Dynamic Viscosity = Energy,Work,Heat Quantity

Length^5 · Frequency^2 · Mass Density = Energy,Work,Heat Quantity

Length^3 · Frequency^2 · Dynamic Viscosity = Power, Radiance Flux

Length^5 · Frequency^3 · Mass Density = Power, Radiance Flux

Length^2 · Frequency · Magnetic Flux Density = Electric Potential Difference

Length^2 · Frequency · Mass Density = Dynamic Viscosity

Length^4 · Pressure,Stress^2 · Absorbed Dose = Power, Radiance Flux^2

Length^3 · Pressure,Stress^2 = Power, Radiance Flux · Dynamic Viscosity

Length · Pressure,Stress · Kinematic Viscosity = Power, Radiance Flux

Length^4 · Pressure,Stress^3 = Power, Radiance Flux^2 · Mass Density

Length^3 · Pressure,Stress = Electric Charge · Electric Potential Difference

Length^3 · Pressure,Stress · Capacitance = Electric Charge^2

Length^3 · Pressure,Stress = Electric Potential Difference^2 · Capacitance

Length · Magnetic Flux Density^2 = Pressure,Stress · Inductance

$Length^3 \cdot Pressure,Stress \cdot Inductance = Magnetic\ Flux^2$

$Length^2 \cdot Pressure,Stress^2 = Absorbed\ Dose \cdot Dynamic\ Viscosity^2$

$Length^2 \cdot Pressure,Stress = Dynamic\ Viscosity \cdot Kinematic\ Viscosity$

$Length^2 \cdot Pressure,Stress \cdot Mass\ Density = Dynamic\ Viscosity^2$

$Length^2 \cdot Pressure,Stress = Kinematic\ Viscosity^2 \cdot Mass\ Density$

$Length^2 \cdot Power,\ Radiance\ Flux^2 = Energy,Work,Heat\ Quantity^2 \cdot Absorbed\ Dose$

$Length^3 \cdot Power,\ Radiance\ Flux \cdot Dynamic\ Viscosity = Energy,Work,Heat\ Quantity^2$

$Length^2 \cdot Power,\ Radiance\ Flux = Energy,Work,Heat\ Quantity \cdot Kinematic\ Viscosity$

$Length^5 \cdot Power,\ Radiance\ Flux^2 \cdot Mass\ Density = Energy,Work,Heat\ Quantity^3$

$Length^4 \cdot Magnetic\ Flux\ Density^2 = Energy,Work,Heat\ Quantity \cdot Inductance$

$Length^4 \cdot Absorbed\ Dose \cdot Dynamic\ Viscosity^2 = Energy,Work,Heat\ Quantity^2$

$Length^3 \cdot Absorbed\ Dose \cdot Mass\ Density = Energy,Work,Heat\ Quantity$

$Length \cdot Dynamic\ Viscosity \cdot Kinematic\ Viscosity = Energy,Work,Heat\ Quantity$

$Length \cdot Dynamic\ Viscosity^2 = Energy,Work,Heat\ Quantity \cdot Mass\ Density$

$Length \cdot Kinematic\ Viscosity^2 \cdot Mass\ Density = Energy,Work,Heat\ Quantity$

$Length \cdot Absorbed\ Dose \cdot Dynamic\ Viscosity = Power,\ Radiance\ Flux$

$Length^4 \cdot Absorbed\ Dose^3 \cdot Mass\ Density^2 = Power,\ Radiance\ Flux^2$

$Length \cdot Power,\ Radiance\ Flux = Dynamic\ Viscosity \cdot Kinematic\ Viscosity^2$

$Length \cdot Power,\ Radiance\ Flux \cdot Mass\ Density^2 = Dynamic\ Viscosity^3$

$Length \cdot Power,\ Radiance\ Flux = Kinematic\ Viscosity^3 \cdot Mass\ Density$

$Length^2 \cdot Magnetic\ Flux\ Density = Electric\ Charge \cdot Electric\ Resistance$

$Length^3 \cdot Dynamic\ Viscosity = Electric\ Charge^2 \cdot Electric\ Resistance$

Length $\cdot$ Dynamic Viscosity = Electric Charge $\cdot$ Magnetic Flux Density

Length^5 $\cdot$ Mass Density = Electric Charge^2 $\cdot$ Inductance

Length^3 $\cdot$ Dynamic Viscosity = Electric Charge $\cdot$ Magnetic Flux

Length^2 $\cdot$ Magnetic Flux Density^2 $\cdot$ Absorbed Dose = Electric Potential Difference^2

Length^2 $\cdot$ Electric Potential Difference^2 = Absorbed Dose $\cdot$ Magnetic Flux^2

Length^2 $\cdot$ Electric Potential Difference = Magnetic Flux $\cdot$ Kinematic Viscosity

Length^2 = Capacitance^2 $\cdot$ Electric Resistance^2 $\cdot$ Absorbed Dose

Length^2 = Capacitance $\cdot$ Electric Resistance $\cdot$ Kinematic Viscosity

Length $\cdot$ Mass Density = Capacitance $\cdot$ Magnetic Flux Density^2

Length^2 = Capacitance $\cdot$ Inductance $\cdot$ Absorbed Dose

Length^4 = Capacitance $\cdot$ Inductance $\cdot$ Kinematic Viscosity^2

Length^5 $\cdot$ Mass Density = Capacitance $\cdot$ Magnetic Flux^2

Length $\cdot$ Magnetic Flux Density^2 = Electric Resistance $\cdot$ Dynamic Viscosity

Length^2 $\cdot$ Electric Resistance^2 = Inductance^2 $\cdot$ Absorbed Dose

Length^2 $\cdot$ Electric Resistance = Inductance $\cdot$ Kinematic Viscosity

Length^3 $\cdot$ Electric Resistance $\cdot$ Dynamic Viscosity = Magnetic Flux^2

Length^2 $\cdot$ Absorbed Dose $\cdot$ Mass Density^2 = Dynamic Viscosity^2

Volumetric Velocity^2 $\cdot$ Dynamic Viscosity^2 = Area^3 $\cdot$ Pressure,Stress^2

Volumetric Velocity^2 $\cdot$ Mass Density = Area^2 $\cdot$ Pressure,Stress

Volumetric Velocity^2 $\cdot$ Energy,Work,Heat Quantity^2 = Area^3 $\cdot$ Power, Radiance Flux^2

Volumetric Velocity^4 $\cdot$ Mass Density^2 = Area $\cdot$ Energy,Work,Heat Quantity^2

Volumetric Velocity^4 $\cdot$ Dynamic Viscosity^2 = Area^3 $\cdot$ Power, Radiance Flux^2

Volumetric Velocity^3 $\cdot$ Mass Density = Area^2 $\cdot$ Power, Radiance Flux

Volumetric Velocity^2 · Magnetic Flux Density^2 = Area · Electric Potential Difference^2

Volumetric Velocity^2 · Magnetic Flux^2 = Area^3 · Electric Potential Difference^2

Volumetric Velocity^2 · Capacitance^2 · Electric Resistance^2 = Area^3

Volumetric Velocity^2 · Capacitance · Inductance = Area^3

Volumetric Velocity^2 · Inductance^2 = Area^3 · Electric Resistance^2

Volumetric Velocity^2 · Mass Density^2 = Area · Dynamic Viscosity^2

Volumetric Velocity^6 = Volume^4 · Specific Volume^3 · Pressure,Stress^3

Volumetric Velocity^6 = Volume · Specific Volume^3 · Energy,Work,Heat Quantity^3

Volumetric Velocity^9 = Volume^4 · Specific Volume^3 · Power, Radiance Flux^3

Volumetric Velocity^3 = Volume · Specific Volume^3 · Dynamic Viscosity^3

Volumetric Velocity^3 · Electric Charge^3 = Volume · Current Density^3

Volumetric Velocity^3 · Electric Charge^3 = Volume^4 · Magnetic Field Strength^3

Volumetric Velocity · Dynamic Viscosity = Volume · Pressure,Stress

Volumetric Velocity^6 · Mass Density^3 = Volume^4 · Pressure,Stress^3

Volumetric Velocity · Energy,Work,Heat Quantity = Volume · Power, Radiance Flux

Volumetric Velocity^6 · Mass Density^3 = Volume · Energy,Work,Heat Quantity^3

Volumetric Velocity^2 · Dynamic Viscosity = Volume · Power, Radiance Flux

Volumetric Velocity^9 · Mass Density^3 = Volume^4 · Power, Radiance Flux^3

Volumetric Velocity^3 · Magnetic Flux Density^3 = Volume · Electric Potential Difference^3

Volumetric Velocity · Magnetic Flux = Volume · Electric Potential Difference

Volumetric Velocity · Capacitance · Electric Resistance = Volume

Volumetric Velocity^2 · Capacitance · Inductance = Volume^2

Volumetric Velocity $\cdot$ Inductance = Volume $\cdot$ Electric Resistance

Volumetric Velocity^3 $\cdot$ Mass Density^3 = Volume $\cdot$ Dynamic Viscosity^3

Volumetric Velocity^2 $\cdot$ Wavenumber^4 = Specific Volume $\cdot$ Pressure,Stress

Volumetric Velocity^2 $\cdot$ Wavenumber = Specific Volume $\cdot$ Energy,Work,Heat Quantity

Volumetric Velocity^3 $\cdot$ Wavenumber^4 = Specific Volume $\cdot$ Power, Radiance Flux

Volumetric Velocity $\cdot$ Wavenumber = Specific Volume $\cdot$ Dynamic Viscosity

Volumetric Velocity $\cdot$ Wavenumber $\cdot$ Electric Charge = Current Density

Volumetric Velocity $\cdot$ Wavenumber^4 $\cdot$ Electric Charge = Magnetic Field Strength

Volumetric Velocity $\cdot$ Wavenumber^3 $\cdot$ Dynamic Viscosity = Pressure,Stress

Volumetric Velocity^2 $\cdot$ Wavenumber^4 $\cdot$ Mass Density = Pressure,Stress

Volumetric Velocity $\cdot$ Wavenumber^3 $\cdot$ Energy,Work,Heat Quantity = Power, Radiance Flux

Volumetric Velocity^2 $\cdot$ Wavenumber $\cdot$ Mass Density = Energy,Work,Heat Quantity

Volumetric Velocity^2 $\cdot$ Wavenumber^3 $\cdot$ Dynamic Viscosity = Power, Radiance Flux

Volumetric Velocity^3 $\cdot$ Wavenumber^4 $\cdot$ Mass Density = Power, Radiance Flux

Volumetric Velocity $\cdot$ Wavenumber $\cdot$ Magnetic Flux Density = Electric Potential Difference

Volumetric Velocity $\cdot$ Wavenumber^3 $\cdot$ Magnetic Flux = Electric Potential Difference

Volumetric Velocity $\cdot$ Wavenumber^3 $\cdot$ Capacitance $\cdot$ Electric Resistance = A Constant

Volumetric Velocity^2 $\cdot$ Wavenumber^6 $\cdot$ Capacitance $\cdot$ Inductance = A Constant

Volumetric Velocity $\cdot$ Wavenumber^3 $\cdot$ Inductance = Electric Resistance

Volumetric Velocity $\cdot$ Wavenumber $\cdot$ Mass Density = Dynamic Viscosity

Volumetric Velocity^3 = Specific Volume $\cdot$ Current Density^2 $\cdot$ Electric Resistance

Volumetric Velocity^2 $\cdot$ Frequency^4 = Specific Volume^3 $\cdot$ Pressure,Stress^3

Volumetric Velocity^5 $\cdot$ Frequency = Specific Volume^3 $\cdot$ Energy,Work,Heat Quantity^3

Volumetric Velocity^5 · Frequency^4 = Specific Volume^3 · Power, Radiance Flux^3

Volumetric Velocity^2 · Frequency = Specific Volume^3 · Dynamic Viscosity^3

Volumetric Velocity^6 · Pressure,Stress = Specific Volume^3 · Energy,Work,Heat Quantity^4

Volumetric Velocity^2 · Pressure,Stress = Specific Volume^3 · Dynamic Viscosity^4

Volumetric Velocity^2 · Specific Volume · Pressure,Stress = Kinematic Viscosity^4

Volumetric Velocity^5 · Power, Radiance Flux = Specific Volume^3 · Energy,Work,Heat Quantity^4

Volumetric Velocity^6 · Absorbed Dose = Specific Volume^4 · Energy,Work,Heat Quantity^4

Volumetric Velocity · Kinematic Viscosity = Specific Volume · Energy,Work,Heat Quantity

Volumetric Velocity · Absorbed Dose = Specific Volume · Power, Radiance Flux

Volumetric Velocity · Power, Radiance Flux = Specific Volume^3 · Dynamic Viscosity^4

Volumetric Velocity · Specific Volume · Power, Radiance Flux = Kinematic Viscosity^4

Volumetric Velocity = Specific Volume · Electric Charge · Magnetic Flux Density

Volumetric Velocity^2 · Absorbed Dose = Specific Volume^4 · Dynamic Viscosity^4

Volumetric Velocity · Magnetic Field Strength = Current Density · Frequency

Volumetric Velocity^3 · Magnetic Field Strength · Electric Charge^3 = Current Density^4

Volumetric Velocity^6 · Magnetic Field Strength^4 = Current Density^4 · Absorbed Dose^3

Volumetric Velocity^3 · Magnetic Field Strength = Current Density · Kinematic Viscosity^3

Volumetric Velocity^2 · Frequency · Electric Charge^3 = Current Density^3

Volumetric Velocity^2 · Electric Charge^4 · Absorbed Dose = Current Density^4

Volumetric Velocity^3 · Mass Density = Current Density^2 · Electric Resistance

Volumetric Velocity · Dynamic Viscosity = Current Density · Magnetic Flux Density

Volumetric Velocity · Magnetic Field Strength^3 = Frequency^4 · Electric Charge^3

213

Volumetric Velocity $\cdot$ Magnetic Field Strength $\cdot$ Magnetic Flux Density = Power, Radiance Flux

Volumetric Velocity $\cdot$ Magnetic Field Strength = Electric Charge $\cdot$ Absorbed Dose

Volumetric Velocity^3 $\cdot$ Magnetic Field Strength = Electric Charge $\cdot$ Kinematic Viscosity^4

Volumetric Velocity $\cdot$ Pressure,Stress = Frequency $\cdot$ Energy,Work,Heat Quantity

Volumetric Velocity^2 $\cdot$ Frequency^4 $\cdot$ Mass Density^3 = Pressure,Stress^3

Volumetric Velocity^5 $\cdot$ Frequency $\cdot$ Mass Density^3 = Energy,Work,Heat Quantity^3

Volumetric Velocity $\cdot$ Frequency $\cdot$ Dynamic Viscosity = Power, Radiance Flux

Volumetric Velocity^5 $\cdot$ Frequency^4 $\cdot$ Mass Density^3 = Power, Radiance Flux^3

Volumetric Velocity^2 $\cdot$ Frequency $\cdot$ Magnetic Flux Density^3 = Electric Potential Difference^3

Volumetric Velocity^2 $\cdot$ Magnetic Flux Density^3 = Frequency^2 $\cdot$ Magnetic Flux^3

Volumetric Velocity^2 $\cdot$ Frequency $\cdot$ Mass Density^3 = Dynamic Viscosity^3

Volumetric Velocity^6 $\cdot$ Pressure,Stress^4 = Energy,Work,Heat Quantity^4 $\cdot$ Absorbed Dose^3

Volumetric Velocity^3 $\cdot$ Pressure,Stress = Energy,Work,Heat Quantity $\cdot$ Kinematic Viscosity^3

Volumetric Velocity^6 $\cdot$ Pressure,Stress $\cdot$ Mass Density^3 = Energy,Work,Heat Quantity^4

Volumetric Velocity $\cdot$ Pressure,Stress $\cdot$ Electric Resistance = Electric Potential Difference^2

Volumetric Velocity^2 $\cdot$ Pressure,Stress^4 = Absorbed Dose^3 $\cdot$ Dynamic Viscosity^4

Volumetric Velocity^2 $\cdot$ Pressure,Stress = Dynamic Viscosity $\cdot$ Kinematic Viscosity^3

Volumetric Velocity^2 $\cdot$ Pressure,Stress $\cdot$ Mass Density^3 = Dynamic Viscosity^4

Volumetric Velocity^2 $\cdot$ Pressure,Stress = Kinematic Viscosity^4 $\cdot$ Mass Density

Volumetric Velocity^2 $\cdot$ Power, Radiance Flux^4 = Energy,Work,Heat Quantity^4 $\cdot$ Absorbed Dose^3

Volumetric Velocity^2 $\cdot$ Power, Radiance Flux = Energy,Work,Heat Quantity $\cdot$ Kinematic Viscosity^3

Volumetric Velocity^5 $\cdot$ Power, Radiance Flux $\cdot$ Mass Density^3 = Energy,Work,Heat Quantity^4

Volumetric Velocity^6 · Absorbed Dose · Mass Density^4 = Energy,Work,Heat Quantity^4

Volumetric Velocity · Kinematic Viscosity · Mass Density = Energy,Work,Heat Quantity

Volumetric Velocity^2 · Absorbed Dose^3 · Dynamic Viscosity^4 = Power, Radiance Flux^4

Volumetric Velocity · Absorbed Dose · Mass Density = Power, Radiance Flux

Volumetric Velocity · Power, Radiance Flux = Dynamic Viscosity · Kinematic Viscosity^3

Volumetric Velocity · Power, Radiance Flux · Mass Density^3 = Dynamic Viscosity^4

Volumetric Velocity · Power, Radiance Flux = Kinematic Viscosity^4 · Mass Density

Volumetric Velocity · Dynamic Viscosity = Electric Charge · Electric Potential Difference

Volumetric Velocity · Capacitance · Dynamic Viscosity = Electric Charge^2

Volumetric Velocity · Mass Density = Electric Charge · Magnetic Flux Density

Volumetric Velocity · Dynamic Viscosity = Electric Potential Difference^2 · Capacitance

Volumetric Velocity^2 · Magnetic Flux Density^4 · Absorbed Dose = Electric Potential Difference^4

Volumetric Velocity^2 · Magnetic Flux Density^3 = Electric Potential Difference^2 · Magnetic Flux

Volumetric Velocity^2 · Electric Potential Difference^4 = Absorbed Dose^3 · Magnetic Flux^4

Volumetric Velocity^2 · Electric Potential Difference = Magnetic Flux · Kinematic Viscosity^3

Volumetric Velocity^2 = Capacitance^4 · Electric Resistance^4 · Absorbed Dose^3

Volumetric Velocity^2 = Capacitance · Electric Resistance · Kinematic Viscosity^3

Volumetric Velocity^2 = Capacitance^2 · Inductance^2 · Absorbed Dose^3

Volumetric Velocity^4 = Capacitance · Inductance · Kinematic Viscosity^6

Volumetric Velocity^2 · Electric Resistance^4 = Inductance^4 · Absorbed Dose^3

Volumetric Velocity^2 · Electric Resistance = Inductance · Kinematic Viscosity^3

Volumetric Velocity^2 · Magnetic Flux Density^2 = Absorbed Dose · Magnetic Flux^2

Volumetric Velocity^2 · Magnetic Flux Density = Magnetic Flux · Kinematic Viscosity^2

215

Volumetric Velocity $\cdot$ Inductance $\cdot$ Dynamic Viscosity = Magnetic Flux^2

Volumetric Velocity^2 $\cdot$ Absorbed Dose $\cdot$ Mass Density^4 = Dynamic Viscosity^4

Force $\cdot$ Velocity^2 = Energy $\cdot$ Acceleration

Force^4 $\cdot$ Specific Volume = Energy^3 $\cdot$ Acceleration

Force $\cdot$ Acceleration = Energy $\cdot$ Frequency^2

Force $\cdot$ Energy $\cdot$ Acceleration = Power, Radiance Flux^2

Force $\cdot$ Absorbed Dose = Energy $\cdot$ Acceleration

Force^5 = Energy^3 $\cdot$ Acceleration $\cdot$ Dynamic Viscosity^2

Force^3 $\cdot$ Kinematic Viscosity^2 = Energy^3 $\cdot$ Acceleration

Force^4 = Energy^3 $\cdot$ Acceleration $\cdot$ Mass Density

Force^3 $\cdot$ Mass $\cdot$ Specific Volume = Energy^3

Force^2 = Energy $\cdot$ Mass $\cdot$ Frequency^2

Force^2 $\cdot$ Energy = Mass $\cdot$ Power, Radiance Flux^2

Force^4 $\cdot$ Mass = Energy^3 $\cdot$ Dynamic Viscosity^2

Force^2 $\cdot$ Mass $\cdot$ Kinematic Viscosity^2 = Energy^3

Force^3 $\cdot$ Mass = Energy^3 $\cdot$ Mass Density

Force^3 $\cdot$ Specific Volume = Energy^2 $\cdot$ Velocity^2

Force $\cdot$ Velocity = Energy $\cdot$ Frequency

Force^2 = Energy $\cdot$ Velocity $\cdot$ Dynamic Viscosity

Force $\cdot$ Kinematic Viscosity = Energy $\cdot$ Velocity

Force^3 = Energy^2 $\cdot$ Velocity^2 $\cdot$ Mass Density

Force^5 $\cdot$ Specific Volume = Energy^4 $\cdot$ Frequency^2

Force^5 · Specific Volume = Energy^2 · Power, Radiance Flux^2

Force^3 · Specific Volume = Energy^2 · Absorbed Dose

Force^3 · Current Density = Energy^3 · Magnetic Field Strength

Force^4 · Current Density^2 · Inductance = Energy^5

Force^2 · Current Density · Magnetic Flux = Energy^3

Force^3 = Energy^2 · Magnetic Field Strength · Magnetic Flux Density

Force^2 = Energy · Magnetic Field Strength^2 · Inductance

Force^2 · Absorbed Dose = Energy^2 · Frequency^2

Force^3 = Energy^2 · Frequency · Dynamic Viscosity

Force^2 · Kinematic Viscosity = Energy^2 · Frequency

Force^5 = Energy^4 · Frequency^2 · Mass Density

Force^3 = Energy · Power, Radiance Flux · Dynamic Viscosity

Force^2 · Kinematic Viscosity = Energy · Power, Radiance Flux

Force^5 = Energy^2 · Power, Radiance Flux^2 · Mass Density

Force^4 · Inductance = Energy^3 · Magnetic Flux Density^2

Force^2 · Magnetic Flux = Energy^2 · Magnetic Flux Density

Force^4 = Energy^2 · Absorbed Dose · Dynamic Viscosity^2

Force^2 · Kinematic Viscosity^2 = Energy^2 · Absorbed Dose

Force^3 = Energy^2 · Absorbed Dose · Mass Density

Force · Acceleration^2 · Specific Volume = Velocity^6

Force · Acceleration^2 = Velocity^4 · Pressure,Stress

Force · Velocity^2 = Acceleration · Energy,Work,Heat Quantity

Force · Acceleration = Velocity^3 · Dynamic Viscosity

$$\text{Force} \cdot \text{Acceleration}^2 = \text{Velocity}^6 \cdot \text{Mass Density}$$

$$\text{Force}^2 \cdot \text{Specific Volume}^2 = \text{Acceleration}^2 \cdot \text{Area}^3$$

$$\text{Force}^4 \cdot \text{Acceleration}^2 \cdot \text{Area} = \text{Power, Radiance Flux}^4$$

$$\text{Force}^4 = \text{Acceleration}^2 \cdot \text{Area}^3 \cdot \text{Dynamic Viscosity}^4$$

$$\text{Force}^2 = \text{Acceleration}^2 \cdot \text{Area}^3 \cdot \text{Mass Density}^2$$

$$\text{Force} \cdot \text{Specific Volume} = \text{Acceleration} \cdot \text{Volume}$$

$$\text{Force}^6 \cdot \text{Acceleration}^3 \cdot \text{Volume} = \text{Power, Radiance Flux}^6$$

$$\text{Force}^2 = \text{Acceleration} \cdot \text{Volume} \cdot \text{Dynamic Viscosity}^2$$

$$\text{Force} = \text{Acceleration} \cdot \text{Volume} \cdot \text{Mass Density}$$

$$\text{Force} \cdot \text{Wavenumber}^3 \cdot \text{Specific Volume} = \text{Acceleration}$$

$$\text{Force}^2 \cdot \text{Acceleration} = \text{Wavenumber} \cdot \text{Power, Radiance Flux}^2$$

$$\text{Force}^2 \cdot \text{Wavenumber}^3 = \text{Acceleration} \cdot \text{Dynamic Viscosity}^2$$

$$\text{Force} \cdot \text{Wavenumber}^3 = \text{Acceleration} \cdot \text{Mass Density}$$

$$\text{Force} \cdot \text{Specific Volume} \cdot \text{Frequency}^6 = \text{Acceleration}^4$$

$$\text{Force} \cdot \text{Acceleration}^2 = \text{Specific Volume}^2 \cdot \text{Pressure,Stress}^3$$

$$\text{Force}^4 \cdot \text{Specific Volume} = \text{Acceleration} \cdot \text{Energy,Work,Heat Quantity}^3$$

$$\text{Force}^7 \cdot \text{Acceleration}^2 \cdot \text{Specific Volume} = \text{Power, Radiance Flux}^6$$

$$\text{Force} \cdot \text{Acceleration}^2 \cdot \text{Specific Volume} = \text{Absorbed Dose}^3$$

$$\text{Force} \cdot \text{Frequency}^4 = \text{Acceleration}^2 \cdot \text{Pressure,Stress}$$

$$\text{Force} \cdot \text{Acceleration} = \text{Frequency}^2 \cdot \text{Energy,Work,Heat Quantity}$$

$$\text{Force} \cdot \text{Acceleration} = \text{Frequency} \cdot \text{Power, Radiance Flux}$$

$$\text{Force} \cdot \text{Frequency}^3 = \text{Acceleration}^2 \cdot \text{Dynamic Viscosity}$$

Force $\cdot$ Frequency^6 = Acceleration^4 $\cdot$ Mass Density

Force^5 $\cdot$ Acceleration^2 = Pressure,Stress $\cdot$ Power, Radiance Flux^4

Force $\cdot$ Acceleration^2 = Pressure,Stress $\cdot$ Absorbed Dose^2

Force $\cdot$ Pressure,Stress^3 = Acceleration^2 $\cdot$ Dynamic Viscosity^4

Force^3 $\cdot$ Acceleration^2 = Pressure,Stress^3 $\cdot$ Kinematic Viscosity^4

Force $\cdot$ Acceleration^2 $\cdot$ Mass Density^2 = Pressure,Stress^3

Force $\cdot$ Acceleration $\cdot$ Energy,Work,Heat Quantity = Power, Radiance Flux^2

Force $\cdot$ Absorbed Dose = Acceleration $\cdot$ Energy,Work,Heat Quantity

Force^5 = Acceleration $\cdot$ Energy,Work,Heat Quantity^3 $\cdot$ Dynamic Viscosity^2

Force^3 $\cdot$ Kinematic Viscosity^2 = Acceleration $\cdot$ Energy,Work,Heat Quantity^3

Force^4 = Acceleration $\cdot$ Energy,Work,Heat Quantity^3 $\cdot$ Mass Density

Force^4 $\cdot$ Acceleration = Power, Radiance Flux^3 $\cdot$ Dynamic Viscosity

Force^3 $\cdot$ Acceleration $\cdot$ Kinematic Viscosity = Power, Radiance Flux^3

Force^7 $\cdot$ Acceleration^2 = Power, Radiance Flux^6 $\cdot$ Mass Density

Force $\cdot$ Acceleration $\cdot$ Inductance = Electric Potential Difference^2

Force^2 $\cdot$ Acceleration^2 = Absorbed Dose^3 $\cdot$ Dynamic Viscosity^2

Force $\cdot$ Acceleration^2 = Absorbed Dose^3 $\cdot$ Mass Density

Force^2 $\cdot$ Area = Mass^2 $\cdot$ Velocity^4

Force^3 $\cdot$ Volume = Mass^3 $\cdot$ Velocity^6

Force = Mass $\cdot$ Velocity^2 $\cdot$ Wavenumber

Force^3 $\cdot$ Specific Volume = Mass^2 $\cdot$ Velocity^6

Force = Mass $\cdot$ Velocity $\cdot$ Frequency

Force^3 = Mass^2 $\cdot$ Velocity^4 $\cdot$ Pressure,Stress

$$\text{Force}^2 = \text{Mass} \cdot \text{Velocity}^3 \cdot \text{Dynamic Viscosity}$$

$$\text{Force} \cdot \text{Kinematic Viscosity} = \text{Mass} \cdot \text{Velocity}^3$$

$$\text{Force}^3 = \text{Mass}^2 \cdot \text{Velocity}^6 \cdot \text{Mass Density}$$

$$\text{Force}^2 = \text{Mass}^2 \cdot \text{Area} \cdot \text{Frequency}^4$$

$$\text{Force}^6 \cdot \text{Area} = \text{Mass}^2 \cdot \text{Power, Radiance Flux}^4$$

$$\text{Force}^2 \cdot \text{Area} = \text{Mass}^2 \cdot \text{Absorbed Dose}^2$$

$$\text{Force}^2 \cdot \text{Mass}^2 = \text{Area}^3 \cdot \text{Dynamic Viscosity}^4$$

$$\text{Force}^2 \cdot \text{Area}^3 = \text{Mass}^2 \cdot \text{Kinematic Viscosity}^4$$

$$\text{Force}^3 = \text{Mass}^3 \cdot \text{Volume} \cdot \text{Frequency}^6$$

$$\text{Force}^9 \cdot \text{Volume} = \text{Mass}^3 \cdot \text{Power, Radiance Flux}^6$$

$$\text{Force}^3 \cdot \text{Volume} = \text{Mass}^3 \cdot \text{Absorbed Dose}^3$$

$$\text{Force} \cdot \text{Mass} = \text{Volume} \cdot \text{Dynamic Viscosity}^2$$

$$\text{Force} \cdot \text{Volume} = \text{Mass} \cdot \text{Kinematic Viscosity}^2$$

$$\text{Force} \cdot \text{Wavenumber} = \text{Mass} \cdot \text{Frequency}^2$$

$$\text{Force}^3 = \text{Mass} \cdot \text{Wavenumber} \cdot \text{Power, Radiance Flux}^2$$

$$\text{Force} = \text{Mass} \cdot \text{Wavenumber} \cdot \text{Absorbed Dose}$$

$$\text{Force} \cdot \text{Mass} \cdot \text{Wavenumber}^3 = \text{Dynamic Viscosity}^2$$

$$\text{Force} = \text{Mass} \cdot \text{Wavenumber}^3 \cdot \text{Kinematic Viscosity}^2$$

$$\text{Force}^3 = \text{Mass}^4 \cdot \text{Specific Volume} \cdot \text{Frequency}^6$$

$$\text{Force}^3 = \text{Mass}^2 \cdot \text{Specific Volume}^2 \cdot \text{Pressure,Stress}^3$$

$$\text{Force}^3 \cdot \text{Mass} \cdot \text{Specific Volume} = \text{Energy,Work,Heat Quantity}^3$$

$$\text{Force}^9 \cdot \text{Specific Volume} = \text{Mass}^2 \cdot \text{Power, Radiance Flux}^6$$

$Force^3 \cdot Specific\ Volume = Mass^2 \cdot Absorbed\ Dose^3$

$Force \cdot Pressure, Stress = Mass^2 \cdot Frequency^4$

$Force^2 = Mass \cdot Frequency^2 \cdot Energy, Work, Heat\ Quantity$

$Force^2 = Mass \cdot Frequency \cdot Power,\ Radiance\ Flux$

$Force^2 = Mass^2 \cdot Frequency^2 \cdot Absorbed\ Dose$

$Force \cdot Dynamic\ Viscosity = Mass^2 \cdot Frequency^3$

$Force^2 = Mass^2 \cdot Frequency^3 \cdot Kinematic\ Viscosity$

$Force^3 \cdot Mass\ Density = Mass^4 \cdot Frequency^6$

$Force^7 = Mass^2 \cdot Pressure, Stress \cdot Power,\ Radiance\ Flux^4$

$Force^3 = Mass^2 \cdot Pressure, Stress \cdot Absorbed\ Dose^2$

$Force \cdot Dynamic\ Viscosity^4 = Mass^2 \cdot Pressure, Stress^3$

$Force^5 = Mass^2 \cdot Pressure, Stress^3 \cdot Kinematic\ Viscosity^4$

$Force^3 \cdot Mass\ Density^2 = Mass^2 \cdot Pressure, Stress^3$

$Force^2 \cdot Energy, Work, Heat\ Quantity = Mass \cdot Power,\ Radiance\ Flux^2$

$Force^4 \cdot Mass = Energy, Work, Heat\ Quantity^3 \cdot Dynamic\ Viscosity^2$

$Force^2 \cdot Mass \cdot Kinematic\ Viscosity^2 = Energy, Work, Heat\ Quantity^3$

$Force^3 \cdot Mass = Energy, Work, Heat\ Quantity^3 \cdot Mass\ Density$

$Force^5 = Mass \cdot Power,\ Radiance\ Flux^3 \cdot Dynamic\ Viscosity$

$Force^4 \cdot Kinematic\ Viscosity = Mass \cdot Power,\ Radiance\ Flux^3$

$Force^9 = Mass^2 \cdot Power,\ Radiance\ Flux^6 \cdot Mass\ Density$

$Force^2 \cdot Inductance = Mass \cdot Electric\ Potential\ Difference^2$

$Force^4 = Mass^2 \cdot Absorbed\ Dose^3 \cdot Dynamic\ Viscosity^2$

$Force^2 \cdot Kinematic\ Viscosity^2 = Mass^2 \cdot Absorbed\ Dose^3$

$$\text{Force}^3 = \text{Mass}^2 \cdot \text{Absorbed Dose}^3 \cdot \text{Mass Density}$$

$$\text{Force} \cdot \text{Specific Volume} = \text{Velocity}^2 \cdot \text{Area}$$

$$\text{Force}^2 = \text{Velocity}^2 \cdot \text{Area} \cdot \text{Dynamic Viscosity}^2$$

$$\text{Force} = \text{Velocity}^2 \cdot \text{Area} \cdot \text{Mass Density}$$

$$\text{Force}^3 \cdot \text{Specific Volume}^3 = \text{Velocity}^6 \cdot \text{Volume}^2$$

$$\text{Force}^3 = \text{Velocity}^3 \cdot \text{Volume} \cdot \text{Dynamic Viscosity}^3$$

$$\text{Force}^3 = \text{Velocity}^6 \cdot \text{Volume}^2 \cdot \text{Mass Density}^3$$

$$\text{Force} \cdot \text{Wavenumber}^2 \cdot \text{Specific Volume} = \text{Velocity}^2$$

$$\text{Force} \cdot \text{Wavenumber} = \text{Velocity} \cdot \text{Dynamic Viscosity}$$

$$\text{Force} \cdot \text{Wavenumber}^2 = \text{Velocity}^2 \cdot \text{Mass Density}$$

$$\text{Force} \cdot \text{Specific Volume} \cdot \text{Frequency}^2 = \text{Velocity}^4$$

$$\text{Force}^3 \cdot \text{Specific Volume} = \text{Velocity}^2 \cdot \text{Energy,Work,Heat Quantity}^2$$

$$\text{Force} \cdot \text{Frequency}^2 = \text{Velocity}^2 \cdot \text{Pressure,Stress}$$

$$\text{Force} \cdot \text{Velocity} = \text{Frequency} \cdot \text{Energy,Work,Heat Quantity}$$

$$\text{Force} \cdot \text{Frequency} = \text{Velocity}^2 \cdot \text{Dynamic Viscosity}$$

$$\text{Force} \cdot \text{Frequency}^2 = \text{Velocity}^4 \cdot \text{Mass Density}$$

$$\text{Force} \cdot \text{Pressure,Stress} = \text{Velocity}^2 \cdot \text{Dynamic Viscosity}^2$$

$$\text{Force} \cdot \text{Velocity}^2 = \text{Pressure,Stress} \cdot \text{Kinematic Viscosity}^2$$

$$\text{Force}^2 = \text{Velocity} \cdot \text{Energy,Work,Heat Quantity} \cdot \text{Dynamic Viscosity}$$

$$\text{Force} \cdot \text{Kinematic Viscosity} = \text{Velocity} \cdot \text{Energy,Work,Heat Quantity}$$

$$\text{Force}^3 = \text{Velocity}^2 \cdot \text{Energy,Work,Heat Quantity}^2 \cdot \text{Mass Density}$$

$$\text{Force} = \text{Velocity} \cdot \text{Electric Charge} \cdot \text{Magnetic Flux Density}$$

Force $\cdot$ Velocity $\cdot$ Electric Resistance = Electric Potential Difference^2

Force $\cdot$ Specific Volume = Area^2 $\cdot$ Frequency^2

Force^3 $\cdot$ Specific Volume = Area $\cdot$ Power, Radiance Flux^2

Force $\cdot$ Specific Volume = Area $\cdot$ Absorbed Dose

Force^2 $\cdot$ Area = Current Density^2 $\cdot$ Magnetic Flux Density^2

Force^2 $\cdot$ Area^5 = Current Density^4 $\cdot$ Inductance^2

Force^2 $\cdot$ Area^3 = Current Density^2 $\cdot$ Magnetic Flux^2

Force = Area $\cdot$ Magnetic Field Strength $\cdot$ Magnetic Flux Density

Force^2 = Area $\cdot$ Magnetic Field Strength^4 $\cdot$ Inductance^2

Force^2 $\cdot$ Area $\cdot$ Frequency^2 = Power, Radiance Flux^2

Force = Area $\cdot$ Frequency $\cdot$ Dynamic Viscosity

Force = Area^2 $\cdot$ Frequency^2 $\cdot$ Mass Density

Force^4 = Area $\cdot$ Power, Radiance Flux^2 $\cdot$ Dynamic Viscosity^2

Force^2 $\cdot$ Kinematic Viscosity^2 = Area $\cdot$ Power, Radiance Flux^2

Force^3 = Area $\cdot$ Power, Radiance Flux^2 $\cdot$ Mass Density

Force^2 $\cdot$ Area = Electric Charge^2 $\cdot$ Electric Potential Difference^2

Force^2 $\cdot$ Area $\cdot$ Capacitance^2 = Electric Charge^4

Force^2 $\cdot$ Area = Electric Potential Difference^4 $\cdot$ Capacitance^2

Force^2 $\cdot$ Inductance^2 = Area^3 $\cdot$ Magnetic Flux Density^4

Force^2 $\cdot$ Area $\cdot$ Inductance^2 = Magnetic Flux^4

Force^2 = Area $\cdot$ Absorbed Dose $\cdot$ Dynamic Viscosity^2

Force = Area $\cdot$ Absorbed Dose $\cdot$ Mass Density

Force^3 $\cdot$ Specific Volume^3 = Volume^4 $\cdot$ Frequency^6

$$\text{Force}^9 \cdot \text{Specific Volume}^3 = \text{Volume}^2 \cdot \text{Power, Radiance Flux}^6$$

$$\text{Force}^3 \cdot \text{Specific Volume}^3 = \text{Volume}^2 \cdot \text{Absorbed Dose}^3$$

$$\text{Force}^3 \cdot \text{Volume} = \text{Current Density}^3 \cdot \text{Magnetic Flux Density}^3$$

$$\text{Force}^3 \cdot \text{Volume}^5 = \text{Current Density}^6 \cdot \text{Inductance}^3$$

$$\text{Force} \cdot \text{Volume} = \text{Current Density} \cdot \text{Magnetic Flux}$$

$$\text{Force}^3 = \text{Volume}^2 \cdot \text{Magnetic Field Strength}^3 \cdot \text{Magnetic Flux Density}^3$$

$$\text{Force}^3 = \text{Volume} \cdot \text{Magnetic Field Strength}^6 \cdot \text{Inductance}^3$$

$$\text{Force}^3 \cdot \text{Volume} \cdot \text{Frequency}^3 = \text{Power, Radiance Flux}^3$$

$$\text{Force}^3 = \text{Volume}^2 \cdot \text{Frequency}^3 \cdot \text{Dynamic Viscosity}^3$$

$$\text{Force}^3 = \text{Volume}^4 \cdot \text{Frequency}^6 \cdot \text{Mass Density}^3$$

$$\text{Force}^6 = \text{Volume} \cdot \text{Power, Radiance Flux}^3 \cdot \text{Dynamic Viscosity}^3$$

$$\text{Force}^3 \cdot \text{Kinematic Viscosity}^3 = \text{Volume} \cdot \text{Power, Radiance Flux}^3$$

$$\text{Force}^9 = \text{Volume}^2 \cdot \text{Power, Radiance Flux}^6 \cdot \text{Mass Density}^3$$

$$\text{Force}^3 \cdot \text{Volume} = \text{Electric Charge}^3 \cdot \text{Electric Potential Difference}^3$$

$$\text{Force}^3 \cdot \text{Volume} \cdot \text{Capacitance}^3 = \text{Electric Charge}^6$$

$$\text{Force}^3 \cdot \text{Volume} = \text{Electric Potential Difference}^6 \cdot \text{Capacitance}^3$$

$$\text{Force} \cdot \text{Inductance} = \text{Volume} \cdot \text{Magnetic Flux Density}^2$$

$$\text{Force}^3 \cdot \text{Volume} \cdot \text{Inductance}^3 = \text{Magnetic Flux}^6$$

$$\text{Force}^6 = \text{Volume}^2 \cdot \text{Absorbed Dose}^3 \cdot \text{Dynamic Viscosity}^6$$

$$\text{Force}^3 = \text{Volume}^2 \cdot \text{Absorbed Dose}^3 \cdot \text{Mass Density}^3$$

$$\text{Force} \cdot \text{Wavenumber}^4 \cdot \text{Specific Volume} = \text{Frequency}^2$$

$$\text{Force}^3 \cdot \text{Wavenumber}^2 \cdot \text{Specific Volume} = \text{Power, Radiance Flux}^2$$

Force $\cdot$ Wavenumber^2 $\cdot$ Specific Volume = Absorbed Dose

Force = Wavenumber $\cdot$ Current Density $\cdot$ Magnetic Flux Density

Force = Wavenumber^5 $\cdot$ Current Density^2 $\cdot$ Inductance

Force = Wavenumber^3 $\cdot$ Current Density $\cdot$ Magnetic Flux

Force $\cdot$ Wavenumber^2 = Magnetic Field Strength $\cdot$ Magnetic Flux Density

Force $\cdot$ Wavenumber = Magnetic Field Strength^2 $\cdot$ Inductance

Force $\cdot$ Frequency = Wavenumber $\cdot$ Power, Radiance Flux

Force $\cdot$ Wavenumber^2 = Frequency $\cdot$ Dynamic Viscosity

Force $\cdot$ Wavenumber^4 = Frequency^2 $\cdot$ Mass Density

Force^2 $\cdot$ Wavenumber = Power, Radiance Flux $\cdot$ Dynamic Viscosity

Force $\cdot$ Wavenumber $\cdot$ Kinematic Viscosity = Power, Radiance Flux

Force^3 $\cdot$ Wavenumber^2 = Power, Radiance Flux^2 $\cdot$ Mass Density

Force = Wavenumber $\cdot$ Electric Charge $\cdot$ Electric Potential Difference

Force $\cdot$ Capacitance = Wavenumber $\cdot$ Electric Charge^2

Force = Wavenumber $\cdot$ Electric Potential Difference^2 $\cdot$ Capacitance

Force $\cdot$ Wavenumber^3 $\cdot$ Inductance = Magnetic Flux Density^2

Force $\cdot$ Inductance = Wavenumber $\cdot$ Magnetic Flux^2

Force^2 $\cdot$ Wavenumber^2 = Absorbed Dose $\cdot$ Dynamic Viscosity^2

Force $\cdot$ Wavenumber^2 = Absorbed Dose $\cdot$ Mass Density

Force $\cdot$ Specific Volume $\cdot$ Electric Charge^2 = Current Density^2

Force $\cdot$ Frequency^2 = Specific Volume $\cdot$ Pressure,Stress^2

Force^5 $\cdot$ Specific Volume = Frequency^2 $\cdot$ Energy,Work,Heat Quantity^4

Force^5 $\cdot$ Specific Volume $\cdot$ Frequency^2 = Power, Radiance Flux^4

$$Force \cdot Specific\ Volume \cdot Frequency^2 = Absorbed\ Dose^2$$

$$Force^2 \cdot Specific\ Volume \cdot Pressure,Stress = Power,\ Radiance\ Flux^2$$

$$Force^5 \cdot Specific\ Volume = Energy,Work,Heat\ Quantity^2 \cdot Power,\ Radiance\ Flux^2$$

$$Force^3 \cdot Specific\ Volume = Energy,Work,Heat\ Quantity^2 \cdot Absorbed\ Dose$$

$$Force \cdot Specific\ Volume \cdot Magnetic\ Flux\ Density^2 = Electric\ Potential\ Difference^2$$

$$Force^3 \cdot Magnetic\ Field\ Strength^2 = Current\ Density^2 \cdot Pressure,Stress^3$$

$$Force^3 \cdot Current\ Density = Magnetic\ Field\ Strength \cdot Energy,Work,Heat\ Quantity^3$$

$$Force^3 = Current\ Density^2 \cdot Magnetic\ Field\ Strength \cdot Magnetic\ Flux\ Density^3$$

$$Force^3 = Current\ Density \cdot Magnetic\ Field\ Strength^5 \cdot Inductance^3$$

$$Force^3 = Current\ Density^2 \cdot Pressure,Stress \cdot Magnetic\ Flux\ Density^2$$

$$Force^7 = Current\ Density^4 \cdot Pressure,Stress^5 \cdot Inductance^2$$

$$Force^5 = Current\ Density^2 \cdot Pressure,Stress^3 \cdot Magnetic\ Flux^2$$

$$Force^4 \cdot Current\ Density^2 \cdot Inductance = Energy,Work,Heat\ Quantity^5$$

$$Force^2 \cdot Current\ Density \cdot Magnetic\ Flux = Energy,Work,Heat\ Quantity^3$$

$$Force \cdot Electric\ Charge = Current\ Density \cdot Dynamic\ Viscosity$$

$$Force \cdot Electric\ Charge^2 = Current\ Density^2 \cdot Mass\ Density$$

$$Force^4 \cdot Inductance = Current\ Density^3 \cdot Magnetic\ Flux\ Density^5$$

$$Force^2 \cdot Magnetic\ Flux = Current\ Density^2 \cdot Magnetic\ Flux\ Density^3$$

$$Force^2 \cdot Current\ Density \cdot Inductance^3 = Magnetic\ Flux^5$$

$$Force \cdot Frequency = Magnetic\ Field\ Strength \cdot Electric\ Potential\ Difference$$

$$Force \cdot Pressure,Stress = Magnetic\ Field\ Strength^4 \cdot Inductance^2$$

$$Force^3 = Magnetic\ Field\ Strength \cdot Energy,Work,Heat\ Quantity^2 \cdot Magnetic\ Flux\ Density$$

$$\text{Force}^2 = \text{Magnetic Field Strength}^2 \cdot \text{Energy,Work,Heat Quantity} \cdot \text{Inductance}$$

$$\text{Force} = \text{Magnetic Field Strength} \cdot \text{Electric Charge} \cdot \text{Electric Resistance}$$

$$\text{Force} \cdot \text{Magnetic Flux Density} = \text{Magnetic Field Strength}^3 \cdot \text{Inductance}^2$$

$$\text{Energy}^2 = \text{Acceleration}^2 \cdot \text{Mass}^2 \cdot \text{Area}$$

$$\text{Energy}^3 = \text{Acceleration}^3 \cdot \text{Mass}^3 \cdot \text{Volume}$$

$$\text{Energy} \cdot \text{Wavenumber} = \text{Acceleration} \cdot \text{Mass}$$

$$\text{Energy}^3 = \text{Acceleration}^3 \cdot \text{Mass}^4 \cdot \text{Specific Volume}$$

$$\text{Energy} \cdot \text{Frequency}^2 = \text{Acceleration}^2 \cdot \text{Mass}$$

$$\text{Energy}^2 \cdot \text{Pressure,Stress} = \text{Acceleration}^3 \cdot \text{Mass}^3$$

$$\text{Energy} \cdot \text{Acceleration}^2 \cdot \text{Mass} = \text{Power, Radiance Flux}^2$$

$$\text{Energy}^3 \cdot \text{Dynamic Viscosity}^2 = \text{Acceleration}^4 \cdot \text{Mass}^5$$

$$\text{Energy}^3 = \text{Acceleration}^2 \cdot \text{Mass}^3 \cdot \text{Kinematic Viscosity}^2$$

$$\text{Energy}^3 \cdot \text{Mass Density} = \text{Acceleration}^3 \cdot \text{Mass}^4$$

$$\text{Energy} \cdot \text{Acceleration}^3 \cdot \text{Specific Volume} = \text{Velocity}^8$$

$$\text{Energy} \cdot \text{Acceleration}^3 = \text{Velocity}^6 \cdot \text{Pressure,Stress}$$

$$\text{Energy} \cdot \text{Acceleration} = \text{Velocity} \cdot \text{Power, Radiance Flux}$$

$$\text{Energy} \cdot \text{Acceleration}^2 = \text{Velocity}^5 \cdot \text{Dynamic Viscosity}$$

$$\text{Energy} \cdot \text{Acceleration}^3 = \text{Velocity}^8 \cdot \text{Mass Density}$$

$$\text{Energy} \cdot \text{Specific Volume} = \text{Acceleration} \cdot \text{Area}^2$$

$$\text{Energy}^4 \cdot \text{Acceleration}^2 = \text{Area} \cdot \text{Power, Radiance Flux}^4$$

$$\text{Energy}^4 = \text{Acceleration}^2 \cdot \text{Area}^5 \cdot \text{Dynamic Viscosity}^4$$

$$\text{Energy} = \text{Acceleration} \cdot \text{Area}^2 \cdot \text{Mass Density}$$

$$\text{Energy}^3 \cdot \text{Specific Volume}^3 = \text{Acceleration}^3 \cdot \text{Volume}^4$$

$$\text{Energy}^6 \cdot \text{Acceleration}^3 = \text{Volume} \cdot \text{Power, Radiance Flux}^6$$

$$\text{Energy}^6 = \text{Acceleration}^3 \cdot \text{Volume}^5 \cdot \text{Dynamic Viscosity}^6$$

$$\text{Energy}^3 = \text{Acceleration}^3 \cdot \text{Volume}^4 \cdot \text{Mass Density}^3$$

$$\text{Energy} \cdot \text{Wavenumber}^4 \cdot \text{Specific Volume} = \text{Acceleration}$$

$$\text{Energy}^2 \cdot \text{Acceleration} \cdot \text{Wavenumber} = \text{Power, Radiance Flux}^2$$

$$\text{Energy}^2 \cdot \text{Wavenumber}^5 = \text{Acceleration} \cdot \text{Dynamic Viscosity}^2$$

$$\text{Energy} \cdot \text{Wavenumber}^4 = \text{Acceleration} \cdot \text{Mass Density}$$

$$\text{Energy} \cdot \text{Specific Volume} \cdot \text{Frequency}^8 = \text{Acceleration}^5$$

$$\text{Energy} \cdot \text{Acceleration}^3 = \text{Specific Volume}^3 \cdot \text{Pressure,Stress}^4$$

$$\text{Energy}^7 \cdot \text{Acceleration}^5 = \text{Specific Volume} \cdot \text{Power, Radiance Flux}^8$$

$$\text{Energy} \cdot \text{Acceleration}^3 \cdot \text{Specific Volume} = \text{Absorbed Dose}^4$$

$$\text{Energy}^3 \cdot \text{Acceleration} = \text{Specific Volume}^5 \cdot \text{Dynamic Viscosity}^8$$

$$\text{Energy}^3 \cdot \text{Acceleration} \cdot \text{Specific Volume}^3 = \text{Kinematic Viscosity}^8$$

$$\text{Energy} \cdot \text{Frequency}^6 = \text{Acceleration}^3 \cdot \text{Pressure,Stress}$$

$$\text{Energy} \cdot \text{Frequency}^5 = \text{Acceleration}^3 \cdot \text{Dynamic Viscosity}$$

$$\text{Energy} \cdot \text{Frequency}^8 = \text{Acceleration}^5 \cdot \text{Mass Density}$$

$$\text{Energy}^5 \cdot \text{Acceleration}^3 \cdot \text{Pressure,Stress} = \text{Power, Radiance Flux}^6$$

$$\text{Energy} \cdot \text{Acceleration}^3 = \text{Pressure,Stress} \cdot \text{Absorbed Dose}^3$$

$$\text{Energy} \cdot \text{Pressure,Stress}^5 = \text{Acceleration}^3 \cdot \text{Dynamic Viscosity}^6$$

$$\text{Energy} \cdot \text{Acceleration} = \text{Pressure,Stress} \cdot \text{Kinematic Viscosity}^2$$

$$\text{Energy} \cdot \text{Acceleration}^3 \cdot \text{Mass Density}^3 = \text{Pressure,Stress}^4$$

$$\text{Energy}^2 \cdot \text{Acceleration}^2 = \text{Power, Radiance Flux}^2 \cdot \text{Absorbed Dose}$$

$\text{Energy}^4 \cdot \text{Acceleration}^3 \cdot \text{Dynamic Viscosity} = \text{Power, Radiance Flux}^5$

$\text{Energy}^3 \cdot \text{Acceleration}^2 = \text{Power, Radiance Flux}^3 \cdot \text{Kinematic Viscosity}$

$\text{Energy}^7 \cdot \text{Acceleration}^5 \cdot \text{Mass Density} = \text{Power, Radiance Flux}^8$

$\text{Energy}^2 \cdot \text{Acceleration}^4 = \text{Absorbed Dose}^5 \cdot \text{Dynamic Viscosity}^2$

$\text{Energy} \cdot \text{Acceleration}^3 = \text{Absorbed Dose}^4 \cdot \text{Mass Density}$

$\text{Energy}^3 \cdot \text{Acceleration} = \text{Dynamic Viscosity}^3 \cdot \text{Kinematic Viscosity}^5$

$\text{Energy}^3 \cdot \text{Acceleration} \cdot \text{Mass Density}^5 = \text{Dynamic Viscosity}^8$

$\text{Energy}^3 \cdot \text{Acceleration} = \text{Kinematic Viscosity}^8 \cdot \text{Mass Density}^3$

$\text{Energy} = \text{Mass} \cdot \text{Area} \cdot \text{Frequency}^2$

$\text{Energy}^3 = \text{Mass} \cdot \text{Area} \cdot \text{Power, Radiance Flux}^2$

$\text{Energy} \cdot \text{Mass} = \text{Area}^2 \cdot \text{Dynamic Viscosity}^2$

$\text{Energy} \cdot \text{Area} = \text{Mass} \cdot \text{Kinematic Viscosity}^2$

$\text{Energy}^3 = \text{Mass}^3 \cdot \text{Volume}^2 \cdot \text{Frequency}^6$

$\text{Energy}^9 = \text{Mass}^3 \cdot \text{Volume}^2 \cdot \text{Power, Radiance Flux}^6$

$\text{Energy}^3 \cdot \text{Mass}^3 = \text{Volume}^4 \cdot \text{Dynamic Viscosity}^6$

$\text{Energy}^3 \cdot \text{Volume}^2 = \text{Mass}^3 \cdot \text{Kinematic Viscosity}^6$

$\text{Energy} \cdot \text{Wavenumber}^2 = \text{Mass} \cdot \text{Frequency}^2$

$\text{Energy}^3 \cdot \text{Wavenumber}^2 = \text{Mass} \cdot \text{Power, Radiance Flux}^2$

$\text{Energy} \cdot \text{Mass} \cdot \text{Wavenumber}^4 = \text{Dynamic Viscosity}^2$

$\text{Energy} = \text{Mass} \cdot \text{Wavenumber}^2 \cdot \text{Kinematic Viscosity}^2$

$\text{Energy}^3 = \text{Mass}^5 \cdot \text{Specific Volume}^2 \cdot \text{Frequency}^6$

$\text{Energy} = \text{Mass} \cdot \text{Specific Volume} \cdot \text{Pressure,Stress}$

$\text{Energy}^9 = \text{Mass}^5 \cdot \text{Specific Volume}^2 \cdot \text{Power, Radiance Flux}^6$

$$\text{Energy}^3 = \text{Mass} \cdot \text{Specific Volume}^4 \cdot \text{Dynamic Viscosity}^6$$

$$\text{Energy}^3 \cdot \text{Specific Volume}^2 = \text{Mass} \cdot \text{Kinematic Viscosity}^6$$

$$\text{Energy} \cdot \text{Pressure,Stress}^2 = \text{Mass}^3 \cdot \text{Frequency}^6$$

$$\text{Energy} \cdot \text{Dynamic Viscosity}^2 = \text{Mass}^3 \cdot \text{Frequency}^4$$

$$\text{Energy} = \text{Mass} \cdot \text{Frequency} \cdot \text{Kinematic Viscosity}$$

$$\text{Energy}^3 \cdot \text{Mass Density}^2 = \text{Mass}^5 \cdot \text{Frequency}^6$$

$$\text{Energy}^7 \cdot \text{Pressure,Stress}^2 = \text{Mass}^3 \cdot \text{Power, Radiance Flux}^6$$

$$\text{Energy} \cdot \text{Dynamic Viscosity}^6 = \text{Mass}^3 \cdot \text{Pressure,Stress}^4$$

$$\text{Energy}^5 = \text{Mass}^3 \cdot \text{Pressure,Stress}^2 \cdot \text{Kinematic Viscosity}^6$$

$$\text{Energy} \cdot \text{Mass Density} = \text{Mass} \cdot \text{Pressure,Stress}$$

$$\text{Energy}^5 \cdot \text{Dynamic Viscosity}^2 = \text{Mass}^3 \cdot \text{Power, Radiance Flux}^4$$

$$\text{Energy}^2 = \text{Mass} \cdot \text{Power, Radiance Flux} \cdot \text{Kinematic Viscosity}$$

$$\text{Energy}^9 \cdot \text{Mass Density}^2 = \text{Mass}^5 \cdot \text{Power, Radiance Flux}^6$$

$$\text{Energy}^3 = \text{Mass} \cdot \text{Dynamic Viscosity}^2 \cdot \text{Kinematic Viscosity}^4$$

$$\text{Energy}^3 \cdot \text{Mass Density}^4 = \text{Mass} \cdot \text{Dynamic Viscosity}^6$$

$$\text{Energy}^3 = \text{Mass} \cdot \text{Kinematic Viscosity}^6 \cdot \text{Mass Density}^2$$

$$\text{Energy}^2 \cdot \text{Specific Volume}^2 = \text{Velocity}^4 \cdot \text{Area}^3$$

$$\text{Energy}^2 \cdot \text{Velocity}^2 = \text{Area} \cdot \text{Power, Radiance Flux}^2$$

$$\text{Energy} = \text{Velocity} \cdot \text{Area} \cdot \text{Dynamic Viscosity}$$

$$\text{Energy}^2 = \text{Velocity}^4 \cdot \text{Area}^3 \cdot \text{Mass Density}^2$$

$$\text{Energy} \cdot \text{Specific Volume} = \text{Velocity}^2 \cdot \text{Volume}$$

$$\text{Energy}^3 \cdot \text{Velocity}^3 = \text{Volume} \cdot \text{Power, Radiance Flux}^3$$

$$\text{Energy}^3 = \text{Velocity}^3 \cdot \text{Volume}^2 \cdot \text{Dynamic Viscosity}^3$$

$$\text{Energy} = \text{Velocity}^2 \cdot \text{Volume} \cdot \text{Mass Density}$$

$$\text{Energy} \cdot \text{Wavenumber}^3 \cdot \text{Specific Volume} = \text{Velocity}^2$$

$$\text{Energy} \cdot \text{Velocity} \cdot \text{Wavenumber} = \text{Power, Radiance Flux}$$

$$\text{Energy} \cdot \text{Wavenumber}^2 = \text{Velocity} \cdot \text{Dynamic Viscosity}$$

$$\text{Energy} \cdot \text{Wavenumber}^3 = \text{Velocity}^2 \cdot \text{Mass Density}$$

$$\text{Energy} \cdot \text{Specific Volume} \cdot \text{Frequency}^3 = \text{Velocity}^5$$

$$\text{Energy}^2 \cdot \text{Velocity}^5 = \text{Specific Volume} \cdot \text{Power, Radiance Flux}^3$$

$$\text{Energy} \cdot \text{Velocity} = \text{Specific Volume}^2 \cdot \text{Dynamic Viscosity}^3$$

$$\text{Energy} \cdot \text{Velocity} \cdot \text{Specific Volume} = \text{Kinematic Viscosity}^3$$

$$\text{Energy} \cdot \text{Frequency}^3 = \text{Velocity}^3 \cdot \text{Pressure,Stress}$$

$$\text{Energy} \cdot \text{Frequency}^2 = \text{Velocity}^3 \cdot \text{Dynamic Viscosity}$$

$$\text{Energy} \cdot \text{Frequency}^3 = \text{Velocity}^5 \cdot \text{Mass Density}$$

$$\text{Energy}^2 \cdot \text{Velocity}^3 \cdot \text{Pressure,Stress} = \text{Power, Radiance Flux}^3$$

$$\text{Energy} \cdot \text{Pressure,Stress}^2 = \text{Velocity}^3 \cdot \text{Dynamic Viscosity}^3$$

$$\text{Energy} \cdot \text{Velocity}^3 = \text{Pressure,Stress} \cdot \text{Kinematic Viscosity}^3$$

$$\text{Energy} \cdot \text{Velocity}^3 \cdot \text{Dynamic Viscosity} = \text{Power, Radiance Flux}^2$$

$$\text{Energy} \cdot \text{Velocity}^2 = \text{Power, Radiance Flux} \cdot \text{Kinematic Viscosity}$$

$$\text{Energy}^2 \cdot \text{Velocity}^5 \cdot \text{Mass Density} = \text{Power, Radiance Flux}^3$$

$$\text{Energy} \cdot \text{Velocity} = \text{Dynamic Viscosity} \cdot \text{Kinematic Viscosity}^2$$

$$\text{Energy} \cdot \text{Velocity} \cdot \text{Mass Density}^2 = \text{Dynamic Viscosity}^3$$

$$\text{Energy} \cdot \text{Velocity} = \text{Kinematic Viscosity}^3 \cdot \text{Mass Density}$$

$$\text{Energy}^2 \cdot \text{Specific Volume}^2 = \text{Area}^5 \cdot \text{Frequency}^4$$

$\text{Energy}^6 \cdot \text{Specific Volume}^2 = \text{Area}^5 \cdot \text{Power, Radiance Flux}^4$

$\text{Energy}^2 \cdot \text{Specific Volume}^2 = \text{Area}^3 \cdot \text{Absorbed Dose}^2$

$\text{Energy}^2 = \text{Area} \cdot \text{Specific Volume}^2 \cdot \text{Dynamic Viscosity}^4$

$\text{Energy}^2 \cdot \text{Specific Volume}^2 = \text{Area} \cdot \text{Kinematic Viscosity}^4$

$\text{Energy} \cdot \text{Area}^2 = \text{Current Density}^2 \cdot \text{Inductance}$

$\text{Energy} \cdot \text{Area} = \text{Current Density} \cdot \text{Magnetic Flux}$

$\text{Energy}^2 = \text{Area}^3 \cdot \text{Magnetic Field Strength}^2 \cdot \text{Magnetic Flux Density}^2$

$\text{Energy} = \text{Area} \cdot \text{Magnetic Field Strength}^2 \cdot \text{Inductance}$

$\text{Energy}^2 = \text{Area} \cdot \text{Magnetic Field Strength}^2 \cdot \text{Magnetic Flux}^2$

$\text{Energy}^2 = \text{Area}^3 \cdot \text{Frequency}^2 \cdot \text{Dynamic Viscosity}^2$

$\text{Energy}^2 = \text{Area}^5 \cdot \text{Frequency}^4 \cdot \text{Mass Density}^2$

$\text{Energy}^2 \cdot \text{Absorbed Dose} = \text{Area} \cdot \text{Power, Radiance Flux}^2$

$\text{Energy}^4 = \text{Area}^3 \cdot \text{Power, Radiance Flux}^2 \cdot \text{Dynamic Viscosity}^2$

$\text{Energy} \cdot \text{Kinematic Viscosity} = \text{Area} \cdot \text{Power, Radiance Flux}$

$\text{Energy}^6 = \text{Area}^5 \cdot \text{Power, Radiance Flux}^4 \cdot \text{Mass Density}^2$

$\text{Energy} \cdot \text{Inductance} = \text{Area}^2 \cdot \text{Magnetic Flux Density}^2$

$\text{Energy}^2 = \text{Area}^2 \cdot \text{Absorbed Dose} \cdot \text{Dynamic Viscosity}^2$

$\text{Energy}^2 = \text{Area}^3 \cdot \text{Absorbed Dose}^2 \cdot \text{Mass Density}^2$

$\text{Energy}^2 = \text{Area} \cdot \text{Dynamic Viscosity}^2 \cdot \text{Kinematic Viscosity}^2$

$\text{Energy}^2 \cdot \text{Mass Density}^2 = \text{Area} \cdot \text{Dynamic Viscosity}^4$

$\text{Energy}^2 = \text{Area} \cdot \text{Kinematic Viscosity}^4 \cdot \text{Mass Density}^2$

$\text{Energy}^3 \cdot \text{Specific Volume}^3 = \text{Volume}^5 \cdot \text{Frequency}^6$

$\text{Energy}^9 \cdot \text{Specific Volume}^3 = \text{Volume}^5 \cdot \text{Power, Radiance Flux}^6$

$\text{Energy} \cdot \text{Specific Volume} = \text{Volume} \cdot \text{Absorbed Dose}$

$\text{Energy}^3 = \text{Volume} \cdot \text{Specific Volume}^3 \cdot \text{Dynamic Viscosity}^6$

$\text{Energy}^3 \cdot \text{Specific Volume}^3 = \text{Volume} \cdot \text{Kinematic Viscosity}^6$

$\text{Energy}^3 \cdot \text{Volume}^4 = \text{Current Density}^6 \cdot \text{Inductance}^3$

$\text{Energy}^3 \cdot \text{Volume}^2 = \text{Current Density}^3 \cdot \text{Magnetic Flux}^3$

$\text{Energy} = \text{Volume} \cdot \text{Magnetic Field Strength} \cdot \text{Magnetic Flux Density}$

$\text{Energy}^3 = \text{Volume}^2 \cdot \text{Magnetic Field Strength}^6 \cdot \text{Inductance}^3$

$\text{Energy}^3 = \text{Volume} \cdot \text{Magnetic Field Strength}^3 \cdot \text{Magnetic Flux}^3$

$\text{Energy} = \text{Volume} \cdot \text{Frequency} \cdot \text{Dynamic Viscosity}$

$\text{Energy}^3 = \text{Volume}^5 \cdot \text{Frequency}^6 \cdot \text{Mass Density}^3$

$\text{Energy}^6 \cdot \text{Absorbed Dose}^3 = \text{Volume}^2 \cdot \text{Power, Radiance Flux}^6$

$\text{Energy}^2 = \text{Volume} \cdot \text{Power, Radiance Flux} \cdot \text{Dynamic Viscosity}$

$\text{Energy}^3 \cdot \text{Kinematic Viscosity}^3 = \text{Volume}^2 \cdot \text{Power, Radiance Flux}^3$

$\text{Energy}^9 = \text{Volume}^5 \cdot \text{Power, Radiance Flux}^6 \cdot \text{Mass Density}^3$

$\text{Energy}^3 \cdot \text{Inductance}^3 = \text{Volume}^4 \cdot \text{Magnetic Flux Density}^6$

$\text{Energy}^6 = \text{Volume}^4 \cdot \text{Absorbed Dose}^3 \cdot \text{Dynamic Viscosity}^6$

$\text{Energy} = \text{Volume} \cdot \text{Absorbed Dose} \cdot \text{Mass Density}$

$\text{Energy}^3 = \text{Volume} \cdot \text{Dynamic Viscosity}^3 \cdot \text{Kinematic Viscosity}^3$

$\text{Energy}^3 \cdot \text{Mass Density}^3 = \text{Volume} \cdot \text{Dynamic Viscosity}^6$

$\text{Energy}^3 = \text{Volume} \cdot \text{Kinematic Viscosity}^6 \cdot \text{Mass Density}^3$

$\text{Energy} \cdot \text{Wavenumber}^5 \cdot \text{Specific Volume} = \text{Frequency}^2$

$\text{Energy}^3 \cdot \text{Wavenumber}^5 \cdot \text{Specific Volume} = \text{Power, Radiance Flux}^2$

Energy $\cdot$ Wavenumber^3 $\cdot$ Specific Volume = Absorbed Dose

Energy $\cdot$ Wavenumber = Specific Volume $\cdot$ Dynamic Viscosity^2

Energy $\cdot$ Wavenumber $\cdot$ Specific Volume = Kinematic Viscosity^2

Energy = Wavenumber^4 $\cdot$ Current Density^2 $\cdot$ Inductance

Energy = Wavenumber^2 $\cdot$ Current Density $\cdot$ Magnetic Flux

Energy $\cdot$ Wavenumber^3 = Magnetic Field Strength $\cdot$ Magnetic Flux Density

Energy $\cdot$ Wavenumber^2 = Magnetic Field Strength^2 $\cdot$ Inductance

Energy $\cdot$ Wavenumber = Magnetic Field Strength $\cdot$ Magnetic Flux

Energy $\cdot$ Wavenumber^3 = Frequency $\cdot$ Dynamic Viscosity

Energy $\cdot$ Wavenumber^5 = Frequency^2 $\cdot$ Mass Density

Energy^2 $\cdot$ Wavenumber^2 $\cdot$ Absorbed Dose = Power, Radiance Flux^2

Energy^2 $\cdot$ Wavenumber^3 = Power, Radiance Flux $\cdot$ Dynamic Viscosity

Energy $\cdot$ Wavenumber^2 $\cdot$ Kinematic Viscosity = Power, Radiance Flux

Energy^3 $\cdot$ Wavenumber^5 = Power, Radiance Flux^2 $\cdot$ Mass Density

Energy $\cdot$ Wavenumber^4 $\cdot$ Inductance = Magnetic Flux Density^2

Energy^2 $\cdot$ Wavenumber^4 = Absorbed Dose $\cdot$ Dynamic Viscosity^2

Energy $\cdot$ Wavenumber^3 = Absorbed Dose $\cdot$ Mass Density

Energy $\cdot$ Wavenumber = Dynamic Viscosity $\cdot$ Kinematic Viscosity

Energy $\cdot$ Wavenumber $\cdot$ Mass Density = Dynamic Viscosity^2

Energy $\cdot$ Wavenumber = Kinematic Viscosity^2 $\cdot$ Mass Density

Energy^2 $\cdot$ Frequency^6 = Specific Volume^3 $\cdot$ Pressure,Stress^5

Energy^2 $\cdot$ Specific Volume^2 $\cdot$ Frequency^6 = Absorbed Dose^5

Energy^2 · Frequency = Specific Volume^3 · Dynamic Viscosity^5

Energy^2 · Specific Volume^2 · Frequency = Kinematic Viscosity^5

Energy^4 · Specific Volume^3 · Pressure,Stress^5 = Power, Radiance Flux^6

Energy^2 · Pressure,Stress = Specific Volume^3 · Dynamic Viscosity^6

Energy^2 · Specific Volume^3 · Pressure,Stress = Kinematic Viscosity^6

Energy^4 · Absorbed Dose^5 = Specific Volume^2 · Power, Radiance Flux^6

Energy · Power, Radiance Flux = Specific Volume^3 · Dynamic Viscosity^5

Energy · Specific Volume^2 · Power, Radiance Flux = Kinematic Viscosity^5

Energy^2 · Absorbed Dose = Specific Volume^4 · Dynamic Viscosity^6

Energy^2 · Specific Volume^2 · Absorbed Dose = Kinematic Viscosity^6

Energy · Magnetic Field Strength = Current Density · Pressure,Stress

Energy^3 = Current Density^2 · Magnetic Field Strength^4 · Inductance^3

Energy^3 = Current Density · Magnetic Field Strength^2 · Magnetic Flux^3

Energy^7 = Current Density^6 · Pressure,Stress^4 · Inductance^3

Energy^5 = Current Density^3 · Pressure,Stress^2 · Magnetic Flux^3

Energy · Kinematic Viscosity = Current Density · Electric Potential Difference

Energy · Capacitance · Kinematic Viscosity^2 = Current Density^2

Acceleration^3 · Mass · Specific Volume = Velocity^6

Acceleration^3 · Mass = Velocity^4 · Pressure,Stress

Acceleration · Mass · Velocity = Power, Radiance Flux

Acceleration^2 · Mass = Velocity^3 · Dynamic Viscosity

Acceleration^3 · Mass = Velocity^6 · Mass Density

Acceleration · Mass = Area · Pressure,Stress

$\text{Acceleration}^2 \cdot \text{Mass}^2 \cdot \text{Area} = \text{Energy,Work,Heat Quantity}^2$

$\text{Acceleration}^6 \cdot \text{Mass}^4 \cdot \text{Area} = \text{Power, Radiance Flux}^4$

$\text{Acceleration}^2 \cdot \text{Mass}^4 = \text{Area}^3 \cdot \text{Dynamic Viscosity}^4$

$\text{Acceleration}^3 \cdot \text{Mass}^3 = \text{Volume}^2 \cdot \text{Pressure,Stress}^3$

$\text{Acceleration}^3 \cdot \text{Mass}^3 \cdot \text{Volume} = \text{Energy,Work,Heat Quantity}^3$

$\text{Acceleration}^9 \cdot \text{Mass}^6 \cdot \text{Volume} = \text{Power, Radiance Flux}^6$

$\text{Acceleration} \cdot \text{Mass}^2 = \text{Volume} \cdot \text{Dynamic Viscosity}^2$

$\text{Acceleration} \cdot \text{Mass} \cdot \text{Wavenumber}^2 = \text{Pressure,Stress}$

$\text{Acceleration} \cdot \text{Mass} = \text{Wavenumber} \cdot \text{Energy,Work,Heat Quantity}$

$\text{Acceleration}^3 \cdot \text{Mass}^2 = \text{Wavenumber} \cdot \text{Power, Radiance Flux}^2$

$\text{Acceleration} \cdot \text{Mass}^2 \cdot \text{Wavenumber}^3 = \text{Dynamic Viscosity}^2$

$\text{Acceleration}^3 = \text{Mass} \cdot \text{Specific Volume} \cdot \text{Frequency}^6$

$\text{Acceleration}^3 \cdot \text{Mass} = \text{Specific Volume}^2 \cdot \text{Pressure,Stress}^3$

$\text{Acceleration}^3 \cdot \text{Mass}^4 \cdot \text{Specific Volume} = \text{Energy,Work,Heat Quantity}^3$

$\text{Acceleration}^9 \cdot \text{Mass}^7 \cdot \text{Specific Volume} = \text{Power, Radiance Flux}^6$

$\text{Acceleration}^3 \cdot \text{Mass} \cdot \text{Specific Volume} = \text{Absorbed Dose}^3$

$\text{Acceleration} \cdot \text{Mass} = \text{Specific Volume} \cdot \text{Dynamic Viscosity}^2$

$\text{Acceleration} \cdot \text{Mass} \cdot \text{Specific Volume} = \text{Kinematic Viscosity}^2$

$\text{Acceleration} \cdot \text{Mass} = \text{Magnetic Field Strength} \cdot \text{Magnetic Flux}$

$\text{Acceleration} \cdot \text{Pressure,Stress} = \text{Mass} \cdot \text{Frequency}^4$

$\text{Acceleration}^2 \cdot \text{Mass} = \text{Frequency}^2 \cdot \text{Energy,Work,Heat Quantity}$

$\text{Acceleration}^2 \cdot \text{Mass} = \text{Frequency} \cdot \text{Power, Radiance Flux}$

Acceleration · Dynamic Viscosity = Mass · Frequency^3

Acceleration^3 · Mass Density = Mass · Frequency^6

Acceleration^3 · Mass^3 = Pressure,Stress · Energy,Work,Heat Quantity^2

Acceleration^7 · Mass^5 = Pressure,Stress · Power, Radiance Flux^4

Acceleration^3 · Mass = Pressure,Stress · Absorbed Dose^2

Acceleration · Dynamic Viscosity^4 = Mass · Pressure,Stress^3

Acceleration^5 · Mass^3 = Pressure,Stress^3 · Kinematic Viscosity^4

Acceleration^3 · Mass · Mass Density^2 = Pressure,Stress^3

Acceleration^2 · Mass · Energy,Work,Heat Quantity = Power, Radiance Flux^2

Acceleration^4 · Mass^5 = Energy,Work,Heat Quantity^3 · Dynamic Viscosity^2

Acceleration^2 · Mass^3 · Kinematic Viscosity^2 = Energy,Work,Heat Quantity^3

Acceleration^3 · Mass^4 = Energy,Work,Heat Quantity^3 · Mass Density

Acceleration^2 · Mass^2 · Absorbed Dose = Power, Radiance Flux^2

Acceleration^5 · Mass^4 = Power, Radiance Flux^3 · Dynamic Viscosity

Acceleration^4 · Mass^3 · Kinematic Viscosity = Power, Radiance Flux^3

Acceleration^9 · Mass^7 = Power, Radiance Flux^6 · Mass Density

Acceleration^2 · Mass · Inductance = Electric Potential Difference^2

Acceleration^4 · Mass^2 = Absorbed Dose^3 · Dynamic Viscosity^2

Acceleration^3 · Mass = Absorbed Dose^3 · Mass Density

Acceleration · Mass = Dynamic Viscosity · Kinematic Viscosity

Acceleration · Mass · Mass Density = Dynamic Viscosity^2

Acceleration · Mass = Kinematic Viscosity^2 · Mass Density

Acceleration^3 · Specific Volume · Energy,Work,Heat Quantity = Velocity^8

237

Acceleration^2 · Specific Volume · Power, Radiance Flux = Velocity^7

Acceleration · Specific Volume · Dynamic Viscosity = Velocity^3

Acceleration^3 · Current Density = Velocity^6 · Magnetic Field Strength

Acceleration · Current Density = Velocity^3 · Electric Charge

Acceleration^2 · Electric Charge = Velocity^3 · Magnetic Field Strength

Acceleration^3 · Energy,Work,Heat Quantity = Velocity^6 · Pressure,Stress

Acceleration^2 · Power, Radiance Flux = Velocity^5 · Pressure,Stress

Acceleration · Dynamic Viscosity = Velocity · Pressure,Stress

Acceleration · Energy,Work,Heat Quantity = Velocity · Power, Radiance Flux

Acceleration^2 · Energy,Work,Heat Quantity = Velocity^5 · Dynamic Viscosity

Acceleration^3 · Energy,Work,Heat Quantity = Velocity^8 · Mass Density

Acceleration · Power, Radiance Flux = Velocity^4 · Dynamic Viscosity

Acceleration^2 · Power, Radiance Flux = Velocity^7 · Mass Density

Acceleration · Electric Potential Difference = Velocity^3 · Magnetic Flux Density

Acceleration · Magnetic Flux = Velocity · Electric Potential Difference

Acceleration · Capacitance · Electric Resistance = Velocity

Acceleration^2 · Capacitance · Inductance = Velocity^2

Acceleration · Inductance = Velocity · Electric Resistance

Acceleration^2 · Magnetic Flux = Velocity^4 · Magnetic Flux Density

Acceleration · Dynamic Viscosity = Velocity^3 · Mass Density

Acceleration^2 · Area = Specific Volume^2 · Pressure,Stress^2

Acceleration · Area^2 = Specific Volume · Energy,Work,Heat Quantity

$Acceleration^6 \cdot Area^7 = Specific\ Volume^4 \cdot Power,\ Radiance\ Flux^4$

$Acceleration^2 \cdot Area^3 = Specific\ Volume^4 \cdot Dynamic\ Viscosity^4$

$Acceleration^2 \cdot Area^3 \cdot Electric\ Charge^4 = Current\ Density^4$

$Acceleration^2 \cdot Electric\ Charge^4 = Area^3 \cdot Magnetic\ Field\ Strength^4$

$Acceleration^2 \cdot Area^5 \cdot Pressure,Stress^4 = Power,\ Radiance\ Flux^4$

$Acceleration^2 \cdot Dynamic\ Viscosity^4 = Area \cdot Pressure,Stress^4$

$Acceleration^2 \cdot Area \cdot Mass\ Density^2 = Pressure,Stress^2$

$Acceleration^2 \cdot Energy,Work,Heat\ Quantity^4 = Area \cdot Power,\ Radiance\ Flux^4$

$Acceleration^2 \cdot Area^5 \cdot Dynamic\ Viscosity^4 = Energy,Work,Heat\ Quantity^4$

$Acceleration \cdot Area^2 \cdot Mass\ Density = Energy,Work,Heat\ Quantity$

$Acceleration \cdot Area \cdot Dynamic\ Viscosity = Power,\ Radiance\ Flux$

$Acceleration^6 \cdot Area^7 \cdot Mass\ Density^4 = Power,\ Radiance\ Flux^4$

$Acceleration^2 \cdot Area^3 \cdot Magnetic\ Flux\ Density^4 = Electric\ Potential\ Difference^4$

$Acceleration^2 \cdot Magnetic\ Flux^4 = Area \cdot Electric\ Potential\ Difference^4$

$Acceleration^2 \cdot Capacitance^4 \cdot Electric\ Resistance^4 = Area$

$Acceleration^2 \cdot Capacitance^2 \cdot Inductance^2 = Area$

$Acceleration^2 \cdot Inductance^4 = Area \cdot Electric\ Resistance^4$

$Acceleration^2 \cdot Area^3 \cdot Mass\ Density^4 = Dynamic\ Viscosity^4$

$Acceleration^3 \cdot Volume = Specific\ Volume^3 \cdot Pressure,Stress^3$

$Acceleration^3 \cdot Volume^4 = Specific\ Volume^3 \cdot Energy,Work,Heat\ Quantity^3$

$Acceleration^9 \cdot Volume^7 = Specific\ Volume^6 \cdot Power,\ Radiance\ Flux^6$

$Acceleration \cdot Volume = Specific\ Volume^2 \cdot Dynamic\ Viscosity^2$

$Acceleration \cdot Volume \cdot Electric\ Charge^2 = Current\ Density^2$

Acceleration $\cdot$ Electric Charge^2 = Volume $\cdot$ Magnetic Field Strength^2

Acceleration^3 $\cdot$ Volume^5 $\cdot$ Pressure,Stress^6 = Power, Radiance Flux^6

Acceleration^3 $\cdot$ Dynamic Viscosity^6 = Volume $\cdot$ Pressure,Stress^6

Acceleration^3 $\cdot$ Volume $\cdot$ Mass Density^3 = Pressure,Stress^3

Acceleration^3 $\cdot$ Energy,Work,Heat Quantity^6 = Volume $\cdot$ Power, Radiance Flux^6

Acceleration^3 $\cdot$ Volume^5 $\cdot$ Dynamic Viscosity^6 = Energy,Work,Heat Quantity^6

Acceleration^3 $\cdot$ Volume^4 $\cdot$ Mass Density^3 = Energy,Work,Heat Quantity^3

Acceleration^3 $\cdot$ Volume^2 $\cdot$ Dynamic Viscosity^3 = Power, Radiance Flux^3

Acceleration^9 $\cdot$ Volume^7 $\cdot$ Mass Density^6 = Power, Radiance Flux^6

Acceleration $\cdot$ Volume $\cdot$ Magnetic Flux Density^2 = Electric Potential Difference^2

Acceleration^3 $\cdot$ Magnetic Flux^6 = Volume $\cdot$ Electric Potential Difference^6

Acceleration^3 $\cdot$ Capacitance^6 $\cdot$ Electric Resistance^6 = Volume

Acceleration^3 $\cdot$ Capacitance^3 $\cdot$ Inductance^3 = Volume

Acceleration^3 $\cdot$ Inductance^6 = Volume $\cdot$ Electric Resistance^6

Acceleration $\cdot$ Volume $\cdot$ Mass Density^2 = Dynamic Viscosity^2

Acceleration = Wavenumber $\cdot$ Specific Volume $\cdot$ Pressure,Stress

Acceleration = Wavenumber^4 $\cdot$ Specific Volume $\cdot$ Energy,Work,Heat Quantity

Acceleration^3 = Wavenumber^7 $\cdot$ Specific Volume^2 $\cdot$ Power, Radiance Flux^2

Acceleration = Wavenumber^3 $\cdot$ Specific Volume^2 $\cdot$ Dynamic Viscosity^2

Acceleration $\cdot$ Electric Charge^2 = Wavenumber^3 $\cdot$ Current Density^2

Acceleration $\cdot$ Wavenumber^3 $\cdot$ Electric Charge^2 = Magnetic Field Strength^2

Acceleration $\cdot$ Pressure,Stress^2 = Wavenumber^5 $\cdot$ Power, Radiance Flux^2

Acceleration $\cdot$ Wavenumber $\cdot$ Dynamic Viscosity^2 = Pressure,Stress^2

Acceleration $\cdot$ Mass Density = Wavenumber $\cdot$ Pressure,Stress

Acceleration $\cdot$ Wavenumber $\cdot$ Energy,Work,Heat Quantity^2 = Power, Radiance Flux^2

Acceleration $\cdot$ Dynamic Viscosity^2 = Wavenumber^5 $\cdot$ Energy,Work,Heat Quantity^2

Acceleration $\cdot$ Mass Density = Wavenumber^4 $\cdot$ Energy,Work,Heat Quantity

Acceleration $\cdot$ Dynamic Viscosity = Wavenumber^2 $\cdot$ Power, Radiance Flux

Acceleration^3 $\cdot$ Mass Density^2 = Wavenumber^7 $\cdot$ Power, Radiance Flux^2

Acceleration $\cdot$ Magnetic Flux Density^2 = Wavenumber^3 $\cdot$ Electric Potential Difference^2

Acceleration $\cdot$ Wavenumber $\cdot$ Magnetic Flux^2 = Electric Potential Difference^2

Acceleration $\cdot$ Wavenumber $\cdot$ Capacitance^2 $\cdot$ Electric Resistance^2 = A Constant

Acceleration $\cdot$ Wavenumber $\cdot$ Capacitance $\cdot$ Inductance = A Constant

Acceleration $\cdot$ Wavenumber $\cdot$ Inductance^2 = Electric Resistance^2

Acceleration $\cdot$ Mass Density^2 = Wavenumber^3 $\cdot$ Dynamic Viscosity^2

Acceleration^2 = Specific Volume $\cdot$ Frequency^2 $\cdot$ Pressure,Stress

Acceleration^5 = Specific Volume $\cdot$ Frequency^8 $\cdot$ Energy,Work,Heat Quantity

Acceleration^5 = Specific Volume $\cdot$ Frequency^7 $\cdot$ Power, Radiance Flux

Acceleration^2 = Specific Volume $\cdot$ Frequency^3 $\cdot$ Dynamic Viscosity

Acceleration^3 $\cdot$ Energy,Work,Heat Quantity = Specific Volume^3 $\cdot$ Pressure,Stress^4

Acceleration^4 $\cdot$ Power, Radiance Flux^2 = Specific Volume^5 $\cdot$ Pressure,Stress^7

Acceleration^2 $\cdot$ Dynamic Viscosity^2 = Specific Volume $\cdot$ Pressure,Stress^3

Acceleration^2 $\cdot$ Kinematic Viscosity^2 = Specific Volume^3 $\cdot$ Pressure,Stress^3

Acceleration^5 $\cdot$ Energy,Work,Heat Quantity^7 = Specific Volume $\cdot$ Power, Radiance Flux^8

Acceleration^3 $\cdot$ Specific Volume $\cdot$ Energy,Work,Heat Quantity = Absorbed Dose^4

Acceleration $\cdot$ Energy,Work,Heat Quantity^3 = Specific Volume^5 $\cdot$ Dynamic Viscosity^8

Acceleration $\cdot$ Specific Volume^3 $\cdot$ Energy,Work,Heat Quantity^3 = Kinematic Viscosity^8

Acceleration^4 $\cdot$ Specific Volume^2 $\cdot$ Power, Radiance Flux^2 = Absorbed Dose^7

Acceleration $\cdot$ Specific Volume^4 $\cdot$ Dynamic Viscosity^7 = Power, Radiance Flux^3

Acceleration $\cdot$ Kinematic Viscosity^7 = Specific Volume^3 $\cdot$ Power, Radiance Flux^3

Acceleration $\cdot$ Inductance = Specific Volume $\cdot$ Magnetic Flux Density^2

Acceleration^2 $\cdot$ Specific Volume^2 $\cdot$ Dynamic Viscosity^2 = Absorbed Dose^3

Acceleration^3 $\cdot$ Magnetic Field Strength = Current Density $\cdot$ Frequency^6

Acceleration $\cdot$ Electric Charge^2 = Current Density $\cdot$ Magnetic Field Strength

Acceleration^3 $\cdot$ Current Density = Magnetic Field Strength $\cdot$ Absorbed Dose^3

Acceleration $\cdot$ Current Density = Magnetic Field Strength $\cdot$ Kinematic Viscosity^2

Acceleration^2 $\cdot$ Electric Charge = Current Density $\cdot$ Frequency^3

Acceleration^2 $\cdot$ Current Density^2 = Electric Charge^2 $\cdot$ Absorbed Dose^3

Acceleration $\cdot$ Magnetic Field Strength = Frequency^3 $\cdot$ Electric Charge

Acceleration $\cdot$ Pressure,Stress $\cdot$ Capacitance = Magnetic Field Strength^2

Acceleration^4 $\cdot$ Electric Charge^2 = Magnetic Field Strength^2 $\cdot$ Absorbed Dose^3

Acceleration $\cdot$ Electric Charge = Magnetic Field Strength $\cdot$ Kinematic Viscosity

Acceleration $\cdot$ Capacitance $\cdot$ Magnetic Flux Density = Magnetic Field Strength

Acceleration $\cdot$ Dynamic Viscosity = Magnetic Field Strength^2 $\cdot$ Electric Resistance

Acceleration^3 $\cdot$ Pressure,Stress = Frequency^6 $\cdot$ Energy,Work,Heat Quantity

Acceleration^3 $\cdot$ Pressure,Stress = Frequency^5 $\cdot$ Power, Radiance Flux

Acceleration^2 $\cdot$ Mass Density = Frequency^2 $\cdot$ Pressure,Stress

$$\text{Acceleration}^3 \cdot \text{Dynamic Viscosity} = \text{Frequency}^5 \cdot \text{Energy,Work,Heat Quantity}$$

$$\text{Acceleration}^5 \cdot \text{Mass Density} = \text{Frequency}^8 \cdot \text{Energy,Work,Heat Quantity}$$

$$\text{Acceleration}^3 \cdot \text{Dynamic Viscosity} = \text{Frequency}^4 \cdot \text{Power, Radiance Flux}$$

$$\text{Acceleration}^5 \cdot \text{Mass Density} = \text{Frequency}^7 \cdot \text{Power, Radiance Flux}$$

$$\text{Acceleration}^2 \cdot \text{Magnetic Flux Density} = \text{Frequency}^3 \cdot \text{Electric Potential Difference}$$

$$\text{Acceleration}^2 \cdot \text{Magnetic Flux Density} = \text{Frequency}^4 \cdot \text{Magnetic Flux}$$

$$\text{Acceleration}^2 \cdot \text{Mass Density} = \text{Frequency}^3 \cdot \text{Dynamic Viscosity}$$

$$\text{Acceleration}^3 \cdot \text{Pressure,Stress} \cdot \text{Energy,Work,Heat Quantity}^5 = \text{Power, Radiance Flux}^6$$

$$\text{Acceleration}^3 \cdot \text{Energy,Work,Heat Quantity} = \text{Pressure,Stress} \cdot \text{Absorbed Dose}^3$$

$$\text{Acceleration}^3 \cdot \text{Dynamic Viscosity}^6 = \text{Pressure,Stress}^5 \cdot \text{Energy,Work,Heat Quantity}$$

$$\text{Acceleration} \cdot \text{Energy,Work,Heat Quantity} = \text{Pressure,Stress} \cdot \text{Kinematic Viscosity}^2$$

$$\text{Acceleration}^3 \cdot \text{Energy,Work,Heat Quantity} \cdot \text{Mass Density}^3 = \text{Pressure,Stress}^4$$

$$\text{Acceleration}^4 \cdot \text{Power, Radiance Flux}^2 = \text{Pressure,Stress}^2 \cdot \text{Absorbed Dose}^5$$

$$\text{Acceleration}^3 \cdot \text{Dynamic Viscosity}^5 = \text{Pressure,Stress}^4 \cdot \text{Power, Radiance Flux}$$

$$\text{Acceleration} \cdot \text{Power, Radiance Flux}^3 = \text{Pressure,Stress}^3 \cdot \text{Kinematic Viscosity}^5$$

$$\text{Acceleration}^4 \cdot \text{Power, Radiance Flux}^2 \cdot \text{Mass Density}^5 = \text{Pressure,Stress}^7$$

$$\text{Acceleration} \cdot \text{Capacitance} \cdot \text{Magnetic Flux Density}^2 = \text{Pressure,Stress}$$

$$\text{Acceleration}^2 \cdot \text{Dynamic Viscosity}^2 = \text{Pressure,Stress}^2 \cdot \text{Absorbed Dose}$$

$$\text{Acceleration}^2 \cdot \text{Dynamic Viscosity}^3 = \text{Pressure,Stress}^3 \cdot \text{Kinematic Viscosity}$$

$$\text{Acceleration}^2 \cdot \text{Dynamic Viscosity}^2 \cdot \text{Mass Density} = \text{Pressure,Stress}^3$$

$$\text{Acceleration}^2 \cdot \text{Kinematic Viscosity}^2 \cdot \text{Mass Density}^3 = \text{Pressure,Stress}^3$$

$$\text{Acceleration}^2 \cdot \text{Energy,Work,Heat Quantity}^2 = \text{Power, Radiance Flux}^2 \cdot \text{Absorbed Dose}$$

$$\text{Acceleration}^3 \cdot \text{Energy,Work,Heat Quantity}^4 \cdot \text{Dynamic Viscosity} = \text{Power, Radiance Flux}^5$$

Acceleration^2 · Energy,Work,Heat Quantity^3 = Power, Radiance Flux^3 · Kinematic Viscosity

Acceleration^5 · Energy,Work,Heat Quantity^7 · Mass Density = Power, Radiance Flux^8

Acceleration^4 · Energy,Work,Heat Quantity^2 = Absorbed Dose^5 · Dynamic Viscosity^2

Acceleration^3 · Energy,Work,Heat Quantity = Absorbed Dose^4 · Mass Density

Acceleration · Energy,Work,Heat Quantity^3 = Dynamic Viscosity^3 · Kinematic Viscosity^5

Acceleration · Energy,Work,Heat Quantity^3 · Mass Density^5 = Dynamic Viscosity^8

Acceleration · Energy,Work,Heat Quantity^3 = Kinematic Viscosity^8 · Mass Density^3

Acceleration · Power, Radiance Flux = Absorbed Dose^2 · Dynamic Viscosity

Acceleration^4 · Power, Radiance Flux^2 = Absorbed Dose^7 · Mass Density^2

Acceleration · Dynamic Viscosity^3 · Kinematic Viscosity^4 = Power, Radiance Flux^3

Acceleration · Dynamic Viscosity^7 = Power, Radiance Flux^3·· Mass Density^4

Acceleration · Kinematic Viscosity^7 · Mass Density^3 = Power, Radiance Flux^3

Mass^2 · Velocity^4 = Area^3 · Pressure,Stress^2

Mass^2 · Velocity^6 = Area · Power, Radiance Flux^2

Mass · Velocity = Area · Dynamic Viscosity

Mass · Velocity^2 = Volume · Pressure,Stress

Mass^3 · Velocity^9 = Volume · Power, Radiance Flux^3

Mass^3 · Velocity^3 = Volume^2 · Dynamic Viscosity^3

Mass · Velocity^2 · Wavenumber^3 = Pressure,Stress

Mass · Velocity^3 · Wavenumber = Power, Radiance Flux

Mass · Velocity · Wavenumber^2 = Dynamic Viscosity

Mass · Specific Volume · Frequency^3 = Velocity^3

$Mass^2 \cdot Velocity^9 = Specific\ Volume \cdot Power, Radiance\ Flux^3$

$Mass \cdot Velocity^3 = Specific\ Volume^2 \cdot Dynamic\ Viscosity^3$

$Mass \cdot Velocity^3 \cdot Specific\ Volume = Kinematic\ Viscosity^3$

$Mass \cdot Velocity^2 = Current\ Density \cdot Magnetic\ Flux\ Density$

$Mass \cdot Frequency^3 = Velocity \cdot Pressure, Stress$

$Mass \cdot Velocity^2 \cdot Frequency = Power, Radiance\ Flux$

$Mass \cdot Frequency^2 = Velocity \cdot Dynamic\ Viscosity$

$Mass \cdot Frequency^3 = Velocity^3 \cdot Mass\ Density$

$Mass^2 \cdot Velocity^7 \cdot Pressure, Stress = Power, Radiance\ Flux^3$

$Mass \cdot Pressure, Stress^2 = Velocity \cdot Dynamic\ Viscosity^3$

$Mass \cdot Velocity^5 = Pressure, Stress \cdot Kinematic\ Viscosity^3$

$Mass \cdot Velocity^5 \cdot Dynamic\ Viscosity = Power, Radiance\ Flux^2$

$Mass \cdot Velocity^4 = Power, Radiance\ Flux \cdot Kinematic\ Viscosity$

$Mass^2 \cdot Velocity^9 \cdot Mass\ Density = Power, Radiance\ Flux^3$

$Mass \cdot Velocity^2 = Electric\ Charge \cdot Electric\ Potential\ Difference$

$Mass \cdot Velocity^2 \cdot Capacitance = Electric\ Charge^2$

$Mass \cdot Velocity^2 = Electric\ Potential\ Difference^2 \cdot Capacitance$

$Mass \cdot Velocity^2 \cdot Inductance = Magnetic\ Flux^2$

$Mass \cdot Velocity^3 = Dynamic\ Viscosity \cdot Kinematic\ Viscosity^2$

$Mass \cdot Velocity^3 \cdot Mass\ Density^2 = Dynamic\ Viscosity^3$

$Mass \cdot Velocity^3 = Kinematic\ Viscosity^3 \cdot Mass\ Density$

$Mass^2 \cdot Frequency^4 = Area \cdot Pressure, Stress^2$

$Mass \cdot Area \cdot Frequency^2 = Energy, Work, Heat\ Quantity$

$$\text{Mass} \cdot \text{Area} \cdot \text{Frequency}^3 = \text{Power, Radiance Flux}$$

$$\text{Mass}^2 \cdot \text{Frequency}^2 = \text{Area} \cdot \text{Dynamic Viscosity}^2$$

$$\text{Mass}^2 \cdot \text{Power, Radiance Flux}^4 = \text{Area}^7 \cdot \text{Pressure,Stress}^6$$

$$\text{Mass}^2 \cdot \text{Absorbed Dose}^2 = \text{Area}^3 \cdot \text{Pressure,Stress}^2$$

$$\text{Mass}^2 \cdot \text{Pressure,Stress}^2 = \text{Area} \cdot \text{Dynamic Viscosity}^4$$

$$\text{Mass}^2 \cdot \text{Kinematic Viscosity}^4 = \text{Area}^5 \cdot \text{Pressure,Stress}^2$$

$$\text{Mass} \cdot \text{Area} \cdot \text{Power, Radiance Flux}^2 = \text{Energy,Work,Heat Quantity}^3$$

$$\text{Mass} \cdot \text{Energy,Work,Heat Quantity} = \text{Area}^2 \cdot \text{Dynamic Viscosity}^2$$

$$\text{Mass} \cdot \text{Kinematic Viscosity}^2 = \text{Area} \cdot \text{Energy,Work,Heat Quantity}$$

$$\text{Mass}^2 \cdot \text{Absorbed Dose}^3 = \text{Area} \cdot \text{Power, Radiance Flux}^2$$

$$\text{Mass}^4 \cdot \text{Power, Radiance Flux}^2 = \text{Area}^5 \cdot \text{Dynamic Viscosity}^6$$

$$\text{Mass} \cdot \text{Kinematic Viscosity}^3 = \text{Area}^2 \cdot \text{Power, Radiance Flux}$$

$$\text{Mass} \cdot \text{Area} = \text{Electric Charge}^2 \cdot \text{Inductance}$$

$$\text{Mass} = \text{Area} \cdot \text{Capacitance} \cdot \text{Magnetic Flux Density}^2$$

$$\text{Mass} \cdot \text{Area} = \text{Capacitance} \cdot \text{Magnetic Flux}^2$$

$$\text{Mass}^2 \cdot \text{Absorbed Dose} = \text{Area}^2 \cdot \text{Dynamic Viscosity}^2$$

$$\text{Mass}^2 \cdot \text{Kinematic Viscosity}^2 = \text{Area}^3 \cdot \text{Dynamic Viscosity}^2$$

$$\text{Mass}^3 \cdot \text{Frequency}^6 = \text{Volume} \cdot \text{Pressure,Stress}^3$$

$$\text{Mass}^3 \cdot \text{Volume}^2 \cdot \text{Frequency}^6 = \text{Energy,Work,Heat Quantity}^3$$

$$\text{Mass}^3 \cdot \text{Volume}^2 \cdot \text{Frequency}^9 = \text{Power, Radiance Flux}^3$$

$$\text{Mass}^3 \cdot \text{Frequency}^3 = \text{Volume} \cdot \text{Dynamic Viscosity}^3$$

$$\text{Mass}^3 \cdot \text{Power, Radiance Flux}^6 = \text{Volume}^7 \cdot \text{Pressure,Stress}^9$$

$Mass \cdot Absorbed\ Dose = Volume \cdot Pressure, Stress$

$Mass^3 \cdot Pressure, Stress^3 = Volume \cdot Dynamic\ Viscosity^6$

$Mass^3 \cdot Kinematic\ Viscosity^6 = Volume^5 \cdot Pressure, Stress^3$

$Mass^3 \cdot Volume^2 \cdot Power, Radiance\ Flux^6 = Energy, Work, Heat\ Quantity^9$

$Mass^3 \cdot Energy, Work, Heat\ Quantity^3 = Volume^4 \cdot Dynamic\ Viscosity^6$

$Mass^3 \cdot Kinematic\ Viscosity^6 = Volume^2 \cdot Energy, Work, Heat\ Quantity^3$

$Mass^6 \cdot Absorbed\ Dose^9 = Volume^2 \cdot Power, Radiance\ Flux^6$

$Mass^6 \cdot Power, Radiance\ Flux^3 = Volume^5 \cdot Dynamic\ Viscosity^9$

$Mass^3 \cdot Kinematic\ Viscosity^9 = Volume^4 \cdot Power, Radiance\ Flux^3$

$Mass^3 \cdot Volume^2 = Electric\ Charge^6 \cdot Inductance^3$

$Mass^3 = Volume^2 \cdot Capacitance^3 \cdot Magnetic\ Flux\ Density^6$

$Mass^3 \cdot Volume^2 = Capacitance^3 \cdot Magnetic\ Flux^6$

$Mass^6 \cdot Absorbed\ Dose^3 = Volume^4 \cdot Dynamic\ Viscosity^6$

$Mass \cdot Kinematic\ Viscosity = Volume \cdot Dynamic\ Viscosity$

$Mass \cdot Wavenumber \cdot Frequency^2 = Pressure, Stress$

$Mass \cdot Frequency^2 = Wavenumber^2 \cdot Energy, Work, Heat\ Quantity$

$Mass \cdot Frequency^3 = Wavenumber^2 \cdot Power, Radiance\ Flux$

$Mass \cdot Wavenumber \cdot Frequency = Dynamic\ Viscosity$

$Mass \cdot Wavenumber^7 \cdot Power, Radiance\ Flux^2 = Pressure, Stress^3$

$G_c \cdot Kinematic\ Viscosity^2 \cdot Mass\ Density = Absorbed\ Dose^2$

$Volume^2 \cdot Frequency^6 = Specific\ Volume^3 \cdot Pressure, Stress^3$

$Volume^5 \cdot Frequency^6 = Specific\ Volume^3 \cdot Energy, Work, Heat\ Quantity^3$

$Volume^5 \cdot Frequency^9 = Specific\ Volume^3 \cdot Power, Radiance\ Flux^3$

$$\text{Volume}^2 \cdot \text{Frequency}^3 = \text{Specific Volume}^3 \cdot \text{Dynamic Viscosity}^3$$

$$\text{Volume}^4 \cdot \text{Specific Volume}^3 \cdot \text{Pressure,Stress}^9 = \text{Power, Radiance Flux}^6$$

$$\text{Volume}^2 \cdot \text{Pressure,Stress}^3 = \text{Specific Volume}^3 \cdot \text{Dynamic Viscosity}^6$$

$$\text{Volume}^2 \cdot \text{Specific Volume}^3 \cdot \text{Pressure,Stress}^3 = \text{Kinematic Viscosity}^6$$

$$\text{Volume}^5 \cdot \text{Power, Radiance Flux}^6 = \text{Specific Volume}^3 \cdot \text{Energy,Work,Heat Quantity}^9$$

$$\text{Volume} \cdot \text{Absorbed Dose} = \text{Specific Volume} \cdot \text{Energy,Work,Heat Quantity}$$

$$\text{Volume} \cdot \text{Specific Volume}^3 \cdot \text{Dynamic Viscosity}^6 = \text{Energy,Work,Heat Quantity}^3$$

$$\text{Volume} \cdot \text{Kinematic Viscosity}^6 = \text{Specific Volume}^3 \cdot \text{Energy,Work,Heat Quantity}^3$$

$$\text{Volume}^4 \cdot \text{Absorbed Dose}^9 = \text{Specific Volume}^6 \cdot \text{Power, Radiance Flux}^6$$

$$\text{Volume} \cdot \text{Power, Radiance Flux}^3 = \text{Specific Volume}^6 \cdot \text{Dynamic Viscosity}^9$$

$$\text{Volume} \cdot \text{Specific Volume}^3 \cdot \text{Power, Radiance Flux}^3 = \text{Kinematic Viscosity}^9$$

$$\text{Volume}^5 = \text{Specific Volume}^3 \cdot \text{Electric Charge}^6 \cdot \text{Inductance}^3$$

$$\text{Volume} = \text{Specific Volume}^3 \cdot \text{Capacitance}^3 \cdot \text{Magnetic Flux Density}^6$$

$$\text{Volume}^5 = \text{Specific Volume}^3 \cdot \text{Capacitance}^3 \cdot \text{Magnetic Flux}^6$$

$$\text{Volume}^2 \cdot \text{Absorbed Dose}^3 = \text{Specific Volume}^6 \cdot \text{Dynamic Viscosity}^6$$

$$\text{Volume}^2 \cdot \text{Frequency}^3 \cdot \text{Electric Charge}^3 = \text{Current Density}^3$$

$$\text{Volume} \cdot \text{Pressure,Stress} = \text{Current Density} \cdot \text{Magnetic Flux Density}$$

$$\text{Volume}^7 \cdot \text{Pressure,Stress}^3 = \text{Current Density}^6 \cdot \text{Inductance}^3$$

$$\text{Volume}^5 \cdot \text{Pressure,Stress}^3 = \text{Current Density}^3 \cdot \text{Magnetic Flux}^3$$

$$\text{Volume}^4 \cdot \text{Energy,Work,Heat Quantity}^3 = \text{Current Density}^6 \cdot \text{Inductance}^3$$

$$\text{Volume}^2 \cdot \text{Energy,Work,Heat Quantity}^3 = \text{Current Density}^3 \cdot \text{Magnetic Flux}^3$$

$$\text{Volume}^2 \cdot \text{Power, Radiance Flux}^3 = \text{Current Density}^3 \cdot \text{Electric Potential Difference}^3$$

Volume^4 · Power, Radiance Flux^3 = Current Density^6 · Electric Resistance^3

Volume^2 · Electric Charge^6 · Absorbed Dose^3 = Current Density^6

Volume^2 · Electric Potential Difference^3 = Current Density^3 · Electric Resistance^3

Volume^4 · Magnetic Flux Density^3 = Current Density^3 · Inductance^3

Volume^2 · Magnetic Flux^3 = Current Density^3 · Inductance^3

Volume · Magnetic Field Strength^3 = Frequency^3 · Electric Charge^3

Volume · Pressure,Stress^3 = Magnetic Field Strength^6 · Inductance^3

Volume^2 · Pressure,Stress^3 = Magnetic Field Strength^3 · Magnetic Flux^3

Volume · Magnetic Field Strength · Magnetic Flux Density = Energy,Work,Heat Quantity

Volume^2 · Magnetic Field Strength^6 · Inductance^3 = Energy,Work,Heat Quantity^3

Volume · Magnetic Field Strength^3 · Magnetic Flux^3 = Energy,Work,Heat Quantity^3

Volume · Magnetic Field Strength^3 · Electric Potential Difference^3 = Power, Radiance Flux^3

Volume^2 · Magnetic Field Strength^6 · Electric Resistance^3 = Power, Radiance Flux^3

Volume^4 · Magnetic Field Strength^6 = Electric Charge^6 · Absorbed Dose^3

Volume · Magnetic Field Strength = Electric Charge · Kinematic Viscosity

Volume · Magnetic Field Strength^3 · Electric Resistance^3 = Electric Potential Difference^3

Volume · Magnetic Flux Density^3 = Magnetic Field Strength^3 · Inductance^3

Volume · Magnetic Field Strength^3 · Inductance^3 = Magnetic Flux^3

Volume · Frequency · Pressure,Stress = Power, Radiance Flux

Volume^2 · Frequency^6 · Mass Density^3 = Pressure,Stress^3

Volume · Frequency · Dynamic Viscosity = Energy,Work,Heat Quantity

Volume^5 · Frequency^6 · Mass Density^3 = Energy,Work,Heat Quantity^3

Volume · Frequency^2 · Dynamic Viscosity = Power, Radiance Flux

$Volume^5 \cdot Frequency^9 \cdot Mass\ Density^3 = Power, Radiance\ Flux^3$

$Volume^2 \cdot Frequency^3 \cdot Magnetic\ Flux\ Density^3 = Electric\ Potential\ Difference^3$

$Volume^2 \cdot Frequency^3 \cdot Mass\ Density^3 = Dynamic\ Viscosity^3$

$Volume^4 \cdot Pressure, Stress^6 \cdot Absorbed\ Dose^3 = Power, Radiance\ Flux^6$

$Volume \cdot Pressure, Stress^2 = Power, Radiance\ Flux \cdot Dynamic\ Viscosity$

$Volume \cdot Pressure, Stress^3 \cdot Kinematic\ Viscosity^3 = Power, Radiance\ Flux^3$

$Volume^4 \cdot Pressure, Stress^9 = Power, Radiance\ Flux^6 \cdot Mass\ Density^3$

$Volume \cdot Pressure, Stress = Electric\ Charge \cdot Electric\ Potential\ Difference$

$Volume \cdot Pressure, Stress \cdot Capacitance = Electric\ Charge^2$

$Volume \cdot Pressure, Stress = Electric\ Potential\ Difference^2 \cdot Capacitance$

$Volume \cdot Magnetic\ Flux\ Density^6 = Pressure, Stress^3 \cdot Inductance^3$

$Volume \cdot Pressure, Stress \cdot Inductance = Magnetic\ Flux^2$

$Volume^2 \cdot Pressure, Stress^6 = Absorbed\ Dose^3 \cdot Dynamic\ Viscosity^6$

$Volume^2 \cdot Pressure, Stress^3 = Dynamic\ Viscosity^3 \cdot Kinematic\ Viscosity^3$

$Volume^2 \cdot Pressure, Stress^3 \cdot Mass\ Density^3 = Dynamic\ Viscosity^6$

$Volume^2 \cdot Pressure, Stress^3 = Kinematic\ Viscosity^6 \cdot Mass\ Density^3$

$Volume^2 \cdot Power, Radiance\ Flux^6 = Energy, Work, Heat\ Quantity^6 \cdot Absorbed\ Dose^3$

$Volume \cdot Power, Radiance\ Flux \cdot Dynamic\ Viscosity = Energy, Work, Heat\ Quantity^2$

$Volume^2 \cdot Power, Radiance\ Flux^3 = Energy, Work, Heat\ Quantity^3 \cdot Kinematic\ Viscosity^3$

$Volume^5 \cdot Power, Radiance\ Flux^6 \cdot Mass\ Density^3 = Energy, Work, Heat\ Quantity^9$

$Volume^4 \cdot Magnetic\ Flux\ Density^6 = Energy, Work, Heat\ Quantity^3 \cdot Inductance^3$

$Volume^4 \cdot Absorbed\ Dose^3 \cdot Dynamic\ Viscosity^6 = Energy, Work, Heat\ Quantity^6$

Volume $\cdot$ Absorbed Dose $\cdot$ Mass Density = Energy,Work,Heat Quantity

Volume $\cdot$ Dynamic Viscosity^3 $\cdot$ Kinematic Viscosity^3 = Energy,Work,Heat Quantity^3

Volume $\cdot$ Dynamic Viscosity^6 = Energy,Work,Heat Quantity^3 $\cdot$ Mass Density^3

Volume $\cdot$ Kinematic Viscosity^6 $\cdot$ Mass Density^3 = Energy,Work,Heat Quantity^3

Volume $\cdot$ Absorbed Dose^3 $\cdot$ Dynamic Viscosity^3 = Power, Radiance Flux^3

Volume^4 $\cdot$ Absorbed Dose^9 $\cdot$ Mass Density^6 = Power, Radiance Flux^6

Volume $\cdot$ Power, Radiance Flux^3 = Dynamic Viscosity^3 $\cdot$ Kinematic Viscosity^6

Volume $\cdot$ Power, Radiance Flux^3 $\cdot$ Mass Density^6 = Dynamic Viscosity^9

Volume $\cdot$ Power, Radiance Flux^3 = Kinematic Viscosity^9 $\cdot$ Mass Density^3

Volume^2 $\cdot$ Magnetic Flux Density^3 = Electric Charge^3 $\cdot$ Electric Resistance^3

Volume $\cdot$ Dynamic Viscosity = Electric Charge^2 $\cdot$ Electric Resistance

Volume $\cdot$ Dynamic Viscosity^3 = Electric Charge^3 $\cdot$ Magnetic Flux Density^3

Volume^5 $\cdot$ Mass Density^3 = Electric Charge^6 $\cdot$ Inductance^3

Volume $\cdot$ Dynamic Viscosity = Electric Charge $\cdot$ Magnetic Flux

Volume^2 $\cdot$ Magnetic Flux Density^6 $\cdot$ Absorbed Dose^3 = Electric Potential Difference^6

Volume^2 $\cdot$ Electric Potential Difference^6 = Absorbed Dose^3 $\cdot$ Magnetic Flux^6

Volume^2 $\cdot$ Electric Potential Difference^3 = Magnetic Flux^3 $\cdot$ Kinematic Viscosity^3

Volume^2 = Capacitance^6 $\cdot$ Electric Resistance^6 $\cdot$ Absorbed Dose^3

Volume^2 = Capacitance^3 $\cdot$ Electric Resistance^3 $\cdot$ Kinematic Viscosity^3

Volume $\cdot$ Mass Density^3 = Capacitance^3 $\cdot$ Magnetic Flux Density^6

Volume^2 = Capacitance^3 $\cdot$ Inductance^3 $\cdot$ Absorbed Dose^3

Volume^4 = Capacitance^3 $\cdot$ Inductance^3 $\cdot$ Kinematic Viscosity^6

Volume^5 $\cdot$ Mass Density^3 = Capacitance^3 $\cdot$ Magnetic Flux^6

251

Volume · Magnetic Flux Density^6 = Electric Resistance^3 · Dynamic Viscosity^3

Volume^2 · Electric Resistance^6 = Inductance^6 · Absorbed Dose^3

Volume^2 · Electric Resistance^3 = Inductance^3 · Kinematic Viscosity^3

Volume · Electric Resistance · Dynamic Viscosity = Magnetic Flux^2

Volume^2 · Absorbed Dose^3 · Mass Density^6 = Dynamic Viscosity^6

Wavenumber^2 · Specific Volume · Pressure,Stress = Frequency^2

Wavenumber^5 · Specific Volume · Energy,Work,Heat Quantity = Frequency^2

Wavenumber^5 · Specific Volume · Power, Radiance Flux = Frequency^3

Wavenumber^2 · Specific Volume · Dynamic Viscosity = Frequency

Wavenumber^4 · Power, Radiance Flux^2 = Specific Volume · Pressure,Stress^3

Wavenumber^2 · Specific Volume · Dynamic Viscosity^2 = Pressure,Stress

Wavenumber^2 · Kinematic Viscosity^2 = Specific Volume · Pressure,Stress

Wavenumber^5 · Specific Volume · Energy,Work,Heat Quantity^3 = Power, Radiance Flux^2

Wavenumber^3 · Specific Volume · Energy,Work,Heat Quantity = Absorbed Dose

Wavenumber · Energy,Work,Heat Quantity = Specific Volume · Dynamic Viscosity^2

Wavenumber · Specific Volume · Energy,Work,Heat Quantity = Kinematic Viscosity^2

Wavenumber^4 · Specific Volume^2 · Power, Radiance Flux^2 = Absorbed Dose^3

Wavenumber · Specific Volume^2 · Dynamic Viscosity^3 = Power, Radiance Flux

Wavenumber · Kinematic Viscosity^3 = Specific Volume · Power, Radiance Flux

Wavenumber^5 · Specific Volume · Electric Charge^2 · Inductance = A Constant

Wavenumber · Specific Volume · Capacitance · Magnetic Flux Density^2 = A Constant

Wavenumber^5 · Specific Volume · Capacitance · Magnetic Flux^2 = A Constant

Wavenumber^2 · Specific Volume^2 · Dynamic Viscosity^2 = Absorbed Dose

Wavenumber^2 · Current Density = Frequency · Electric Charge

Wavenumber^3 · Current Density · Magnetic Flux Density = Pressure,Stress

Wavenumber^7 · Current Density^2 · Inductance = Pressure,Stress

Wavenumber^5 · Current Density · Magnetic Flux = Pressure,Stress

Wavenumber^4 · Current Density^2 · Inductance = Energy,Work,Heat Quantity

Wavenumber^2 · Current Density · Magnetic Flux = Energy,Work,Heat Quantity

Wavenumber^2 · Current Density · Electric Potential Difference = Power, Radiance Flux

Wavenumber^4 · Current Density^2 · Electric Resistance = Power, Radiance Flux

Wavenumber^2 · Current Density^2 = Electric Charge^2 · Absorbed Dose

Wavenumber^2 · Current Density · Electric Resistance = Electric Potential Difference

Wavenumber^4 · Current Density · Inductance = Magnetic Flux Density

Wavenumber^2 · Current Density · Inductance = Magnetic Flux

Wavenumber · Frequency · Electric Charge = Magnetic Field Strength

Wavenumber · Magnetic Field Strength^2 · Inductance = Pressure,Stress

Wavenumber^2 · Magnetic Field Strength · Magnetic Flux = Pressure,Stress

Wavenumber^3 · Energy,Work,Heat Quantity = Magnetic Field Strength · Magnetic Flux Density

Wavenumber^2 · Energy,Work,Heat Quantity = Magnetic Field Strength^2 · Inductance

Wavenumber · Energy,Work,Heat Quantity = Magnetic Field Strength · Magnetic Flux

Wavenumber · Power, Radiance Flux = Magnetic Field Strength · Electric Potential Difference

Wavenumber^2 · Power, Radiance Flux = Magnetic Field Strength^2 · Electric Resistance

Wavenumber^4 · Electric Charge^2 · Absorbed Dose = Magnetic Field Strength^2

Wavenumber^3 · Electric Charge · Kinematic Viscosity = Magnetic Field Strength

Wavenumber · Electric Potential Difference = Magnetic Field Strength · Electric Resistance

Wavenumber · Magnetic Field Strength · Inductance = Magnetic Flux Density

Wavenumber · Magnetic Flux = Magnetic Field Strength · Inductance

Wavenumber^3 · Power, Radiance Flux = Frequency · Pressure,Stress

Wavenumber^2 · Pressure,Stress = Frequency^2 · Mass Density

Wavenumber^3 · Energy,Work,Heat Quantity = Frequency · Dynamic Viscosity

Wavenumber^5 · Energy,Work,Heat Quantity = Frequency^2 · Mass Density

Wavenumber^3 · Power, Radiance Flux = Frequency^2 · Dynamic Viscosity

Wavenumber^5 · Power, Radiance Flux = Frequency^3 · Mass Density

Wavenumber^2 · Electric Potential Difference = Frequency · Magnetic Flux Density

Wavenumber^2 · Dynamic Viscosity = Frequency · Mass Density

Wavenumber^4 · Power, Radiance Flux^2 = Pressure,Stress^2 · Absorbed Dose

Wavenumber^3 · Power, Radiance Flux · Dynamic Viscosity = Pressure,Stress^2

Wavenumber · Power, Radiance Flux = Pressure,Stress · Kinematic Viscosity

Wavenumber^4 · Power, Radiance Flux^2 · Mass Density = Pressure,Stress^3

Wavenumber^3 · Electric Charge · Electric Potential Difference = Pressure,Stress

Wavenumber^3 · Electric Charge^2 = Pressure,Stress · Capacitance

Wavenumber^3 · Electric Potential Difference^2 · Capacitance = Pressure,Stress

Wavenumber · Pressure,Stress · Inductance = Magnetic Flux Density^2

Wavenumber^3 · Magnetic Flux^2 = Pressure,Stress · Inductance

Wavenumber^2 · Absorbed Dose · Dynamic Viscosity^2 = Pressure,Stress^2

Wavenumber^2 · Dynamic Viscosity · Kinematic Viscosity = Pressure,Stress

Wavenumber^2 · Dynamic Viscosity^2 = Pressure,Stress · Mass Density

Wavenumber^2 · Kinematic Viscosity^2 · Mass Density = Pressure,Stress

Wavenumber^2 · Energy,Work,Heat Quantity^2 · Absorbed Dose = Power, Radiance Flux^2

Wavenumber^3 · Energy,Work,Heat Quantity^2 = Power, Radiance Flux · Dynamic Viscosity

Wavenumber^2 · Energy,Work,Heat Quantity · Kinematic Viscosity = Power, Radiance Flux

Wavenumber^5 · Energy,Work,Heat Quantity^3 = Power, Radiance Flux^2 · Mass Density

Wavenumber^4 · Energy,Work,Heat Quantity · Inductance = Magnetic Flux Density^2

Wavenumber^4 · Energy,Work,Heat Quantity^2 = Absorbed Dose · Dynamic Viscosity^2

Wavenumber^3 · Energy,Work,Heat Quantity = Absorbed Dose · Mass Density

Wavenumber · Energy,Work,Heat Quantity = Dynamic Viscosity · Kinematic Viscosity

Wavenumber · Energy,Work,Heat Quantity · Mass Density = Dynamic Viscosity^2

Wavenumber · Energy,Work,Heat Quantity = Kinematic Viscosity^2 · Mass Density

Wavenumber · Power, Radiance Flux = Absorbed Dose · Dynamic Viscosity

Wavenumber^4 · Power, Radiance Flux^2 = Absorbed Dose^3 · Mass Density^2

Wavenumber · Dynamic Viscosity · Kinematic Viscosity^2 = Power, Radiance Flux

Wavenumber · Dynamic Viscosity^3 = Power, Radiance Flux · Mass Density^2

Wavenumber · Kinematic Viscosity^3 · Mass Density = Power, Radiance Flux

Wavenumber^2 · Electric Charge · Electric Resistance = Magnetic Flux Density

Wavenumber^3 · Electric Charge^2 · Electric Resistance = Dynamic Viscosity

Wavenumber · Electric Charge · Magnetic Flux Density = Dynamic Viscosity

Wavenumber^5 · Electric Charge^2 · Inductance = Mass Density

Wavenumber^3 · Electric Charge · Magnetic Flux = Dynamic Viscosity

Wavenumber^2 · Electric Potential Difference^2 = Magnetic Flux Density^2 · Absorbed Dose

Wavenumber^2 · Absorbed Dose · Magnetic Flux^2 = Electric Potential Difference^2

Wavenumber^2 · Magnetic Flux · Kinematic Viscosity = Electric Potential Difference

Wavenumber^2 · Capacitance^2 · Electric Resistance^2 · Absorbed Dose = A Constant

Wavenumber^2 · Capacitance · Electric Resistance · Kinematic Viscosity = A Constant

Wavenumber · Capacitance · Magnetic Flux Density^2 = Mass Density

Wavenumber^2 · Capacitance · Inductance · Absorbed Dose = A Constant

Wavenumber^4 · Capacitance · Inductance · Kinematic Viscosity^2 = A Constant

Wavenumber^5 · Capacitance · Magnetic Flux^2 = Mass Density

Wavenumber · Electric Resistance · Dynamic Viscosity = Magnetic Flux Density^2

Wavenumber^2 · Inductance^2 · Absorbed Dose = Electric Resistance^2

Wavenumber^2 · Inductance · Kinematic Viscosity = Electric Resistance

Wavenumber^3 · Magnetic Flux^2 = Electric Resistance · Dynamic Viscosity

Wavenumber^2 · Dynamic Viscosity^2 = Absorbed Dose · Mass Density^2

Specific Volume · Electric Charge · Dynamic Viscosity = Current Density

Specific Volume · Pressure,Stress^3 = Magnetic Field Strength^4 · Electric Resistance^2

Specific Volume · Magnetic Flux Density^3 = Magnetic Field Strength · Electric Resistance^2

Specific Volume^2 · Magnetic Field Strength^4 · Electric Resistance^2 = Absorbed Dose^3

Specific Volume · Magnetic Field Strength · Magnetic Flux Density = Absorbed Dose

Specific Volume · Dynamic Viscosity^2 = Magnetic Field Strength · Magnetic Flux

Specific Volume · Magnetic Field Strength · Magnetic Flux = Kinematic Viscosity^2

Specific Volume^3 · Pressure,Stress^5 = Frequency^6 · Energy,Work,Heat Quantity^2

Specific Volume^3 · Pressure,Stress^5 = Frequency^4 · Power, Radiance Flux^2

Specific Volume · Pressure,Stress = Frequency · Kinematic Viscosity

Specific Volume^2 · Frequency^6 · Energy,Work,Heat Quantity^2 = Absorbed Dose^5

Specific Volume^3 · Dynamic Viscosity^5 = Frequency · Energy,Work,Heat Quantity^2

Specific Volume^2 · Frequency · Energy,Work,Heat Quantity^2 = Kinematic Viscosity^5

Specific Volume^2 · Frequency^4 · Power, Radiance Flux^2 = Absorbed Dose^5

Specific Volume^3 · Frequency · Dynamic Viscosity^5 = Power, Radiance Flux^2

Specific Volume^2 · Power, Radiance Flux^2 = Frequency · Kinematic Viscosity^5

Specific Volume · Frequency · Dynamic Viscosity = Absorbed Dose

Specific Volume^3 · Pressure,Stress^5 · Energy,Work,Heat Quantity^4 = Power, Radiance Flux^6

Specific Volume^3 · Dynamic Viscosity^6 = Pressure,Stress · Energy,Work,Heat Quantity^2

Specific Volume^3 · Pressure,Stress · Energy,Work,Heat Quantity^2 = Kinematic Viscosity^6

Specific Volume^3 · Pressure,Stress · Dynamic Viscosity^4 = Power, Radiance Flux^2

Specific Volume · Power, Radiance Flux^2 = Pressure,Stress · Kinematic Viscosity^4

Specific Volume · Magnetic Flux Density^4 = Pressure,Stress · Electric Resistance^2

Specific Volume^2 · Power, Radiance Flux^6 = Energy,Work,Heat Quantity^4 · Absorbed Dose^5

Specific Volume^3 · Dynamic Viscosity^5 = Energy,Work,Heat Quantity · Power, Radiance Flux

Specific Volume^2 · Energy,Work,Heat Quantity · Power, Radiance Flux = Kinematic Viscosity^5

Specific Volume^4 · Dynamic Viscosity^6 = Energy,Work,Heat Quantity^2 · Absorbed Dose

Specific Volume^2 · Energy,Work,Heat Quantity^2 · Absorbed Dose = Kinematic Viscosity^6

Specific Volume^2 · Absorbed Dose · Dynamic Viscosity^4 = Power, Radiance Flux^2

Specific Volume^2 · Power, Radiance Flux^2 = Absorbed Dose · Kinematic Viscosity^4

Specific Volume · Magnetic Flux Density · Dynamic Viscosity = Electric Potential Difference

Specific Volume^2 · Capacitance^2 · Magnetic Flux Density^5 = Magnetic Flux

Specific Volume^2 · Magnetic Flux Density^4 = Electric Resistance^2 · Absorbed Dose

Current Density · Magnetic Field Strength^2 = Frequency^3 · Electric Charge^3

Current Density^2 · Frequency^6 = Magnetic Field Strength^2 · Absorbed Dose^3

Current Density^2 · Frequency^3 = Magnetic Field Strength^2 · Kinematic Viscosity^3

Current Density · Pressure,Stress = Magnetic Field Strength · Energy,Work,Heat Quantity

Current Density · Pressure,Stress^3 = Magnetic Field Strength^7 · Inductance^3

Current Density^2 · Pressure,Stress^3 = Magnetic Field Strength^5 · Magnetic Flux^3

Current Density^2 · Magnetic Field Strength^4 · Inductance^3 = Energy,Work,Heat Quantity^3

Current Density · Magnetic Field Strength^2 · Magnetic Flux^3 = Energy,Work,Heat Quantity^3

Current Density · Magnetic Field Strength^2 · Electric Potential Difference^3 = Power, Radiance Flux^3

Current Density^2 · Magnetic Field Strength^4 · Electric Resistance^3 = Power, Radiance Flux^3

Current Density^4 · Magnetic Field Strength^2 = Electric Charge^6 · Absorbed Dose^3

Current Density · Magnetic Field Strength^2 · Electric Resistance^3 = Electric Potential Difference^3

Current Density · Magnetic Flux Density^3 = Magnetic Field Strength^4 · Inductance^3

Current Density^2 · Magnetic Flux Density^3 = Magnetic Field Strength^2 · Magnetic Flux^3

Current Density · Magnetic Field Strength^2 · Inductance^3 = Magnetic Flux^3

Current Density^2 · Absorbed Dose^3 = Magnetic Field Strength^2 · Kinematic Viscosity^6

Current Density · Frequency · Magnetic Flux Density = Power, Radiance Flux

Current Density · Frequency = Electric Charge · Absorbed Dose

Current Density^6 · Pressure,Stress^4 · Inductance^3 = Energy,Work,Heat Quantity^7

Current Density^3 · Pressure,Stress^2 · Magnetic Flux^3 = Energy,Work,Heat Quantity^5

Current Density · Magnetic Flux Density^7 = Pressure,Stress^4 · Inductance^3

Current Density^2 · Magnetic Flux Density^5 = Pressure,Stress^2 · Magnetic Flux^3

Current Density^3 · Pressure,Stress^2 · Inductance^5 = Magnetic Flux^7

Current Density · Electric Potential Difference = Energy,Work,Heat Quantity · Kinematic Viscosity

Current Density^2 = Energy,Work,Heat Quantity · Capacitance · Kinematic Viscosity^2

Current Density · Magnetic Flux Density = Electric Charge · Electric Potential Difference

Current Density · Capacitance · Magnetic Flux Density = Electric Charge^2

Current Density · Mass Density = Electric Charge · Dynamic Viscosity

Current Density · Magnetic Flux Density = Electric Potential Difference^2 · Capacitance

Current Density = Electric Potential Difference · Capacitance · Kinematic Viscosity

Current Density = Capacitance · Magnetic Flux Density · Kinematic Viscosity^2

Current Density = Capacitance · Absorbed Dose · Magnetic Flux

Current Density · Electric Resistance = Magnetic Flux · Kinematic Viscosity

Current Density · Magnetic Flux Density · Inductance = Magnetic Flux^2

Magnetic Field Strength^2 · Absorbed Dose = Frequency^4 · Electric Charge^2

Magnetic Field Strength^2 · Kinematic Viscosity = Frequency^3 · Electric Charge^2

Magnetic Field Strength · Magnetic Flux Density = Frequency · Dynamic Viscosity

Magnetic Field Strength^6 · Inductance^3 = Pressure,Stress^2 · Energy,Work,Heat Quantity

Magnetic Field Strength^3 · Magnetic Flux^3 = Pressure,Stress · Energy,Work,Heat Quantity^2

Magnetic Field Strength · Electric Potential Difference = Pressure,Stress · Kinematic Viscosity

Magnetic Field Strength^4 · Electric Resistance^2 = Pressure,Stress^2 · Absorbed Dose

Magnetic Field Strength^4 · Electric Resistance^2 · Mass Density = Pressure,Stress^3

Magnetic Field Strength^3 · Inductance^2 = Pressure,Stress · Magnetic Flux

Magnetic Field Strength^4 · Inductance^3 = Energy,Work,Heat Quantity · Magnetic Flux Density^2

Magnetic Field Strength^2 · Magnetic Flux^3 = Energy,Work,Heat Quantity^2 · Magnetic Flux Density

Magnetic Field Strength^2 · Inductance · Kinematic Viscosity = Power, Radiance Flux

Magnetic Field Strength^2 · Absorbed Dose · Magnetic Flux^2 = Power, Radiance Flux^2

Magnetic Field Strength^2 · Kinematic Viscosity^4 = Electric Charge^2 · Absorbed Dose^3

Magnetic Field Strength^2 · Inductance^2 · Absorbed Dose = Electric Potential Difference^2

Magnetic Field Strength^3 · Inductance^2 = Electric Potential Difference · Dynamic Viscosity

Magnetic Field Strength · Electric Potential Difference = Absorbed Dose · Dynamic Viscosity

Magnetic Field Strength^2 · Electric Resistance^2 = Magnetic Flux Density^2 · Absorbed Dose

Magnetic Field Strength · Electric Resistance^2 · Mass Density = Magnetic Flux Density^3

Magnetic Field Strength^4 · Electric Resistance^2 = Absorbed Dose^3 · Mass Density^2

Magnetic Field Strength^2 · Inductance^2 = Magnetic Flux Density · Magnetic Flux

Magnetic Field Strength · Magnetic Flux Density = Absorbed Dose · Mass Density

Magnetic Field Strength^4 · Inductance^2 = Absorbed Dose · Dynamic Viscosity^2

Magnetic Field Strength · Magnetic Flux = Dynamic Viscosity · Kinematic Viscosity

Magnetic Field Strength · Magnetic Flux · Mass Density = Dynamic Viscosity^2

Magnetic Field Strength · Magnetic Flux = Kinematic Viscosity^2 · Mass Density

Frequency^6 · Energy,Work,Heat Quantity^2 = Pressure,Stress^2 · Absorbed Dose^3

Frequency^3 · Energy,Work,Heat Quantity^2 = Pressure,Stress^2 · Kinematic Viscosity^3

Frequency^6 · Energy,Work,Heat Quantity^2 · Mass Density^3 = Pressure,Stress^5

Frequency^4 · Power, Radiance Flux^2 = Pressure,Stress^2 · Absorbed Dose^3

Frequency · Power, Radiance Flux^2 = Pressure,Stress^2 · Kinematic Viscosity^3

Frequency^4 · Power, Radiance Flux^2 · Mass Density^3 = Pressure,Stress^5

Frequency · Kinematic Viscosity · Mass Density = Pressure,Stress

Frequency · Electric Charge^2 · Electric Resistance = Energy,Work,Heat Quantity

Frequency^2 · Electric Charge^2 · Inductance = Energy,Work,Heat Quantity

Frequency · Electric Charge · Magnetic Flux = Energy,Work,Heat Quantity

Frequency · Energy,Work,Heat Quantity · Electric Resistance = Electric Potential Difference^2

Frequency^2 · Energy,Work,Heat Quantity · Inductance = Electric Potential Difference^2

Frequency^2 · Capacitance · Magnetic Flux^2 = Energy,Work,Heat Quantity

Frequency · Magnetic Flux^2 = Energy,Work,Heat Quantity · Electric Resistance

Frequency^4 · Energy,Work,Heat Quantity^2 = Absorbed Dose^3 · Dynamic Viscosity^2

Frequency^6 · Energy,Work,Heat Quantity^2 = Absorbed Dose^5 · Mass Density^2

Frequency · Energy,Work,Heat Quantity^2 = Dynamic Viscosity^2 · Kinematic Viscosity^3

Frequency · Energy,Work,Heat Quantity^2 · Mass Density^3 = Dynamic Viscosity^5

Frequency · Energy,Work,Heat Quantity^2 = Kinematic Viscosity^5 · Mass Density^2

Frequency · Electric Charge · Electric Potential Difference = Power, Radiance Flux

Frequency · Electric Charge^2 = Power, Radiance Flux · Capacitance

Frequency^2 · Electric Charge^2 · Electric Resistance = Power, Radiance Flux

Frequency^3 · Electric Charge^2 · Inductance = Power, Radiance Flux

Frequency^2 · Electric Charge · Magnetic Flux = Power, Radiance Flux

Frequency · Electric Potential Difference^2 · Capacitance = Power, Radiance Flux

Frequency · Power, Radiance Flux · Inductance = Electric Potential Difference^2

Frequency^3 · Capacitance · Magnetic Flux^2 = Power, Radiance Flux

$\text{Frequency}^2 \cdot \text{Magnetic Flux}^2 = \text{Power, Radiance Flux} \cdot \text{Electric Resistance}$

$\text{Frequency} \cdot \text{Magnetic Flux}^2 = \text{Power, Radiance Flux} \cdot \text{Inductance}$

$\text{Frequency}^2 \cdot \text{Power, Radiance Flux}^2 = \text{Absorbed Dose}^3 \cdot \text{Dynamic Viscosity}^2$

$\text{Frequency}^4 \cdot \text{Power, Radiance Flux}^2 = \text{Absorbed Dose}^5 \cdot \text{Mass Density}^2$

$\text{Frequency} \cdot \text{Dynamic Viscosity}^2 \cdot \text{Kinematic Viscosity}^3 = \text{Power, Radiance Flux}^2$

$\text{Frequency} \cdot \text{Dynamic Viscosity}^5 = \text{Power, Radiance Flux}^2 \cdot \text{Mass Density}^3$

$\text{Frequency} \cdot \text{Kinematic Viscosity}^5 \cdot \text{Mass Density}^2 = \text{Power, Radiance Flux}^2$

$\text{Frequency} \cdot \text{Electric Charge} \cdot \text{Electric Resistance} = \text{Electric Potential Difference}$

$\text{Frequency}^2 \cdot \text{Electric Charge} \cdot \text{Inductance} = \text{Electric Potential Difference}$

$\text{Frequency} \cdot \text{Capacitance} \cdot \text{Magnetic Flux} = \text{Electric Charge}$

$\text{Frequency} \cdot \text{Electric Charge} \cdot \text{Inductance} = \text{Magnetic Flux}$

$\text{Frequency} \cdot \text{Electric Potential Difference} = \text{Magnetic Flux Density} \cdot \text{Absorbed Dose}$

$\text{Frequency}^2 \cdot \text{Magnetic Flux} = \text{Magnetic Flux Density} \cdot \text{Absorbed Dose}$

$\text{Frequency} \cdot \text{Magnetic Flux} = \text{Magnetic Flux Density} \cdot \text{Kinematic Viscosity}$

$\text{Frequency} \cdot \text{Dynamic Viscosity} = \text{Absorbed Dose} \cdot \text{Mass Density}$

$\text{Pressure,Stress}^2 \cdot \text{Energy,Work,Heat Quantity}^4 \cdot \text{Absorbed Dose}^3 = \text{Power, Radiance Flux}^6$

$\text{Pressure,Stress} \cdot \text{Energy,Work,Heat Quantity} = \text{Power, Radiance Flux} \cdot \text{Dynamic Viscosity}$

$\text{Pressure,Stress}^2 \cdot \text{Energy,Work,Heat Quantity} \cdot \text{Kinematic Viscosity}^3 = \text{Power, Radiance Flux}^3$

$\text{Pressure,Stress}^5 \cdot \text{Energy,Work,Heat Quantity}^4 = \text{Power, Radiance Flux}^6 \cdot \text{Mass Density}^3$

$\text{Pressure,Stress}^4 \cdot \text{Inductance}^3 = \text{Energy,Work,Heat Quantity} \cdot \text{Magnetic Flux Density}^6$

$\text{Pressure,Stress}^2 \cdot \text{Magnetic Flux}^3 = \text{Energy,Work,Heat Quantity}^2 \cdot \text{Magnetic Flux Density}^3$

$\text{Pressure,Stress}^4 \cdot \text{Energy,Work,Heat Quantity}^2 = \text{Absorbed Dose}^3 \cdot \text{Dynamic Viscosity}^6$

$\text{Pressure,Stress}^2 \cdot \text{Kinematic Viscosity}^6 = \text{Energy,Work,Heat Quantity}^2 \cdot \text{Absorbed Dose}^3$

Pressure,Stress $\cdot$ Energy,Work,Heat Quantity^2 = Dynamic Viscosity^3 $\cdot$ Kinematic Viscosity^3

Pressure,Stress $\cdot$ Energy,Work,Heat Quantity^2 $\cdot$ Mass Density^3 = Dynamic Viscosity^6

Pressure,Stress $\cdot$ Energy,Work,Heat Quantity^2 = Kinematic Viscosity^6 $\cdot$ Mass Density^3

Pressure,Stress^2 $\cdot$ Power, Radiance Flux^2 = Absorbed Dose^3 $\cdot$ Dynamic Viscosity^4

Pressure,Stress^2 $\cdot$ Kinematic Viscosity^4 = Power, Radiance Flux^2 $\cdot$ Absorbed Dose

Pressure,Stress $\cdot$ Dynamic Viscosity $\cdot$ Kinematic Viscosity^3 = Power, Radiance Flux^2

Pressure,Stress $\cdot$ Dynamic Viscosity^4 = Power, Radiance Flux^2 $\cdot$ Mass Density^3

Pressure,Stress $\cdot$ Kinematic Viscosity^4 $\cdot$ Mass Density = Power, Radiance Flux^2

Pressure,Stress $\cdot$ Magnetic Flux = Electric Potential Difference $\cdot$ Dynamic Viscosity

Pressure,Stress $\cdot$ Capacitance $\cdot$ Electric Resistance = Dynamic Viscosity

Pressure,Stress^2 $\cdot$ Capacitance $\cdot$ Inductance = Dynamic Viscosity^2

Pressure,Stress^2 $\cdot$ Electric Resistance^2 = Magnetic Flux Density^4 $\cdot$ Absorbed Dose

Pressure,Stress $\cdot$ Electric Resistance^2 $\cdot$ Mass Density = Magnetic Flux Density^4

Pressure,Stress $\cdot$ Inductance = Electric Resistance $\cdot$ Dynamic Viscosity

Pressure,Stress^2 $\cdot$ Inductance^2 = Magnetic Flux Density^3 $\cdot$ Magnetic Flux

Pressure,Stress $\cdot$ Kinematic Viscosity = Absorbed Dose $\cdot$ Dynamic Viscosity

Energy,Work,Heat Quantity^2 = Power, Radiance Flux $\cdot$ Electric Charge^2 $\cdot$ Electric Resistance

Energy,Work,Heat Quantity^3 = Power, Radiance Flux^2 $\cdot$ Electric Charge^2 $\cdot$ Inductance

Energy,Work,Heat Quantity^2 = Power, Radiance Flux $\cdot$ Electric Charge $\cdot$ Magnetic Flux

Energy,Work,Heat Quantity $\cdot$ Electric Potential Difference^2 = Power, Radiance Flux^2 $\cdot$ Inductance

Energy,Work,Heat Quantity $\cdot$ Electric Potential Difference = Power, Radiance Flux $\cdot$ Magnetic Flux

Energy,Work,Heat Quantity = Power, Radiance Flux $\cdot$ Capacitance $\cdot$ Electric Resistance

Energy,Work,Heat Quantity^2 = Power, Radiance Flux^2 $\cdot$ Capacitance $\cdot$ Inductance

Energy,Work,Heat Quantity^3 = Power, Radiance Flux^2 $\cdot$ Capacitance $\cdot$ Magnetic Flux^2

Energy,Work,Heat Quantity $\cdot$ Electric Resistance = Power, Radiance Flux $\cdot$ Inductance

Energy,Work,Heat Quantity^2 $\cdot$ Electric Resistance = Power, Radiance Flux $\cdot$ Magnetic Flux^2

Energy,Work,Heat Quantity^2 $\cdot$ Absorbed Dose^3 $\cdot$ Dynamic Viscosity^2 = Power, Radiance Flux^4

Energy,Work,Heat Quantity $\cdot$ Absorbed Dose = Power, Radiance Flux $\cdot$ Kinematic Viscosity

Energy,Work,Heat Quantity^4 $\cdot$ Absorbed Dose^5 $\cdot$ Mass Density^2 = Power, Radiance Flux^6

Energy,Work,Heat Quantity $\cdot$ Power, Radiance Flux = Dynamic Viscosity^2 $\cdot$ Kinematic Viscosity^3

Energy,Work,Heat Quantity $\cdot$ Power, Radiance Flux $\cdot$ Mass Density^3 = Dynamic Viscosity^5

Energy,Work,Heat Quantity $\cdot$ Power, Radiance Flux = Kinematic Viscosity^5 $\cdot$ Mass Density^2

Energy,Work,Heat Quantity $\cdot$ Inductance = Electric Charge^2 $\cdot$ Electric Resistance^2

Energy,Work,Heat Quantity = Electric Charge $\cdot$ Magnetic Flux Density $\cdot$ Kinematic Viscosity

Energy,Work,Heat Quantity $\cdot$ Electric Resistance^2 = Electric Potential Difference^2 $\cdot$ Inductance

Energy,Work,Heat Quantity $\cdot$ Electric Resistance = Electric Potential Difference $\cdot$ Magnetic Flux

Energy,Work,Heat Quantity $\cdot$ Capacitance $\cdot$ Electric Resistance^2 = Magnetic Flux^2

Energy,Work,Heat Quantity = Capacitance $\cdot$ Magnetic Flux Density^2 $\cdot$ Kinematic Viscosity^2

Energy,Work,Heat Quantity^2 $\cdot$ Absorbed Dose = Dynamic Viscosity^2 $\cdot$ Kinematic Viscosity^4

Energy,Work,Heat Quantity^2 $\cdot$ Absorbed Dose $\cdot$ Mass Density^4 = Dynamic Viscosity^6

Energy,Work,Heat Quantity^2 $\cdot$ Absorbed Dose = Kinematic Viscosity^6 $\cdot$ Mass Density^2

Power, Radiance Flux^2 $\cdot$ Inductance = Electric Charge $\cdot$ Electric Potential Difference^3

Power, Radiance Flux $\cdot$ Magnetic Flux = Electric Charge $\cdot$ Electric Potential Difference^2

Power, Radiance Flux $\cdot$ Capacitance^2 $\cdot$ Electric Resistance = Electric Charge^2

Power, Radiance Flux^2 $\cdot$ Capacitance^3 $\cdot$ Inductance = Electric Charge^4

Power, Radiance Flux $\cdot$ Capacitance^2 $\cdot$ Magnetic Flux = Electric Charge^3

Power, Radiance Flux $\cdot$ Inductance^2 = Electric Charge^2 $\cdot$ Electric Resistance^3

Power, Radiance Flux = Electric Charge $\cdot$ Magnetic Flux Density $\cdot$ Absorbed Dose

Power, Radiance Flux $\cdot$ Electric Charge $\cdot$ Inductance^2 = Magnetic Flux^3

Power, Radiance Flux^2 $\cdot$ Inductance = Electric Potential Difference^4 $\cdot$ Capacitance

Power, Radiance Flux $\cdot$ Magnetic Flux = Electric Potential Difference^3 $\cdot$ Capacitance

Power, Radiance Flux $\cdot$ Inductance = Electric Potential Difference $\cdot$ Magnetic Flux

Power, Radiance Flux $\cdot$ Capacitance^2 $\cdot$ Electric Resistance^3 = Magnetic Flux^2

Power, Radiance Flux^2 $\cdot$ Capacitance $\cdot$ Inductance^3 = Magnetic Flux^4

Power, Radiance Flux $\cdot$ Electric Resistance = Magnetic Flux Density^2 $\cdot$ Kinematic Viscosity^2

Power, Radiance Flux $\cdot$ Inductance^2 = Electric Resistance $\cdot$ Magnetic Flux^2

Power, Radiance Flux^2 = Absorbed Dose $\cdot$ Dynamic Viscosity^2 $\cdot$ Kinematic Viscosity^2

Power, Radiance Flux^2 $\cdot$ Mass Density^2 = Absorbed Dose $\cdot$ Dynamic Viscosity^4

Power, Radiance Flux^2 = Absorbed Dose $\cdot$ Kinematic Viscosity^4 $\cdot$ Mass Density^2

Electric Charge $\cdot$ Electric Resistance^2 = Electric Potential Difference $\cdot$ Inductance

Force^3 $\cdot$ Frequency^2 = Pressure,Stress $\cdot$ Power, Radiance Flux^2

Force $\cdot$ Frequency^2 = Pressure,Stress $\cdot$ Absorbed Dose

Force $\cdot$ Frequency = Pressure,Stress $\cdot$ Kinematic Viscosity

Force $\cdot$ Frequency^2 $\cdot$ Mass Density = Pressure,Stress^2

Force^2 $\cdot$ Absorbed Dose = Frequency^2 $\cdot$ Energy,Work,Heat Quantity^2

Force^3 = Frequency $\cdot$ Energy,Work,Heat Quantity^2 $\cdot$ Dynamic Viscosity

$$\text{Force}^2 \cdot \text{Kinematic Viscosity} = \text{Frequency} \cdot \text{Energy,Work,Heat Quantity}^2$$

$$\text{Force}^5 = \text{Frequency}^2 \cdot \text{Energy,Work,Heat Quantity}^4 \cdot \text{Mass Density}$$

$$\text{Force}^3 \cdot \text{Frequency} = \text{Power, Radiance Flux}^2 \cdot \text{Dynamic Viscosity}$$

$$\text{Force}^2 \cdot \text{Frequency} \cdot \text{Kinematic Viscosity} = \text{Power, Radiance Flux}^2$$

$$\text{Force}^5 \cdot \text{Frequency}^2 = \text{Power, Radiance Flux}^4 \cdot \text{Mass Density}$$

$$\text{Force} \cdot \text{Frequency} = \text{Absorbed Dose} \cdot \text{Dynamic Viscosity}$$

$$\text{Force} \cdot \text{Frequency}^2 = \text{Absorbed Dose}^2 \cdot \text{Mass Density}$$

$$\text{Force}^3 \cdot \text{Pressure,Stress} = \text{Power, Radiance Flux}^2 \cdot \text{Dynamic Viscosity}^2$$

$$\text{Force} \cdot \text{Pressure,Stress} \cdot \text{Kinematic Viscosity}^2 = \text{Power, Radiance Flux}^2$$

$$\text{Force}^2 \cdot \text{Pressure,Stress} = \text{Power, Radiance Flux}^2 \cdot \text{Mass Density}$$

$$\text{Force}^3 = \text{Pressure,Stress} \cdot \text{Electric Charge}^2 \cdot \text{Electric Potential Difference}^2$$

$$\text{Force}^3 \cdot \text{Capacitance}^2 = \text{Pressure,Stress} \cdot \text{Electric Charge}^4$$

$$\text{Force}^3 = \text{Pressure,Stress} \cdot \text{Electric Potential Difference}^4 \cdot \text{Capacitance}^2$$

$$\text{Force} \cdot \text{Magnetic Flux Density}^4 = \text{Pressure,Stress}^3 \cdot \text{Inductance}^2$$

$$\text{Force} \cdot \text{Magnetic Flux Density} = \text{Pressure,Stress} \cdot \text{Magnetic Flux}$$

$$\text{Force}^3 \cdot \text{Inductance}^2 = \text{Pressure,Stress} \cdot \text{Magnetic Flux}^4$$

$$\text{Force} \cdot \text{Pressure,Stress} = \text{Absorbed Dose} \cdot \text{Dynamic Viscosity}^2$$

$$\text{Force} \cdot \text{Absorbed Dose} = \text{Pressure,Stress} \cdot \text{Kinematic Viscosity}^2$$

$$\text{Force}^3 = \text{Energy,Work,Heat Quantity} \cdot \text{Power, Radiance Flux} \cdot \text{Dynamic Viscosity}$$

$$\text{Force}^2 \cdot \text{Kinematic Viscosity} = \text{Energy,Work,Heat Quantity} \cdot \text{Power, Radiance Flux}$$

$$\text{Force}^5 = \text{Energy,Work,Heat Quantity}^2 \cdot \text{Power, Radiance Flux}^2 \cdot \text{Mass Density}$$

$$\text{Force}^4 \cdot \text{Inductance} = \text{Energy,Work,Heat Quantity}^3 \cdot \text{Magnetic Flux Density}^2$$

$\text{Force}^2 \cdot \text{Magnetic Flux} = \text{Energy,Work,Heat Quantity}^2 \cdot \text{Magnetic Flux Density}$

$\text{Force}^4 = \text{Energy,Work,Heat Quantity}^2 \cdot \text{Absorbed Dose} \cdot \text{Dynamic Viscosity}^2$

$\text{Force}^2 \cdot \text{Kinematic Viscosity}^2 = \text{Energy,Work,Heat Quantity}^2 \cdot \text{Absorbed Dose}$

$\text{Force}^3 = \text{Energy,Work,Heat Quantity}^2 \cdot \text{Absorbed Dose} \cdot \text{Mass Density}$

$\text{Force}^2 = \text{Power, Radiance Flux} \cdot \text{Electric Charge} \cdot \text{Magnetic Flux Density}$

$\text{Force}^2 = \text{Electric Charge}^2 \cdot \text{Magnetic Flux Density}^2 \cdot \text{Absorbed Dose}$

$\text{Force}^2 \cdot \text{Electric Resistance}^2 \cdot \text{Absorbed Dose} = \text{Electric Potential Difference}^4$

$\text{Force} \cdot \text{Magnetic Flux Density} = \text{Electric Potential Difference} \cdot \text{Dynamic Viscosity}$

$\text{Force} \cdot \text{Magnetic Flux Density}^2 = \text{Electric Potential Difference}^2 \cdot \text{Mass Density}$

$\text{Force}^2 \cdot \text{Inductance}^2 = \text{Magnetic Flux Density} \cdot \text{Magnetic Flux}^3$

$\text{Energy} \cdot \text{Pressure,Stress}^2 = \text{Magnetic Field Strength}^6 \cdot \text{Inductance}^3$

$\text{Energy}^2 \cdot \text{Pressure,Stress} = \text{Magnetic Field Strength}^3 \cdot \text{Magnetic Flux}^3$

$\text{Energy} \cdot \text{Magnetic Flux Density}^2 = \text{Magnetic Field Strength}^4 \cdot \text{Inductance}^3$

$\text{Energy}^2 \cdot \text{Magnetic Flux Density} = \text{Magnetic Field Strength}^2 \cdot \text{Magnetic Flux}^3$

$\text{Energy}^2 \cdot \text{Frequency}^6 = \text{Pressure,Stress}^2 \cdot \text{Absorbed Dose}^3$

$\text{Energy}^2 \cdot \text{Frequency}^3 = \text{Pressure,Stress}^2 \cdot \text{Kinematic Viscosity}^3$

$\text{Energy}^2 \cdot \text{Frequency}^6 \cdot \text{Mass Density}^3 = \text{Pressure,Stress}^5$

$\text{Energy} = \text{Frequency} \cdot \text{Electric Charge}^2 \cdot \text{Electric Resistance}$

$\text{Energy} = \text{Frequency}^2 \cdot \text{Electric Charge}^2 \cdot \text{Inductance}$

$\text{Energy} = \text{Frequency} \cdot \text{Electric Charge} \cdot \text{Magnetic Flux}$

$\text{Energy} \cdot \text{Frequency} \cdot \text{Electric Resistance} = \text{Electric Potential Difference}^2$

$\text{Energy} \cdot \text{Frequency}^2 \cdot \text{Inductance} = \text{Electric Potential Difference}^2$

$\text{Energy} = \text{Frequency}^2 \cdot \text{Capacitance} \cdot \text{Magnetic Flux}^2$

Energy $\cdot$ Electric Resistance = Frequency $\cdot$ Magnetic Flux^2

Energy^2 $\cdot$ Frequency^4 = Absorbed Dose^3 $\cdot$ Dynamic Viscosity^2

Energy^2 $\cdot$ Frequency^6 = Absorbed Dose^5 $\cdot$ Mass Density^2

Energy^2 $\cdot$ Frequency = Dynamic Viscosity^2 $\cdot$ Kinematic Viscosity^3

Energy^2 $\cdot$ Frequency $\cdot$ Mass Density^3 = Dynamic Viscosity^5

Energy^2 $\cdot$ Frequency = Kinematic Viscosity^5 $\cdot$ Mass Density^2

Energy^4 $\cdot$ Pressure,Stress^2 $\cdot$ Absorbed Dose^3 = Power, Radiance Flux^6

Energy $\cdot$ Pressure,Stress = Power, Radiance Flux $\cdot$ Dynamic Viscosity

Energy $\cdot$ Pressure,Stress^2 $\cdot$ Kinematic Viscosity^3 = Power, Radiance Flux^3

Energy^4 $\cdot$ Pressure,Stress^5 = Power, Radiance Flux^6 $\cdot$ Mass Density^3

Energy $\cdot$ Magnetic Flux Density^6 = Pressure,Stress^4 $\cdot$ Inductance^3

Energy^2 $\cdot$ Magnetic Flux Density^3 = Pressure,Stress^2 $\cdot$ Magnetic Flux^3

Energy^2 $\cdot$ Pressure,Stress^4 = Absorbed Dose^3 $\cdot$ Dynamic Viscosity^6

Energy^2 $\cdot$ Absorbed Dose^3 = Pressure,Stress^2 $\cdot$ Kinematic Viscosity^6

Energy^2 $\cdot$ Pressure,Stress = Dynamic Viscosity^3 $\cdot$ Kinematic Viscosity^3

Energy^2 $\cdot$ Pressure,Stress $\cdot$ Mass Density^3 = Dynamic Viscosity^6

Energy^2 $\cdot$ Pressure,Stress = Kinematic Viscosity^6 $\cdot$ Mass Density^3

Energy^2 = Power, Radiance Flux $\cdot$ Electric Charge^2 $\cdot$ Electric Resistance

Energy^3 = Power, Radiance Flux^2 $\cdot$ Electric Charge^2 $\cdot$ Inductance

Energy^2 = Power, Radiance Flux $\cdot$ Electric Charge $\cdot$ Magnetic Flux

Energy $\cdot$ Electric Potential Difference^2 = Power, Radiance Flux^2 $\cdot$ Inductance

Energy $\cdot$ Electric Potential Difference = Power, Radiance Flux $\cdot$ Magnetic Flux

Energy = Power, Radiance Flux $\cdot$ Capacitance $\cdot$ Electric Resistance

Energy^2 = Power, Radiance Flux^2 $\cdot$ Capacitance $\cdot$ Inductance

Energy^3 = Power, Radiance Flux^2 $\cdot$ Capacitance $\cdot$ Magnetic Flux^2

Energy $\cdot$ Electric Resistance = Power, Radiance Flux $\cdot$ Inductance

Energy^2 $\cdot$ Electric Resistance = Power, Radiance Flux $\cdot$ Magnetic Flux^2

Energy^2 $\cdot$ Absorbed Dose^3 $\cdot$ Dynamic Viscosity^2 = Power, Radiance Flux^4

Energy $\cdot$ Absorbed Dose = Power, Radiance Flux $\cdot$ Kinematic Viscosity

Energy^4 $\cdot$ Absorbed Dose^5 $\cdot$ Mass Density^2 = Power, Radiance Flux^6

Energy $\cdot$ Power, Radiance Flux = Dynamic Viscosity^2 $\cdot$ Kinematic Viscosity^3

Energy $\cdot$ Power, Radiance Flux $\cdot$ Mass Density^3 = Dynamic Viscosity^5

Energy $\cdot$ Power, Radiance Flux = Kinematic Viscosity^5 $\cdot$ Mass Density^2

Energy $\cdot$ Inductance = Electric Charge^2 $\cdot$ Electric Resistance^2

Energy = Electric Charge $\cdot$ Magnetic Flux Density $\cdot$ Kinematic Viscosity

Energy $\cdot$ Electric Resistance^2 = Electric Potential Difference^2 $\cdot$ Inductance

Energy $\cdot$ Electric Resistance = Electric Potential Difference $\cdot$ Magnetic Flux

Energy $\cdot$ Capacitance $\cdot$ Electric Resistance^2 = Magnetic Flux^2

Energy = Capacitance $\cdot$ Magnetic Flux Density^2 $\cdot$ Kinematic Viscosity^2

Energy^2 $\cdot$ Absorbed Dose = Dynamic Viscosity^2 $\cdot$ Kinematic Viscosity^4

Energy^2 $\cdot$ Absorbed Dose $\cdot$ Mass Density^4 = Dynamic Viscosity^6

Energy^2 $\cdot$ Absorbed Dose = Kinematic Viscosity^6 $\cdot$ Mass Density^2

Acceleration^2 $\cdot$ Electric Potential Difference^2 = Magnetic Flux Density^2 $\cdot$ Absorbed Dose^3

Acceleration^2 $\cdot$ Magnetic Flux Density $\cdot$ Magnetic Flux^3 = Electric Potential Difference^4

Acceleration^2 $\cdot$ Magnetic Flux^2 = Electric Potential Difference^2 $\cdot$ Absorbed Dose

$Acceleration^2 \cdot Magnetic\ Flux^3 = Electric\ Potential\ Difference^3 \cdot Kinematic\ Viscosity$

$Acceleration^2 \cdot Capacitance^2 \cdot Electric\ Resistance^2 = Absorbed\ Dose$

$Acceleration^2 \cdot Capacitance^3 \cdot Electric\ Resistance^3 = Kinematic\ Viscosity$

$Acceleration^2 \cdot Capacitance \cdot Inductance = Absorbed\ Dose$

$Acceleration^4 \cdot Capacitance^3 \cdot Inductance^3 = Kinematic\ Viscosity^2$

$Acceleration^2 \cdot Inductance^2 = Electric\ Resistance^2 \cdot Absorbed\ Dose$

$Acceleration^2 \cdot Inductance^3 = Electric\ Resistance^3 \cdot Kinematic\ Viscosity$

$Acceleration \cdot Inductance \cdot Mass\ Density = Magnetic\ Flux\ Density^2$

$Acceleration^2 \cdot Magnetic\ Flux = Magnetic\ Flux\ Density \cdot Absorbed\ Dose^2$

$Acceleration^2 \cdot Magnetic\ Flux^3 = Magnetic\ Flux\ Density^3 \cdot Kinematic\ Viscosity^4$

$Acceleration^2 \cdot Dynamic\ Viscosity^2 = Absorbed\ Dose^3 \cdot Mass\ Density^2$

$Mass \cdot Power, Radiance\ Flux^2 = Wavenumber^2 \cdot Energy, Work, Heat\ Quantity^3$

$Mass \cdot Wavenumber^4 \cdot Energy, Work, Heat\ Quantity = Dynamic\ Viscosity^2$

$Mass \cdot Wavenumber^2 \cdot Kinematic\ Viscosity^2 = Energy, Work, Heat\ Quantity$

$Mass^2 \cdot Wavenumber^2 \cdot Absorbed\ Dose^3 = Power, Radiance\ Flux^2$

$Mass^2 \cdot Wavenumber^5 \cdot Power, Radiance\ Flux = Dynamic\ Viscosity^3$

$Mass \cdot Wavenumber^4 \cdot Kinematic\ Viscosity^3 = Power, Radiance\ Flux$

$Mass = Wavenumber^2 \cdot Electric\ Charge^2 \cdot Inductance$

$Mass \cdot Wavenumber^2 = Capacitance \cdot Magnetic\ Flux\ Density^2$

$Mass = Wavenumber^2 \cdot Capacitance \cdot Magnetic\ Flux^2$

$Mass^2 \cdot Wavenumber^4 \cdot Absorbed\ Dose = Dynamic\ Viscosity^2$

$Mass \cdot Wavenumber^3 \cdot Kinematic\ Viscosity = Dynamic\ Viscosity$

Mass $\cdot$ Specific Volume $\cdot$ Magnetic Field Strength = Current Density

Mass^2 $\cdot$ Frequency^6 = Specific Volume $\cdot$ Pressure,Stress^3

Mass^5 $\cdot$ Specific Volume^2 $\cdot$ Frequency^6 = Energy,Work,Heat Quantity^3

Mass^5 $\cdot$ Specific Volume^2 $\cdot$ Frequency^9 = Power, Radiance Flux^3

Mass^2 $\cdot$ Specific Volume^2 $\cdot$ Frequency^6 = Absorbed Dose^3

Mass^2 $\cdot$ Frequency^3 = Specific Volume $\cdot$ Dynamic Viscosity^3

Mass^2 $\cdot$ Specific Volume^2 $\cdot$ Frequency^3 = Kinematic Viscosity^3

Mass $\cdot$ Specific Volume $\cdot$ Pressure,Stress = Energy,Work,Heat Quantity

Mass^4 $\cdot$ Specific Volume^7 $\cdot$ Pressure,Stress^9 = Power, Radiance Flux^6

Mass^2 $\cdot$ Pressure,Stress^3 = Specific Volume $\cdot$ Dynamic Viscosity^6

Mass^2 $\cdot$ Specific Volume^5 $\cdot$ Pressure,Stress^3 = Kinematic Viscosity^6

Mass^5 $\cdot$ Specific Volume^2 $\cdot$ Power, Radiance Flux^6 = Energy,Work,Heat Quantity^9

Mass $\cdot$ Specific Volume^4 $\cdot$ Dynamic Viscosity^6 = Energy,Work,Heat Quantity^3

Mass $\cdot$ Kinematic Viscosity^6 = Specific Volume^2 $\cdot$ Energy,Work,Heat Quantity^3

Mass^4 $\cdot$ Absorbed Dose^9 = Specific Volume^2 $\cdot$ Power, Radiance Flux^6

Mass $\cdot$ Power, Radiance Flux^3 = Specific Volume^5 $\cdot$ Dynamic Viscosity^9

Mass $\cdot$ Specific Volume^4 $\cdot$ Power, Radiance Flux^3 = Kinematic Viscosity^9

Mass^5 $\cdot$ Specific Volume^2 = Electric Charge^6 $\cdot$ Inductance^3

Mass = Specific Volume^2 $\cdot$ Capacitance^3 $\cdot$ Magnetic Flux Density^6

Mass^5 $\cdot$ Specific Volume^2 = Capacitance^3 $\cdot$ Magnetic Flux^6

Mass^2 $\cdot$ Specific Volume^2 $\cdot$ Magnetic Flux Density^3 = Magnetic Flux^3

Mass^2 $\cdot$ Absorbed Dose^3 = Specific Volume^4 $\cdot$ Dynamic Viscosity^6

Mass^2 $\cdot$ Specific Volume^2 $\cdot$ Absorbed Dose^3 = Kinematic Viscosity^6

Mass · Magnetic Field Strength = Current Density · Mass Density

Mass · Current Density = Electric Charge^3 · Electric Resistance

Mass · Current Density = Electric Charge^2 · Magnetic Flux

Mass · Current Density · Electric Resistance^2 = Magnetic Flux^3

Mass · Kinematic Viscosity^3 = Current Density^2 · Electric Resistance

Mass · Absorbed Dose = Current Density · Magnetic Flux Density

Mass · Kinematic Viscosity^2 = Current Density · Magnetic Flux

Mass · Frequency^4 · Capacitance = Magnetic Field Strength^2

Mass · Frequency^3 = Magnetic Field Strength^2 · Electric Resistance

Mass · Frequency^2 = Magnetic Field Strength^2 · Inductance

Mass · Magnetic Field Strength = Electric Charge · Dynamic Viscosity

Mass = Magnetic Field Strength^2 · Capacitance^3 · Electric Resistance^4

Mass = Magnetic Field Strength^2 · Capacitance · Inductance^2

Mass · Electric Resistance^2 = Magnetic Field Strength^2 · Inductance^3

Mass^3 · Frequency^6 = Pressure,Stress^2 · Energy,Work,Heat Quantity

Mass^3 · Frequency^7 = Pressure,Stress^2 · Power, Radiance Flux

Mass^2 · Frequency^6 = Pressure,Stress^2 · Absorbed Dose

Mass^2 · Frequency^5 = Pressure,Stress^2 · Kinematic Viscosity

Mass^2 · Frequency^6 · Mass Density = Pressure,Stress^3

Mass^3 · Frequency^4 = Energy,Work,Heat Quantity · Dynamic Viscosity^2

Mass · Frequency · Kinematic Viscosity = Energy,Work,Heat Quantity

Mass^5 · Frequency^6 = Energy,Work,Heat Quantity^3 · Mass Density^2

$Mass \cdot Frequency \cdot Absorbed\ Dose = Power, Radiance\ Flux$

$Mass^3 \cdot Frequency^5 = Power, Radiance\ Flux \cdot Dynamic\ Viscosity^2$

$Mass \cdot Frequency^2 \cdot Kinematic\ Viscosity = Power, Radiance\ Flux$

$Mass^5 \cdot Frequency^9 = Power, Radiance\ Flux^3 \cdot Mass\ Density^2$

$Mass \cdot Frequency = Electric\ Charge \cdot Magnetic\ Flux\ Density$

$Mass^2 \cdot Frequency^4 = Absorbed\ Dose \cdot Dynamic\ Viscosity^2$

$Mass^2 \cdot Frequency^6 = Absorbed\ Dose^3 \cdot Mass\ Density^2$

$Mass^2 \cdot Frequency^3 = Dynamic\ Viscosity^2 \cdot Kinematic\ Viscosity$

$Mass^2 \cdot Frequency^3 \cdot Mass\ Density = Dynamic\ Viscosity^3$

$Mass^2 \cdot Frequency^3 = Kinematic\ Viscosity^3 \cdot Mass\ Density^2$

$Mass^3 \cdot Power, Radiance\ Flux^6 = Pressure, Stress^2 \cdot Energy, Work, Heat\ Quantity^7$

$Mass^3 \cdot Pressure, Stress^4 = Energy, Work, Heat\ Quantity \cdot Dynamic\ Viscosity^6$

$Mass^3 \cdot Pressure, Stress^2 \cdot Kinematic\ Viscosity^6 = Energy, Work, Heat\ Quantity^5$

$Mass \cdot Pressure, Stress = Energy, Work, Heat\ Quantity \cdot Mass\ Density$

$Mass^4 \cdot Pressure, Stress^2 \cdot Absorbed\ Dose^7 = Power, Radiance\ Flux^6$

$Mass^3 \cdot Pressure, Stress^5 = Power, Radiance\ Flux \cdot Dynamic\ Viscosity^7$

$Mass \cdot Pressure, Stress^4 \cdot Kinematic\ Viscosity^7 = Power, Radiance\ Flux^5$

$Mass^4 \cdot Pressure, Stress^9 = Power, Radiance\ Flux^6 \cdot Mass\ Density^7$

$Mass^2 \cdot Pressure, Stress^4 = Absorbed\ Dose \cdot Dynamic\ Viscosity^6$

$Mass^2 \cdot Absorbed\ Dose^5 = Pressure, Stress^2 \cdot Kinematic\ Viscosity^6$

$Mass^2 \cdot Pressure, Stress^3 = Dynamic\ Viscosity^5 \cdot Kinematic\ Viscosity$

$Mass^2 \cdot Pressure, Stress^3 \cdot Mass\ Density = Dynamic\ Viscosity^6$

$Mass^2 \cdot Pressure, Stress^3 = Kinematic\ Viscosity^6 \cdot Mass\ Density^5$

$Mass^3 \cdot Power, Radiance\ Flux^4 = Energy, Work, Heat\ Quantity^5 \cdot Dynamic\ Viscosity^2$

$Mass \cdot Power, Radiance\ Flux \cdot Kinematic\ Viscosity = Energy, Work, Heat\ Quantity^2$

$Mass^5 \cdot Power, Radiance\ Flux^6 = Energy, Work, Heat\ Quantity^9 \cdot Mass\ Density^2$

$Mass \cdot Dynamic\ Viscosity^2 \cdot Kinematic\ Viscosity^4 = Energy, Work, Heat\ Quantity^3$

$Mass \cdot Dynamic\ Viscosity^6 = Energy, Work, Heat\ Quantity^3 \cdot Mass\ Density^4$

$Mass \cdot Kinematic\ Viscosity^6 \cdot Mass\ Density^2 = Energy, Work, Heat\ Quantity^3$

$Mass^2 \cdot Absorbed\ Dose^5 \cdot Dynamic\ Viscosity^2 = Power, Radiance\ Flux^4$

$Mass \cdot Absorbed\ Dose^2 = Power, Radiance\ Flux \cdot Kinematic\ Viscosity$

$Mass^4 \cdot Absorbed\ Dose^9 \cdot Mass\ Density^2 = Power, Radiance\ Flux^6$

$Mass \cdot Power, Radiance\ Flux^3 = Dynamic\ Viscosity^4 \cdot Kinematic\ Viscosity^5$

$Mass \cdot Power, Radiance\ Flux^3 \cdot Mass\ Density^5 = Dynamic\ Viscosity^9$

$Mass \cdot Power, Radiance\ Flux^3 = Kinematic\ Viscosity^9 \cdot Mass\ Density^4$

$Mass \cdot Absorbed\ Dose = Electric\ Charge \cdot Electric\ Potential\ Difference$

$Mass \cdot Capacitance \cdot Absorbed\ Dose = Electric\ Charge^2$

$Mass \cdot Kinematic\ Viscosity = Electric\ Charge^2 \cdot Electric\ Resistance$

$Mass^5 = Electric\ Charge^6 \cdot Inductance^3 \cdot Mass\ Density^2$

$Mass \cdot Kinematic\ Viscosity = Electric\ Charge \cdot Magnetic\ Flux$

$Mass \cdot Absorbed\ Dose = Electric\ Potential\ Difference^2 \cdot Capacitance$

$Mass = Capacitance \cdot Magnetic\ Flux\ Density \cdot Magnetic\ Flux$

$Mass \cdot Mass\ Density^2 = Capacitance^3 \cdot Magnetic\ Flux\ Density^6$

$Mass^5 = Capacitance^3 \cdot Magnetic\ Flux^6 \cdot Mass\ Density^2$

$Mass \cdot Electric\ Resistance \cdot Kinematic\ Viscosity = Magnetic\ Flux^2$

Mass $\cdot$ Magnetic Flux Density^2 = Inductance $\cdot$ Dynamic Viscosity^2

Mass^2 $\cdot$ Magnetic Flux Density^3 = Magnetic Flux^3 $\cdot$ Mass Density^2

Mass $\cdot$ Inductance $\cdot$ Absorbed Dose = Magnetic Flux^2

Mass^2 $\cdot$ Absorbed Dose^3 = Dynamic Viscosity^2 $\cdot$ Kinematic Viscosity^4

Mass^2 $\cdot$ Absorbed Dose^3 $\cdot$ Mass Density^4 = Dynamic Viscosity^6

Mass^2 $\cdot$ Absorbed Dose^3 = Kinematic Viscosity^6 $\cdot$ Mass Density^2

Velocity^4 $\cdot$ Area^3 = Specific Volume^2 $\cdot$ Energy,Work,Heat Quantity^2

Velocity^3 $\cdot$ Area = Specific Volume $\cdot$ Power, Radiance Flux

Velocity^2 $\cdot$ Area = Specific Volume^2 $\cdot$ Dynamic Viscosity^2

Velocity^2 $\cdot$ Area $\cdot$ Electric Charge^2 = Current Density^2

Velocity $\cdot$ Electric Charge = Area $\cdot$ Magnetic Field Strength

Velocity $\cdot$ Area $\cdot$ Pressure,Stress = Power, Radiance Flux

Velocity^2 $\cdot$ Dynamic Viscosity^2 = Area $\cdot$ Pressure,Stress^2

Velocity^2 $\cdot$ Energy,Work,Heat Quantity^2 = Area $\cdot$ Power, Radiance Flux^2

Velocity $\cdot$ Area $\cdot$ Dynamic Viscosity = Energy,Work,Heat Quantity

Velocity^4 $\cdot$ Area^3 $\cdot$ Mass Density^2 = Energy,Work,Heat Quantity^2

Velocity^4 $\cdot$ Area $\cdot$ Dynamic Viscosity^2 = Power, Radiance Flux^2

Velocity^3 $\cdot$ Area $\cdot$ Mass Density = Power, Radiance Flux

Velocity^2 $\cdot$ Area $\cdot$ Magnetic Flux Density^2 = Electric Potential Difference^2

Velocity^2 $\cdot$ Magnetic Flux^2 = Area $\cdot$ Electric Potential Difference^2

Velocity^2 $\cdot$ Capacitance^2 $\cdot$ Electric Resistance^2 = Area

Velocity^2 $\cdot$ Capacitance $\cdot$ Inductance = Area

Velocity^2 $\cdot$ Inductance^2 = Area $\cdot$ Electric Resistance^2

$\text{Velocity}^2 \cdot \text{Area} \cdot \text{Mass Density}^2 = \text{Dynamic Viscosity}^2$

$\text{Velocity}^2 \cdot \text{Volume} = \text{Specific Volume} \cdot \text{Energy,Work,Heat Quantity}$

$\text{Velocity}^9 \cdot \text{Volume}^2 = \text{Specific Volume}^3 \cdot \text{Power, Radiance Flux}^3$

$\text{Velocity}^3 \cdot \text{Volume} = \text{Specific Volume}^3 \cdot \text{Dynamic Viscosity}^3$

$\text{Velocity}^3 \cdot \text{Volume} \cdot \text{Electric Charge}^3 = \text{Current Density}^3$

$\text{Velocity}^3 \cdot \text{Electric Charge}^3 = \text{Volume}^2 \cdot \text{Magnetic Field Strength}^3$

$\text{Velocity}^3 \cdot \text{Volume}^2 \cdot \text{Pressure,Stress}^3 = \text{Power, Radiance Flux}^3$

$\text{Velocity}^3 \cdot \text{Dynamic Viscosity}^3 = \text{Volume} \cdot \text{Pressure,Stress}^3$

$\text{Velocity}^3 \cdot \text{Energy,Work,Heat Quantity}^3 = \text{Volume} \cdot \text{Power, Radiance Flux}^3$

$\text{Velocity}^3 \cdot \text{Volume}^2 \cdot \text{Dynamic Viscosity}^3 = \text{Energy,Work,Heat Quantity}^3$

$\text{Velocity}^2 \cdot \text{Volume} \cdot \text{Mass Density} = \text{Energy,Work,Heat Quantity}$

$\text{Velocity}^6 \cdot \text{Volume} \cdot \text{Dynamic Viscosity}^3 = \text{Power, Radiance Flux}^3$

$\text{Velocity}^9 \cdot \text{Volume}^2 \cdot \text{Mass Density}^3 = \text{Power, Radiance Flux}^3$

$\text{Velocity}^3 \cdot \text{Volume} \cdot \text{Magnetic Flux Density}^3 = \text{Electric Potential Difference}^3$

$\text{Velocity}^3 \cdot \text{Magnetic Flux}^3 = \text{Volume} \cdot \text{Electric Potential Difference}^3$

$\text{Velocity}^3 \cdot \text{Capacitance}^3 \cdot \text{Electric Resistance}^3 = \text{Volume}$

$\text{Velocity}^6 \cdot \text{Capacitance}^3 \cdot \text{Inductance}^3 = \text{Volume}^2$

$\text{Velocity}^3 \cdot \text{Inductance}^3 = \text{Volume} \cdot \text{Electric Resistance}^3$

$\text{Velocity}^3 \cdot \text{Volume} \cdot \text{Mass Density}^3 = \text{Dynamic Viscosity}^3$

$\text{Velocity}^2 = \text{Wavenumber}^3 \cdot \text{Specific Volume} \cdot \text{Energy,Work,Heat Quantity}$

$\text{Velocity}^3 = \text{Wavenumber}^2 \cdot \text{Specific Volume} \cdot \text{Power, Radiance Flux}$

$\text{Velocity} = \text{Wavenumber} \cdot \text{Specific Volume} \cdot \text{Dynamic Viscosity}$

Velocity $\cdot$ Electric Charge = Wavenumber $\cdot$ Current Density

Velocity $\cdot$ Wavenumber^2 $\cdot$ Electric Charge = Magnetic Field Strength

Velocity $\cdot$ Pressure,Stress = Wavenumber^2 $\cdot$ Power, Radiance Flux

Velocity $\cdot$ Wavenumber $\cdot$ Dynamic Viscosity = Pressure,Stress

Velocity $\cdot$ Wavenumber $\cdot$ Energy,Work,Heat Quantity = Power, Radiance Flux

Velocity $\cdot$ Dynamic Viscosity = Wavenumber^2 $\cdot$ Energy,Work,Heat Quantity

Velocity^2 $\cdot$ Mass Density = Wavenumber^3 $\cdot$ Energy,Work,Heat Quantity

Velocity^2 $\cdot$ Dynamic Viscosity = Wavenumber $\cdot$ Power, Radiance Flux

Velocity^3 $\cdot$ Mass Density = Wavenumber^2 $\cdot$ Power, Radiance Flux

Velocity $\cdot$ Magnetic Flux Density = Wavenumber $\cdot$ Electric Potential Difference

Velocity $\cdot$ Wavenumber $\cdot$ Magnetic Flux = Electric Potential Difference

Velocity $\cdot$ Wavenumber $\cdot$ Capacitance $\cdot$ Electric Resistance =

Velocity^2 $\cdot$ Wavenumber^2 $\cdot$ Capacitance $\cdot$ Inductance =

Velocity $\cdot$ Wavenumber $\cdot$ Inductance = Electric Resistance

Velocity $\cdot$ Mass Density = Wavenumber $\cdot$ Dynamic Viscosity

Velocity^3 = Specific Volume $\cdot$ Magnetic Field Strength^2 $\cdot$ Electric Resistance

Velocity^2 = Specific Volume $\cdot$ Magnetic Field Strength $\cdot$ Magnetic Flux Density

Velocity^5 = Specific Volume $\cdot$ Frequency^3 $\cdot$ Energy,Work,Heat Quantity

Velocity^5 = Specific Volume $\cdot$ Frequency^2 $\cdot$ Power, Radiance Flux

Velocity^2 = Specific Volume $\cdot$ Frequency $\cdot$ Dynamic Viscosity

Velocity^5 $\cdot$ Energy,Work,Heat Quantity^2 = Specific Volume $\cdot$ Power, Radiance Flux^3

Velocity $\cdot$ Energy,Work,Heat Quantity = Specific Volume^2 $\cdot$ Dynamic Viscosity^3

Velocity $\cdot$ Specific Volume $\cdot$ Energy,Work,Heat Quantity = Kinematic Viscosity^3

Velocity · Specific Volume · Dynamic Viscosity^2 = Power, Radiance Flux

Velocity · Kinematic Viscosity^2 = Specific Volume · Power, Radiance Flux

Velocity · Electric Resistance = Specific Volume · Magnetic Flux Density^2

Velocity^3 · Magnetic Field Strength = Current Density · Frequency^3

Velocity^3 · Electric Charge^3 = Current Density^2 · Magnetic Field Strength

Velocity^3 · Current Density = Magnetic Field Strength · Kinematic Viscosity^3

Velocity^2 · Electric Charge = Current Density · Frequency

Velocity^2 · Capacitance · Magnetic Flux = Current Density

Velocity · Magnetic Field Strength = Frequency^2 · Electric Charge

Velocity · Pressure,Stress = Magnetic Field Strength^2 · Electric Resistance

Velocity · Magnetic Field Strength · Magnetic Flux = Power, Radiance Flux

Velocity^3 · Electric Charge = Magnetic Field Strength · Kinematic Viscosity^2

Velocity · Magnetic Field Strength · Inductance = Electric Potential Difference

Velocity^2 · Dynamic Viscosity = Magnetic Field Strength · Electric Potential Difference

Velocity · Magnetic Flux Density = Magnetic Field Strength · Electric Resistance

Velocity^3 · Mass Density = Magnetic Field Strength^2 · Electric Resistance

Velocity^2 · Mass Density = Magnetic Field Strength · Magnetic Flux Density

Velocity · Dynamic Viscosity = Magnetic Field Strength^2 · Inductance

Velocity^3 · Pressure,Stress = Frequency^3 · Energy,Work,Heat Quantity

Velocity^3 · Pressure,Stress = Frequency^2 · Power, Radiance Flux

Velocity^3 · Dynamic Viscosity = Frequency^2 · Energy,Work,Heat Quantity

Velocity^5 · Mass Density = Frequency^3 · Energy,Work,Heat Quantity

Velocity^3 · Dynamic Viscosity = Frequency · Power, Radiance Flux

Velocity^5 · Mass Density = Frequency^2 · Power, Radiance Flux

Velocity^2 · Magnetic Flux Density = Frequency · Electric Potential Difference

Velocity^2 · Magnetic Flux Density = Frequency^2 · Magnetic Flux

Velocity^2 · Mass Density = Frequency · Dynamic Viscosity

Velocity^3 · Pressure,Stress · Energy,Work,Heat Quantity^2 = Power, Radiance Flux^3

Velocity^3 · Dynamic Viscosity^3 = Pressure,Stress^2 · Energy,Work,Heat Quantity

Velocity^3 · Energy,Work,Heat Quantity = Pressure,Stress · Kinematic Viscosity^3

Velocity^3 · Dynamic Viscosity^2 = Pressure,Stress · Power, Radiance Flux

Velocity · Power, Radiance Flux = Pressure,Stress · Kinematic Viscosity^2

Velocity · Magnetic Flux Density^2 = Pressure,Stress · Electric Resistance

Velocity^2 · Dynamic Viscosity = Pressure,Stress · Kinematic Viscosity

Velocity^3 · Energy,Work,Heat Quantity · Dynamic Viscosity = Power, Radiance Flux^2

Velocity^2 · Energy,Work,Heat Quantity = Power, Radiance Flux · Kinematic Viscosity

Velocity^5 · Energy,Work,Heat Quantity^2 · Mass Density = Power, Radiance Flux^3

Velocity · Energy,Work,Heat Quantity = Dynamic Viscosity · Kinematic Viscosity^2

Velocity · Energy,Work,Heat Quantity · Mass Density^2 = Dynamic Viscosity^3

Velocity · Energy,Work,Heat Quantity = Kinematic Viscosity^3 · Mass Density

Velocity^2 · Electric Charge · Magnetic Flux Density = Power, Radiance Flux

Velocity · Dynamic Viscosity · Kinematic Viscosity = Power, Radiance Flux

Velocity · Dynamic Viscosity^2 = Power, Radiance Flux · Mass Density

Velocity · Kinematic Viscosity^2 · Mass Density = Power, Radiance Flux

Velocity^2 · Magnetic Flux Density · Magnetic Flux = Electric Potential Difference^2

$Velocity^3 \cdot Inductance \cdot Dynamic\ Viscosity = Electric\ Potential\ Difference^2$

$Velocity^2 \cdot Magnetic\ Flux = Electric\ Potential\ Difference \cdot Kinematic\ Viscosity$

$Velocity^2 \cdot Capacitance \cdot Electric\ Resistance = Kinematic\ Viscosity$

$Velocity \cdot Capacitance \cdot Magnetic\ Flux\ Density^2 = Dynamic\ Viscosity$

$Velocity^4 \cdot Capacitance \cdot Inductance = Kinematic\ Viscosity^2$

$Velocity \cdot Electric\ Resistance \cdot Mass\ Density = Magnetic\ Flux\ Density^2$

$Velocity^2 \cdot Inductance = Electric\ Resistance \cdot Kinematic\ Viscosity$

$Velocity^2 \cdot Magnetic\ Flux = Magnetic\ Flux\ Density \cdot Kinematic\ Viscosity^2$

$Area \cdot Frequency^2 = Specific\ Volume \cdot Pressure,Stress$

$Area^5 \cdot Frequency^4 = Specific\ Volume^2 \cdot Energy,Work,Heat\ Quantity^2$

$Area^5 \cdot Frequency^6 = Specific\ Volume^2 \cdot Power,\ Radiance\ Flux^2$

$Area \cdot Frequency = Specific\ Volume \cdot Dynamic\ Viscosity$

$Area^2 \cdot Specific\ Volume \cdot Pressure,Stress^3 = Power,\ Radiance\ Flux^2$

$Area \cdot Pressure,Stress = Specific\ Volume \cdot Dynamic\ Viscosity^2$

$Area \cdot Specific\ Volume \cdot Pressure,Stress = Kinematic\ Viscosity^2$

$Area^5 \cdot Power,\ Radiance\ Flux^4 = Specific\ Volume^2 \cdot Energy,Work,Heat\ Quantity^6$

$Area^3 \cdot Absorbed\ Dose^2 = Specific\ Volume^2 \cdot Energy,Work,Heat\ Quantity^2$

$Area \cdot Specific\ Volume^2 \cdot Dynamic\ Viscosity^4 = Energy,Work,Heat\ Quantity^2$

$Area \cdot Kinematic\ Viscosity^4 = Specific\ Volume^2 \cdot Energy,Work,Heat\ Quantity^2$

$Area^2 \cdot Absorbed\ Dose^3 = Specific\ Volume^2 \cdot Power,\ Radiance\ Flux^2$

$Area \cdot Power,\ Radiance\ Flux^2 = Specific\ Volume^4 \cdot Dynamic\ Viscosity^6$

$Area \cdot Specific\ Volume^2 \cdot Power,\ Radiance\ Flux^2 = Kinematic\ Viscosity^6$

$Area^5 = Specific\ Volume^2 \cdot Electric\ Charge^4 \cdot Inductance^2$

$Area = Specific\ Volume^2 \cdot Capacitance^2 \cdot Magnetic\ Flux\ Density^4$

$Area^5 = Specific\ Volume^2 \cdot Capacitance^2 \cdot Magnetic\ Flux^4$

$Area \cdot Absorbed\ Dose = Specific\ Volume^2 \cdot Dynamic\ Viscosity^2$

$Area \cdot Frequency \cdot Electric\ Charge = Current\ Density$

$Area^3 \cdot Pressure,Stress^2 = Current\ Density^2 \cdot Magnetic\ Flux\ Density^2$

$Area^7 \cdot Pressure,Stress^2 = Current\ Density^4 \cdot Inductance^2$

$Area^5 \cdot Pressure,Stress^2 = Current\ Density^2 \cdot Magnetic\ Flux^2$

$Area^2 \cdot Energy,Work,Heat\ Quantity = Current\ Density^2 \cdot Inductance$

$Area \cdot Energy,Work,Heat\ Quantity = Current\ Density \cdot Magnetic\ Flux$

$Area \cdot Power,\ Radiance\ Flux = Current\ Density \cdot Electric\ Potential\ Difference$

$Area^2 \cdot Power,\ Radiance\ Flux = Current\ Density^2 \cdot Electric\ Resistance$

$Area \cdot Electric\ Charge^2 \cdot Absorbed\ Dose = Current\ Density^2$

$Area \cdot Electric\ Potential\ Difference = Current\ Density \cdot Electric\ Resistance$

$Area^2 \cdot Magnetic\ Flux\ Density = Current\ Density \cdot Inductance$

$Area \cdot Magnetic\ Flux = Current\ Density \cdot Inductance$

$Area \cdot Magnetic\ Field\ Strength^2 = Frequency^2 \cdot Electric\ Charge^2$

$Area \cdot Pressure,Stress^2 = Magnetic\ Field\ Strength^4 \cdot Inductance^2$

$Area \cdot Pressure,Stress = Magnetic\ Field\ Strength \cdot Magnetic\ Flux$

$Area^3 \cdot Magnetic\ Field\ Strength^2 \cdot Magnetic\ Flux\ Density^2 = Energy,Work,Heat\ Quantity^2$

$Area \cdot Magnetic\ Field\ Strength^2 \cdot Inductance = Energy,Work,Heat\ Quantity$

$Area \cdot Magnetic\ Field\ Strength^2 \cdot Magnetic\ Flux^2 = Energy,Work,Heat\ Quantity^2$

$Area \cdot Magnetic\ Field\ Strength^2 \cdot Electric\ Potential\ Difference^2 = Power,\ Radiance\ Flux^2$

Area $\cdot$ Magnetic Field Strength^2 $\cdot$ Electric Resistance = Power, Radiance Flux

Area^2 $\cdot$ Magnetic Field Strength^2 = Electric Charge^2 $\cdot$ Absorbed Dose

Area^3 $\cdot$ Magnetic Field Strength^2 = Electric Charge^2 $\cdot$ Kinematic Viscosity^2

Area $\cdot$ Magnetic Field Strength^2 $\cdot$ Electric Resistance^2 = Electric Potential Difference^2

Area $\cdot$ Magnetic Flux Density^2 = Magnetic Field Strength^2 $\cdot$ Inductance^2

Area $\cdot$ Magnetic Field Strength^2 $\cdot$ Inductance^2 = Magnetic Flux^2

Area^3 $\cdot$ Frequency^2 $\cdot$ Pressure,Stress^2 = Power, Radiance Flux^2

Area $\cdot$ Frequency^2 $\cdot$ Mass Density = Pressure,Stress

Area^3 $\cdot$ Frequency^2 $\cdot$ Dynamic Viscosity^2 = Energy,Work,Heat Quantity^2

Area^5 $\cdot$ Frequency^4 $\cdot$ Mass Density^2 = Energy,Work,Heat Quantity^2

Area^3 $\cdot$ Frequency^4 $\cdot$ Dynamic Viscosity^2 = Power, Radiance Flux^2

Area^5 $\cdot$ Frequency^6 $\cdot$ Mass Density^2 = Power, Radiance Flux^2

Area $\cdot$ Frequency $\cdot$ Magnetic Flux Density = Electric Potential Difference

Area $\cdot$ Frequency $\cdot$ Mass Density = Dynamic Viscosity

Area^2 $\cdot$ Pressure,Stress^2 $\cdot$ Absorbed Dose = Power, Radiance Flux^2

Area^3 $\cdot$ Pressure,Stress^4 = Power, Radiance Flux^2 $\cdot$ Dynamic Viscosity^2

Area $\cdot$ Pressure,Stress^2 $\cdot$ Kinematic Viscosity^2 = Power, Radiance Flux^2

Area^2 $\cdot$ Pressure,Stress^3 = Power, Radiance Flux^2 $\cdot$ Mass Density

Area^3 $\cdot$ Pressure,Stress^2 = Electric Charge^2 $\cdot$ Electric Potential Difference^2

Area^3 $\cdot$ Pressure,Stress^2 $\cdot$ Capacitance^2 = Electric Charge^4

Area^3 $\cdot$ Pressure,Stress^2 = Electric Potential Difference^4 $\cdot$ Capacitance^2

Area $\cdot$ Magnetic Flux Density^4 = Pressure,Stress^2 $\cdot$ Inductance^2

$$\text{Area}^3 \cdot \text{Pressure,Stress}^2 \cdot \text{Inductance}^2 = \text{Magnetic Flux}^4$$

$$\text{Area} \cdot \text{Pressure,Stress}^2 = \text{Absorbed Dose} \cdot \text{Dynamic Viscosity}^2$$

$$\text{Area} \cdot \text{Pressure,Stress} = \text{Dynamic Viscosity} \cdot \text{Kinematic Viscosity}$$

$$\text{Area} \cdot \text{Pressure,Stress} \cdot \text{Mass Density} = \text{Dynamic Viscosity}^2$$

$$\text{Area} \cdot \text{Pressure,Stress} = \text{Kinematic Viscosity}^2 \cdot \text{Mass Density}$$

$$\text{Area} \cdot \text{Power, Radiance Flux}^2 = \text{Energy,Work,Heat Quantity}^2 \cdot \text{Absorbed Dose}$$

$$\text{Area}^3 \cdot \text{Power, Radiance Flux}^2 \cdot \text{Dynamic Viscosity}^2 = \text{Energy,Work,Heat Quantity}^4$$

$$\text{Area} \cdot \text{Power, Radiance Flux} = \text{Energy,Work,Heat Quantity} \cdot \text{Kinematic Viscosity}$$

$$\text{Area}^5 \cdot \text{Power, Radiance Flux}^4 \cdot \text{Mass Density}^2 = \text{Energy,Work,Heat Quantity}^6$$

$$\text{Area}^2 \cdot \text{Pressure,Stress}^2 \cdot \text{Absorbed Dose} = \text{Power, Radiance Flux}^2$$

$$\text{Area}^3 \cdot \text{Pressure,Stress}^4 = \text{Power, Radiance Flux}^2 \cdot \text{Dynamic Viscosity}^2$$

$$\text{Area} \cdot \text{Pressure,Stress}^2 \cdot \text{Kinematic Viscosity}^2 = \text{Power, Radiance Flux}^2$$

$$\text{Area}^2 \cdot \text{Pressure,Stress}^3 = \text{Power, Radiance Flux}^2 \cdot \text{Mass Density}$$

$$\text{Area}^3 \cdot \text{Pressure,Stress}^2 = \text{Electric Charge}^2 \cdot \text{Electric Potential Difference}^2$$

$$\text{Area}^3 \cdot \text{Pressure,Stress}^2 \cdot \text{Capacitance}^2 = \text{Electric Charge}^4$$

$$\text{Area}^3 \cdot \text{Pressure,Stress}^2 = \text{Electric Potential Difference}^4 \cdot \text{Capacitance}^2$$

$$\text{Area} \cdot \text{Magnetic Flux Density}^4 = \text{Pressure,Stress}^2 \cdot \text{Inductance}^2$$

$$\text{Area}^3 \cdot \text{Pressure,Stress}^2 \cdot \text{Inductance}^2 = \text{Magnetic Flux}^4$$

$$\text{Area} \cdot \text{Pressure,Stress}^2 = \text{Absorbed Dose} \cdot \text{Dynamic Viscosity}^2$$

$$\text{Area} \cdot \text{Pressure,Stress} = \text{Dynamic Viscosity} \cdot \text{Kinematic Viscosity}$$

$$\text{Area} \cdot \text{Pressure,Stress} \cdot \text{Mass Density} = \text{Dynamic Viscosity}^2$$

$$\text{Area} \cdot \text{Pressure,Stress} = \text{Kinematic Viscosity}^2 \cdot \text{Mass Density}$$

$$\text{Area} \cdot \text{Power, Radiance Flux}^2 = \text{Energy,Work,Heat Quantity}^2 \cdot \text{Absorbed Dose}$$

$Area^3 \cdot Power, Radiance\ Flux^2 \cdot Dynamic\ Viscosity^2 = Energy, Work, Heat\ Quantity^4$

$Area \cdot Power, Radiance\ Flux = Energy, Work, Heat\ Quantity \cdot Kinematic\ Viscosity$

$Area^5 \cdot Power, Radiance\ Flux^4 \cdot Mass\ Density^2 = Energy, Work, Heat\ Quantity^6$

$Area^2 \cdot Magnetic\ Flux\ Density^2 = Energy, Work, Heat\ Quantity \cdot Inductance$

$Area^2 \cdot Absorbed\ Dose \cdot Dynamic\ Viscosity^2 = Energy, Work, Heat\ Quantity^2$

$Area^3 \cdot Absorbed\ Dose^2 \cdot Mass\ Density^2 = Energy, Work, Heat\ Quantity^2$

$Area \cdot Dynamic\ Viscosity^2 \cdot Kinematic\ Viscosity^2 = Energy, Work, Heat\ Quantity^2$

$Area \cdot Dynamic\ Viscosity^4 = Energy, Work, Heat\ Quantity^2 \cdot Mass\ Density^2$

$Area \cdot Kinematic\ Viscosity^4 \cdot Mass\ Density^2 = Energy, Work, Heat\ Quantity^2$

$Area \cdot Absorbed\ Dose^2 \cdot Dynamic\ Viscosity^2 = Power, Radiance\ Flux^2$

$Area^2 \cdot Absorbed\ Dose^3 \cdot Mass\ Density^2 = Power, Radiance\ Flux^2$

$Area \cdot Power, Radiance\ Flux^2 = Dynamic\ Viscosity^2 \cdot Kinematic\ Viscosity^4$

$Area \cdot Power, Radiance\ Flux^2 \cdot Mass\ Density^4 = Dynamic\ Viscosity^6$

$Area \cdot Power, Radiance\ Flux^2 = Kinematic\ Viscosity^6 \cdot Mass\ Density^2$

$Area \cdot Magnetic\ Flux\ Density = Electric\ Charge \cdot Electric\ Resistance$

$Area^3 \cdot Dynamic\ Viscosity^2 = Electric\ Charge^4 \cdot Electric\ Resistance^2$

$Area \cdot Dynamic\ Viscosity^2 = Electric\ Charge^2 \cdot Magnetic\ Flux\ Density^2$

$Area^5 \cdot Mass\ Density^2 = Electric\ Charge^4 \cdot Inductance^2$

$Area^3 \cdot Dynamic\ Viscosity^2 = Electric\ Charge^2 \cdot Magnetic\ Flux^2$

$Area \cdot Magnetic\ Flux\ Density^2 \cdot Absorbed\ Dose = Electric\ Potential\ Difference^2$

$Area \cdot Electric\ Potential\ Difference^2 = Absorbed\ Dose \cdot Magnetic\ Flux^2$

$Area \cdot Electric\ Potential\ Difference = Magnetic\ Flux \cdot Kinematic\ Viscosity$

Area = Capacitance^2 · Electric Resistance^2 · Absorbed Dose

Area = Capacitance · Electric Resistance · Kinematic Viscosity

Area · Mass Density^2 = Capacitance^2 · Magnetic Flux Density^4

Area = Capacitance · Inductance · Absorbed Dose

Area^2 = Capacitance · Inductance · Kinematic Viscosity^2

Area^5 · Mass Density^2 = Capacitance^2 · Magnetic Flux^4

Area · Magnetic Flux Density^4 = Electric Resistance^2 · Dynamic Viscosity^2

Area · Electric Resistance^2 = Inductance^2 · Absorbed Dose

Area · Electric Resistance = Inductance · Kinematic Viscosity

Area^3 · Electric Resistance^2 · Dynamic Viscosity^2 = Magnetic Flux^4

Area · Absorbed Dose · Mass Density^2 = Dynamic Viscosity^2

Electric Potential Difference · Capacitance · Electric Resistance = Magnetic Flux

Electric Potential Difference^2 · Capacitance · Inductance = Magnetic Flux^2

Electric Potential Difference · Inductance = Electric Resistance · Magnetic Flux

Electric Potential Difference^2 = Magnetic Flux Density · Absorbed Dose · Magnetic Flux

Electric Potential Difference · Mass Density = Magnetic Flux Density · Dynamic Viscosity

Electric Potential Difference^4 = Inductance^2 · Absorbed Dose^3 · Dynamic Viscosity^2

Electric Potential Difference · Kinematic Viscosity = Absorbed Dose · Magnetic Flux

Capacitance · Electric Resistance · Absorbed Dose = Kinematic Viscosity

Capacitance^2 · Magnetic Flux Density^4 · Absorbed Dose = Dynamic Viscosity^2

Capacitance^2 · Magnetic Flux Density^5 = Magnetic Flux · Mass Density^2

Capacitance · Inductance · Absorbed Dose^2 = Kinematic Viscosity^2

Electric Resistance^2 · Absorbed Dose · Mass Density^2 = Magnetic Flux Density^4

285

Electric Resistance$^2 \cdot$ Dynamic Viscosity^2 = Magnetic Flux Density$^3 \cdot$ Magnetic Flux

Electric Resistance $\cdot$ Kinematic Viscosity = Inductance $\cdot$ Absorbed Dose

Magnetic Flux Density $\cdot$ Kinematic Viscosity^2 = Absorbed Dose $\cdot$ Magnetic Flux

APPENDIX B: SEQUENTIALLY INFERRED SET COUNTING WITH ELEMENT RESTRICTIONS AND THE UNIFICATION OF LINEAR, TRIANGULAR, TETRAHEDRAL AND HIGHER ORDER POLYTOPE NUMBER SEQUENCES

B.1. Counting with Rigid Rules

Counting through sets is a common computer problem. Keeping track of how many operations are completed compared to the predicted number of operations becomes an ordeal if appropriate math is not used. Imagine that the following rules are applied to a particular set of numbers:

1.) There is a number of slots a and a maximum number m which is allowed.

2.) The last slot is the only slot which can contain m, while the second to last slot has a maximum value of m-1, the third to last slot m-2 and so on.

3.) The first slot's maximum value is m-a+1

4.) Any number in a slot may not be equal to or larger than any number to its right.

5.) If any slot reaches its maximum value, the next count increases the slot to its left by 1, with numbers right of that numbered sequentially.

6.) The counting stops if all maximum values are reached.

For example, let us try this with a = 3 and m = 5. The sets can then be defined as follows:

$$1\ 2\ 3$$

$$1\ 2\ 4$$

$$1\ 2\ 5$$

$$1\ 3\ 4$$

$$1\ 3\ 5$$

$$1\ 4\ 5$$

$$2\ 3\ 4$$

288

2 3 5

2 4 5

3 4 5

In a couple of minutes, keeping a = 3 and increasing m from 3 to 10 in this manner yields the following counts as a sequence of numbers.

1,4,10,20,35,56,84,120…

This sequence is easily resolved into a function.   However, when a increases, the answer becomes a little more complicated.   For instance, imagine that a = 40 and m = 87, finding that the number of acceptable sets is 9,988,677,302,355,003,038,019,660 is problematic for the above method.

The sequence for a = 4:   1,5,15,35,70,126,210,330,495,715…

The sequence for a = 5:   1,6,21,56,126,252,462,792,1287,2002…

With a little work, a particular pattern seems to appear.   This seems to be described adequately by the following product series.

$$Z(m,a) = \frac{\prod\limits_{n=0}^{a-1} m - n}{a!} , \ m \geq a \ldots \infty$$

Computationally, a progress tracker can be easily implemented in a system if you know the number of steps needed to complete the task at hand and the code contains a counter determining the number of steps already taken. This was the original intent of the work. The expansion was too easy not to generalize the formula. In fact, with a little reduction, a well known combination formula appears.

A quick check will show that the sequences of a=4 and a=5 are strongly (if not exactly) related to the binomial coefficients mentioned by Sloan [10] and may be presented in a more familiar notation by Kim [8]. For a = 3 the numbers are related to tetrahedral numbers, defined by Chou and Deng [9]. For a = 2, the numbers match triangle numbers [7]. Of course, for a = 1, the sequence is simple linear counting of integers n=1...∞. So the general equation given seems to provide formulae for all of these categories. These counting concepts are illustrated as follows

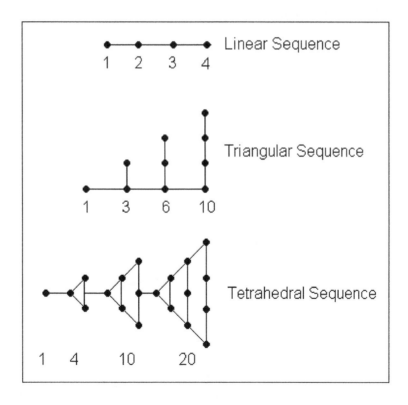

It appears that the general equation discussed may be the unifying formula for all polytope number sequences of this kind. As this was generally stumbled upon, the author leaves it to more versed mathematicians to prove this properly or disprove it by counterexample.

B.2. Counting with Relaxed Rules

Things get interesting when you eliminate certain restrictions, take the following rule set.

1.) There is a number of slots a and a maximum number m which is allowed.

2.) Any number in a slot may not be larger than any number to its right.

3.) If any slot reaches its maximum value, the next count increases the slot to its left by 1, with numbers right of that numbered sequentially.

4.) The counting stops when all slots are at a maximum of m.

For example, let us try this with a = 3 and m = 5. The sets can then be defined as follows:

1 2 3

1 2 4

1 2 5

1 3 4

1 3 5

1 4 5

1 5 5

2 3 4

2 3 5

2 4 5

291

$$2\ 5\ 5$$

$$3\ 4\ 5$$

$$3\ 5\ 5$$

$$4\ 5\ 5$$

$$5\ 5\ 5$$

Keeping a = 3 and increasing m from 3 to 10 in this manner yields the following counts as a sequence of numbers.

4,8,15,26,42,64,93,130…

For a = 1: 1,2,3,4,5,6,7,8,9,10…

For a = 2: 2,4,7,11,16,22,29,37,46…

For a = 3: 4,8,15,26,42,64,93,130…

For a = 4: 8,16,31,57,99,163,256,386,562,794…

.

.

.

This leads to a recursive formula for any a.   Let Z be the number of possible combinations using the above rules.   Math purists beware:

$$Z_a(m) = 1 + \sum_{m=1}^{m'-1} Z_{a-1}(m) \ , \ m \geq a \ldots \infty$$

Here, $Z_1(m) = m$, and the rest is simply a matter of plugging it in. Replace the dummy variable m' with m upon each iteration.

Some of the patterns found in the above formula are interesting to note. For instance, take any of the answers and multiply it by a!, summate the coefficients and the answer seems to consistently be a! again.   For example, take a = 3:

$$Z_3(m) = m^3/6 - m^2/2 + 4m/3$$

$$(3!)(Z_3) = m^3 - 3m^2 + 8m$$

$$1-3+8 = 6 = 3!$$

As another example, a = 5:

$$Z_5(m) = m^5/120 - m^4/12 + 11m^3/24 - 11m^2/12 + 23m/15$$

$$(5!)(Z_5) = m^5 - 10m^4 + 55m^3 - 110m^2 + 184m$$

$$1-10+55-110+184 = 120 = 5!$$

Determining the number of counts in a loop construct such as the ones dealt with here make progress tracking easier to deal with, especially in a program which may be expanded to include more than one term in such specialized loops.

The first sequences discussed seem to provide a unifying formula for all polytope number sequences. While there was no specific mention of the second group of sequences in the Sloan et al. online sequence database [1] or in any research the author has done.   The second sequences were a side effect of the formula discovered in the fist part and included for completeness.

APPENDIX C: EMPIRICAL EVIDENCE FOR A NEW LINEAR RELATIONSHIP BETWEEN FORCE AND ENERGY FROM 10-50 NEWTONS

The following is an example of an experimental setup performed at the University of Central Florida to come up with some evidence that a relationship may exist within the Permutanomicon that has not been thouroughly investigated before now.

The following persons (besides the author) were involved with the experiment either as participants or interested advisors: Ari Litwin, Ray Ramotar, Gabriel Braunstein

---

Using a force sensor the authors shows that there is a correlation between the variables of an unfamiliar energy formula from 10-50N. The variables of force and energy are linearly related through the area of interaction in a classical manner. The experiment used to investigate this relationship is simple and suitable for verification by most stocked laboratories.

Expert systems in physics have come a long way in recent years. Recently, a method was developed to exhaustively search permutations of physical quantities[1]. In order to determine the potential of this expert system, a formula was selected to test. The formula selected was

$$E \quad \sqrt{F^2 A} \tag{1}$$

where E is the energy, F is the force and A is the area. Validating by dimensional analysis is an important step before testing the results. The authors decided to use a common method employed in the literature[2]. It is easily verified that this equation is unitarily correct but lacks a correction constant. Buckingham's theorem assures that this formula's form is independent of the unit system used[3,4].

Using a PASCO force sensor with a range from 10-50N, the authors constructed an experiment to test the validity of the formula. A cart with the force sensor was allowed to roll down an inclined ramp and collide with a solid object. The area of the force sensor's head remained constant, and the height of the elevated end of the ramp was adjustable. During the collision a force is measured across the sensor head as in the classical problem of the inclined ramp, shown in figure 1.

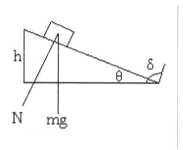

Figure 1: The general experimental setup. A cart travels down a ramp and collides with a solid object. The force sensor has an area witch is constant throughout repeated collisions.

The energy of the system during a collision is the classical formula E=mgh, which is the potential energy of the cart and force sensor at rest at the top of the ramp. This potential energy is converted into kinetic energy upon reaching the bottom of the ramp.

Plugging this into equation 1 shows that there should be a linear relationship between the force and the height that the ramp is inclined to. The data is then expected to pass through the origin.

$$mgh \quad F\sqrt{A} \tag{2}$$

To adjust the energy of the system, the height is changed. Two available force sensor heads were used, a rubber (compressible) head (area = 0.94 cm^2) and a metallic (non-compressible) head (area = 0.28 cm²) with diameters measured by dial caliper. The reason for this was to test the validity of the formula in both elastic and inelastic interactions.

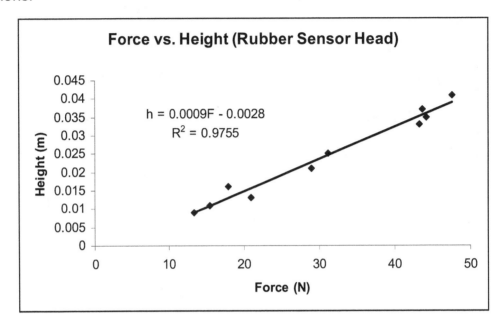

Figure 2: The force versus the height of the impact measurements. The height was modified to create different energies in the impact. The linear trend in the range is apparent on the rubber force sensor head. The intercept is nearly at the origin.

The rubber force sensor displays linear behavior. There was visibly noticeable (and unmeasured) distortion of the rubber on impact. A linear behavior is still observed, leading to the belief that energy dissipation was not significant enough to distort the results. Lowering the height could only be done until friction was enough of a factor to skew the results of the collision at the end of the track. Lowering the track beyond this point requires altering the expected energy to a function in terms of friction.

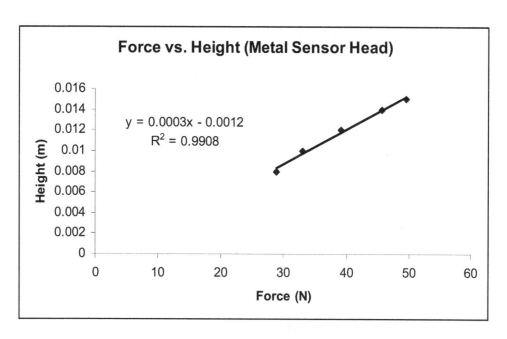

Figure 3: The force versus the height of the impact measurements. The linear trend in the range is apparent on the metal force sensor head. The intercept is nearly at the origin as well.

The metal force sensor also displays linear behavior. There was no noticeable distortion of the metal on impact. The linear trend is also observed. The intensity of the impact was such that the range of the equipment (-50 N to 50 N) was reached in short order for this force sensor head. This seriously limited the amount of measurements that could be done with what was available. Both trends had intercepts that closely crossed the origin.

This experiment was performed with equipment that has limited range in Newtons. Future work should be done with a force sensor that has a larger range. The ramp should be mounted to a wall to avoid excess vibrations that was not completely isolated from the system used in this experiment. The height modifications were done with a bolt to lock the ramp in place. Using a dial mechanism to adjust the height would allow for higher resolution to be obtained. To reduce friction; switching from a track using wheels to a more stable air track setup would be an improvement.

In conclusion, it has been shown that there may be a new relation between energy to force and area. The relationship is shown to be linear from 10-50 Newtons. The results are presented here to encourage further investigation with better equipment.

References for this work:

[1] This manual.

[2] Donald H. Menzel. "Mathematical Physics" Dover 1953.

[3] Buckingham, E. "On Physically Similar Systems: Illustrations of the Use of Dimensional Equations." Phys. Rev. 4, 345-376, 1914.

[4] Buckingham, E. "Model Experiments and the Form of Empirical Equations." Trans. ASME 37, 263, 1915.

APPENDIX D: EXCERPT FROM AN EMAIL BETWEEN MYSELF AND ANOTHER PHYSICIST CONCERNING REVERSE TIME, THE MAXIMUM UNIVERSAL FORCE CONCEPT AND AN ABUNDANCE OF POSSIBLE SOLUTIONS TO SOME COSMOLOGICAL PROBLEMS

First a little history into what inspired me. I was working on an idea of mine over the Christmas Holiday 2004 on a program (you probably remember that I am completely enslaved by computers) meant to derive physics independently of human interaction. I shall put a feather in my cap here and say that I was moderately successful and am now in possession of 6000+ physics formulae my laptop discovered. Some are well known ($E=mc^2$, $F=ma$) and others are not so well known ($F = c^4/G$).

It is this last equation ($F=c^4/G$) that grabbed my attention. It looked like a mighty large force, a maximum force so to speak. I had no clue what it was and so began a literature search and found the following sources.

G.W. Gibbons. 2002, Foundations of Physics, 32, 12, 1891

and

Christoph Schiller. 2005, Motion Mountain: The Physics Textbook, http://www.motionmountain.net/C19-LIMI.pdf

They both mention this force, and call it something like the Universal Maximum Force. So here I had my computer find something that was current in the literature, so I was very excited.

Gibbons and Schiller both mention that this is an upper limit on force but barely do anything with it, like they are afraid to delve too far into the matter. But Schiller mentions something about this force possibly being part of the surface area of a black hole. So I start looking up more information and find nothing.

My interest is peaked, and so I start playing with it on paper with the constant provided.

$$F_{max} = \frac{c^4}{4G} = 3.25 \times 10^{43} N \qquad (1)$$

This can be derived by substituting the Schwarzschild radius into Newton's gravitational formula. Likely, this is what inspires Schiller's comment. But they practically stop there. So I start playing with a classical problem, a particle acted upon by F_{max}. The results are mind candy but really uninteresting to me. I write them here for completion.

$$a(t) = \frac{c^4}{4Gm} \qquad (2)$$

$$v(t) = \frac{c^4 t}{4Gm} + \alpha \qquad (3)$$

$$r(t) = \frac{c^4 t^2}{8Gm} + \beta \qquad (4)$$

Of course, any mass acted upon by a force of this magnitude will exceed the speed of light. To ensure this does not occur, F_{max} may only act on the mass for a maximum time.

$$t_{max} = \frac{4Gm}{c^3} \qquad (5)$$

As m→0, t_{max}→0, as expected. Schiller also mentions that Newton's second law of motion may come into play.

$$g_{bh} = \frac{c^4}{4Gm_{bh}} \qquad (6)$$

This can be derived two ways, one mentioned before.

$$F = \frac{Gm^2}{r_s^2} \text{ with } r_s = \frac{2Gm}{c^2} \qquad (7)$$

$$F = \frac{mc^2}{2r_s} \text{ with } r_s = \frac{2Gm}{c^2} \qquad (8)$$

But something seems seriously wrong with equation 6. As you add more mass, the surface gravity goes down. So my first thought was "huh?" My PI just had dinner with Fernando Atrio Barandela, who mentioned that surface gravity on a black hole is not well defined because it is a Newtonian construct.

But even if undefined, a black hole should still have a surface gravity. Otherwise there would really be no attraction to the surface. Even if the gravity can be described in terms of the curvature of space, it should still have a value with the units of m s^{-2}. This is probably relative to the observer, but still it should exist.

So considering this, I still have a problem, gravity goes down as mass goes up. I call this the "missing g" paradox. Taken to the next logical conclusion, if the mass goes to infinite, gravity goes to 0.

At about this point, I gave up on the theory being sound at all. But morbid curiosity compelled me to continue the calculations as if this were not a problem. Those mind-numbing problems I remember in your class instilled this habit in me which compels me to explore all possibilities before tossing a couple of forests into the waste bin.

So I thought, would this change if I took a test particle and considered relative speeds approaching a black hole. So I applied the Lorentz transformation to the test mass and assumed the black hole was standing still. This is probably simplistic, but was my initial attempt to consider relativity.

$$F = \frac{Gm_t m_{bh}\sqrt{1-(v/c)^2}}{r^2}\bigg| r = r_s = \frac{2Gm}{c^2} \tag{9}$$

$$F = \frac{m_t\sqrt{1-(v/c)^2}}{4m_{bh}}\frac{c^4}{G} \tag{10}$$

$$g_{bh} = \frac{\sqrt{1-(v/c)^2}}{4m_{bh}}\frac{c^4}{G} \tag{11}$$

Here, the form remains the same as equation 6. But after the test particle enters the event horizon, I assumed the velocity would go to zero relative to the black hole and then obviously equation 6 returns.

But this still does not help. If mass goes to infinity, the gravity still plummets to zero. This means as the black hole accretes more matter, it eventually cannot hold itself together if it is spinning and would tear apart.

So then I decide something is missing, and I pour over my list of computer generated formulae for a clue to my problem. After a little searching I find this.

$$F = \frac{kmc^2}{V^{1/3}} \qquad \text{where k is a constant.} \tag{12}$$

$$F = \frac{kmv^2}{V^{1/3}} \qquad \text{where k is a constant.} \tag{13}$$

Intuitively, equations 12 and 13 make sense. Imagine an object being struck by a mass m going a velocity v. The object should experience less of a force if m is spread out over a larger volume. While a brick may make a substantial dent in the ground if dropped from above, if turned into powder and dropped on the same ground spread out over a few meters, the effect would be substantially less.

So from here, I derived the maximum force constant in the following way. I first take Newton's equation.

303

$$F = \frac{GmM}{r^2} \tag{14}$$

One can square equation 12.

$$F^2 = \frac{km_\psi^2 c^4}{V^{2/3}} \text{ where k is a constant.} \tag{15}$$

Now, dividing equation 15 by equation 14, the maximum force principle can be extracted with all extraneous units being cancelled out. m_ψ is denoted only for the purposes of keeping track of each mass and the formulae from which it came.

$$F = \frac{kr^2 m_\psi^2}{mMV^{2/3}} \frac{c^4}{G} = \Omega \frac{c^4}{G} \tag{16}$$

Ω is a unit-less variable and would need to be experimentally determined, as it depends on the system being investigated. Now the task of cleaning up the mass in Ω begins. This helped me realize that there may be more to the maximum force principle than Gibbons or Schiller wrote. So I returned to the black hole case.

Assuming there is only one object in a system undergoing a maximum force, as in a black hole; all of the masses cancel out. If the object is spherical, Ω assumes the following form. (Note that if k = 1, the 1/4 constant of Shiller and Gibbons appears, but with another constant.)

$$\Omega = \frac{k}{4}\left(\frac{6}{\pi}\right)^{2/3} \tag{17}$$

$$F = \frac{k}{4}\left(\frac{6}{\pi}\right)^{2/3} \frac{c^4}{G} \tag{18}$$

Setting $F_{max} = m_{bh}g_{bh}$, one can now extract a number of interesting relations. Solving for m_{bh} or G yields a formula that, when coupled with the Schwarzschild radius gives the radius in terms of the surface gravity of a black hole. Setting r_s to the Schwarzschild radius, m_s in terms of the surface gravity can also be found.

$$r_s = \frac{kc^2}{2g_s}\left(\frac{6}{\pi}\right)^{2/3} \tag{19}$$

$$m_s = \frac{kc^4}{4Gg_s}\left(\frac{6}{\pi}\right)^{2/3} \tag{20}$$

I was not certain how important these would be, but I kept them. I am also aware that the volume of a black hole is not as simple as I presumed according to the literature, but here is a justification of my methods.

Let's start with the area of a sphere: $4/3\,\pi\,r^3$
If you take the derivative with respect to r: $4\,\pi\,r^2$

This is the area of a sphere. Now if you take the Bekenstein-Hawking area of a black hole, it is obviously of the form $8\,\pi\,r^2$. Integrating with respect to r should return the volume, $8/3\,\pi\,r^3$. Here, r is the Schwarzschild radius. I assumed that the volume was that of a sphere, not two spheres.

But assuming that equation 16 is divided by the Bekenstein-Hawking formula, one should obtain the pressure on the event horizon of a Schwarzschild black hole.

$$P_s = \frac{k}{32 m_s^2}\left(\frac{6^{2/3}}{\pi^{5/3}}\right)\frac{c^8}{G^3} \tag{21}$$

Naturally, when equations 16 and 17 are solved for g_s, set equal and solved for r_s, the Schwarzschild radius reappears. Setting $k = (6/\pi)^{2/3}$ returns equations for the Schwarzschild black hole in terms of the surface gravity.

$$r_s = \frac{c^2}{2 g_s} \tag{22}$$

$$m_s = \frac{c^4}{4 G g_s} \tag{23}$$

Notice that equation 23 is equivalent to equation 6. So things have come full circle as one would expect. This is a good thing, but highly annoying again because stubbornly, the black hole "missing g" paradox has returned to haunt me.

Look closely at equation 20 and switch out the roles of mass and gravity.

$$g_s = \frac{kc^4}{4 G m_s}\left(\frac{6}{\pi}\right)^{2/3} \tag{24}$$

So at about this point I'm thinking, either I'm completely wrong or this is the way things work. So what could I assume to make this paradox go away or somehow become useful. The first thing I came up with was that internal to the event horizon (I know, nobody knows what's happening in there, but bear with me. The story gets fun from this point) the laws of physics must be reversed and equally opposite to that of the laws outside of the event horizon.

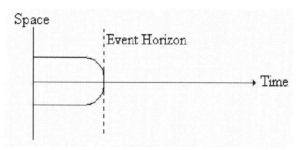

Figure 3: Internal to the event horizon, time reverses.

At the initial onset of a black hole, at the Oppenheimer-Volkoff limit, there would be a boundary where the laws external to the black hole would perfectly balance the laws internally. This mass is what is required to create the minimum event horizon. As energy enters into the black hole (as either light or mass), it crosses the event horizon and enters into a quickly spinning body. In book knowledge physics, whenever an object is on a quickly spinning body, the object flies off and escapes if gravity is not capable of holding it. However, time is reversed, and so the body reacts, as it should, only backwards. This body falls into the black hole, not away from it.

So now, as mass enters into a black hole, it is expected to fall into the black hole with this reverse time taken into account. It would be expected to lift off as the diminishing gravity claims it would in equations 21 and 6. However, the opposite of lifting off is landing, and so the object continues to gather mass centrally. What happens as this mass accumulates? Since everything is attempting to fall out of the black hole due to centripetal acceleration and ever decreasing gravity, everything falls inward, and an object forms instead of tearing apart, a seed object, the singularity, shown in figure 4. This fits in well with reverse time, as it appears the black hole is an object that tore itself apart. Since it is traveling backwards in time, this disruption turns into formation.

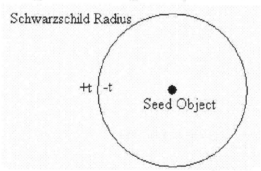

Figure 4: The "seed object" which attempts to form in reverse time.

Here is what's interesting about this concept. Internal to the event horizon, everything reverses in time, and that includes particles. Radioactive elements return to a parent state, Deuterium forms, everything gets younger as time goes on and eventually, should become primordial matter.

There is no reason to believe that there are not other limits where time reverses itself once more, and so another event horizon of maximum force $-F_{max}$ appears. Time flows forward, but the even horizons keep everything shielded.

Now time warp layers are formed, shown in figure 5.

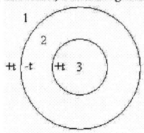

Figure 5: A black hole with two time layers. The event horizon, the outer circle is the boundary between our (forward) time and the first internal (reverse) time layer. The second circle represents another reversal into forward time.

How many of these time layers form? I have no clue, maybe only two, maybe an infinite amount. But the point is that in this model, each layer has a force which is reversed to the layer before it. Objects which pass though the event horizons are eventually forced inwards to pass through all layers.

As each layer is formed, new Schwarzschild radii are formed, which stay far enough apart that they do not cancel each other out. This is assumed because of Penrose's idea "no naked singularities."

$$r_s = \frac{2GM}{c^2} \text{ and } r_{s2} = \frac{2kGM}{c^2} \text{ where k} < 1 \tag{25}$$

These two radii can only be equal when m→∞. When these two radii are equal, the two event horizons cancel out, as $F_{max} - F_{max} = 0$. In fact, all of the sub-Schwarzschild black holes radii are now equal, and so all forces cancel out from the outside in as the final bits of matter from the universe cross each event horizon.

$$\sum_N F_{max}(-1)^N = 0 \tag{26}$$

This requires the number of shells (N) in the black hole to be even, or else a complete cancellation of the event horizons is not possible. Now we return to the "missing-g" paradox in equations 24 or 6. The mass is in this tiny space with a surface element spinning at c in a circle, and all time curves have vanished, since the event horizon forces have canceled out. Time is flowing forward (forward is a relative word of course) again, and according to equations 24 and 6, the surface gravity of this black hole is zero. So now we have a super massive point in space, with no gravity to hold it

307

together, that is spinning its outer radius on the order of velocity c. The mass would naturally rip itself apart, ala the Big Bang.

This struck me as being important. Somehow I had this paradox which turned out to be very useful in creating the universe to begin with.

As it turns out, in this model, the "missing g" paradox is not a paradox at all, but an essential phase in the universe, the Big Bang. The universe appears to recycle itself over and over again in a closed system in the following steps. The complete time model looks like an endless loop, shown in figure 6.

- The universe expands rapidly and throws all matter in all directions.
- This matter coalesces after some time, forming clumps of matter.
- *Something happens to accelerate the expansion of the universe? Dark Energy?*
- This matter reaches a critical value and collapses into black holes.
- These black holes sweep up all of the mass jettisoned from the initial expansion.
- These black holes eventually coalesce themselves, forming larger and larger black holes.
- Two black holes are left and they orbit each other and eventually turn into one black hole.
- All event horizons meet and collapse, and gravity is non-existent. Time begins to flow forward.
- The cycle starts over.

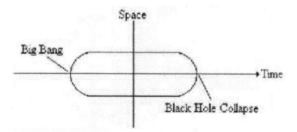

Figure 6: The universal cycle, the universe begins at the Big Bang, goes through a contraction until the Big Crunch and the Big Band happens again.

This fits in well with certain principle physical ideals. A closed temporal model of the universe explains the conservation of energy principle. Otherwise, energy would simply escape out of the universe. Another universe using this energy and matter before our own collapsed thus resulting in The Big Bang. The layering helps with the maximum force principle, as the maximum force being a constant mandates the time reversals. The universe works on one conservation law, the conservation of everything. Therefore in this model, the universe is a perpetual motion machine. We are in but a single iteration of this cycle, with an unknown number of cycles behind us and countless cycles before us.

The idea that the universe may contract, turn into a singularity (The Big Crunch) and bounce has been discussed and explored (albeit in a bit more complicated manner) by others such as Khoury et al. (2002), Seiberg (2002), Elitzur et al. (2002), Tolley and Turok (2002), and Hartle and Hawking (1983). Khoury (2004) gives a nice explanation of the Ekpyrotic/Cyclic universe.

So in my model, the geometry of the universe is open in the beginning, flattens out and then closes. Overall, it must be a closed universe. But there are times when observations will determine it to be any of them.

Now, let's consider a mass m outside of the black hole at some distance r. Let the mass of the black hole be M. Now, Newton's Equation of Gravity (equation 14) may be applied to determine what the attractive force will be. However, something strange happens with the information that comes from a black hole of this type. The information must pass through a zone of negative time (see figure 7). If this is true, then the gravitational influence that m would feel now would be due to the M of the black hole in the future!

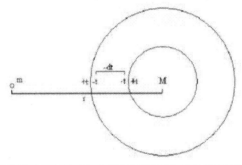

Figure 7: Information of mass M, passes through a negative time layer, traveling back –dt amount of time and reemerges into forward time beyond the event horizon. From here, the information propagates in forward time.

Let us examine this in a thought experiment. Let the distance r be 150 light years. If the information of the mass M travels outwards at a velocity c, it arrives at the negative time zone, and continues traveling outwards to communicate with the mass m. If the negative time interval is say, 20 years, then the information emerges from the black hole 20 years before leaving. If the black hole had a mass B twenty years ago, and a mass A now, then the M = A would be sent to the mass m instead of M = B. 130 years later, m reacts to M = A, but the mass should actually have been M = B. The information M = B was delivered further in the past still. A-B is the missing mass.

$$F = \frac{GmA}{r^2} \qquad \text{even if M was B 130 years ago at the black hole.} \qquad (27)$$

$$A - B = \Delta M \qquad (28)$$

So there is no missing mass! Objects are not just reacting to present masses, but future masses as well. Now it is easy to see how the gravitational signatures of one object may be delivered twice. Imagine a black hole with a time kickback of 20 years, and another mass that will be absorbed by this black hole in 20 years. This mass will transmit its mass m and because the black hole's information is 20 years early, the black hole's mass is M + m and is received at the same time. Thus a stronger reaction to the black hole would be observed.

This nearly reminds me of interference patterns and I had this romantic notion that somehow this may lead to another formulation of gravity waves, but I do not have the knowledge for such a derivation at the moment. So I press forward, disappointed in my own shortcomings here.

But there should be a way to calculate the intensity of the negative time zone, right. So I went back to my program and asked it to derive an equation relating time, the gravitational constant of the universe and the density of matter. Here is what I got.

$$t = -a \sqrt{\frac{1}{G_c \rho}} \qquad \text{where 'a' is a constant.} \qquad (29)$$

Here, I got really disappointed and thought, "That's it, density dropping to zero is a bit much to ask since black holes collect matter." So I went home that day and played a few video games, Halo 2 being one of them. I was sitting there thinking the model just unraveled. Then I thought, "Well, if the density function is a local one, it should drop to zero at all places internal to the event horizon except the singularity, where the density should be infinity. So maybe…"

Once again, I pressed on, assuming that things would clear themselves up. I decided to move towards a density in the simplest way I knew. Mass per volume of course! So imagine a black hole of radius r_s whose volume is the standard sphere. Plug in the Schwarzschild radius and the volume in terms of mass is obtained for a Schwarzschild black hole.

$$V_s = \frac{32\pi m^3 G_c^3}{3c^6} \qquad (30)$$

$$\rho_s = \frac{3c^6}{32\pi m^2 G_c^3} \qquad (31)$$

310

It is now easily seen how the density of a black hole goes to zero as the mass increases, just like the "missing g" paradox, there is a density paradox as well. Plugging this into equation 29, the time period per unit length is given in equation 32.

$$t = -a\sqrt{\frac{32\pi n^2 G_c^2}{3c^6}} \qquad \text{where 'a' is a constant.} \qquad (32)$$

Let us assume that the constant of equation 32 is equal to unity. Then does this system work? Let's test it. If the theory holds, the larger the black hole, the further back in time gravitational information propagates before leaving the event horizon. Let us start with a black hole of 2.5 solar masses. $t = 0.000072s$ per meter of the Schwarzschild black hole. The Schwarzschild radius of such an object would be around 7.4km. Thus the time delay would be around half a second. Let us now examine a super massive black hole that is one million solar masses. $t = 28.6716s$ per meter. The Schwarzschild radius is $2.9 \cdot 10^6$km. The time delay is therefore 2698.17 years. So the information of the mass of a super massive black hole is two thousand seven hundred years early.

At this point, it may be wise to consider the ramifications of equation 32 in the context of the age of the universe and the Hubble constant H_0 (Hubble 1929). The time kickback of the currently seen universe collapsed into a single black hole should match (or come close) to the constrained age or something is wrong. If it matches, it means the age of the universe can be approximated with just the visible matter. Assuming the mass of the universe is $3 \cdot 10^{52}$ kg (I picked this up from NASA's web page), the time kickback is approximately 13.6 Ga. This corresponds to $H_0 = 71.75$ km·s^{-1}/Mpc I think (I'm not familiar with the Hubble constant, I just did a conversion and hoped it would work).

This matches with current approximations summarized nicely by Krauss (2003). The Hubble Space Telescope Key Project discussed by Freedman et al (2000) also returned similar values. So about here, I'm getting really really happy with this model.

Another interesting thing happens when one considers plummeting into a black hole. Imagine that a black hole has a time kickback of 25 years, and that in 20 years one will plunge into the black hole. Effectively 25 years before falling in, one "knows" that the inevitable is coming and actually assists the black hole in sealing one's own fate.

How does this time kickback apply to the rotation curve of the galaxy? We expect the sun to be moving at about 160 km/s, but instead it moves at around 220 km/s. In order for this model to be correct, it must predict the Sun's velocity to be 220 km/s or greater for the observable mass in the galaxy (about $2 \cdot 10^{11}$ solar masses). So let us assume the sun is in a circular orbit about the center of the black hole. When all of the mass of the galaxy collapses into a black hole, the Schwarzschild radius is 3970.25 AU and the time kickback (equation 29 relative to the radius with $a = 1$) is about $1.08 \cdot 10^{14}$ years.

So $1.08 \cdot 10^{14}$ years before the galaxy collapses into one black hole, the sun would feel the effects centrally. Using the classic velocity equation and assuming the sun is about 8000pc from the black hole,

$$v = \sqrt{\frac{mG_c}{r}} = 328.82 \frac{km}{s} \tag{33}$$

Thus the visible disk alone, assuming a time kickback, can reasonably create the velocity we observe. The gap has been closed so to speak, and evoking dark matter in this model is no longer needed.

If these calculations were reversed taking into account the current velocity of the sun, we are feeling the effects of a black hole that is about 45 giga-solar masses that will not exist until approximately 5.5×10^{12} years from now. Also, a large portion of that mass is still sending us information now, and so these numbers are just an estimate. In this model, a black hole is basically a localized time machine. We feel the effects of what we see and what will be.

Now certain questions should arise. Is there a limit to how far in space/time does a mass react to the black hole signatures from the future? In space, there should be no limit, but equation 14 goes as $1/r^2$, and so the effects diminish quickly. However, time wise, there should be limits. To determine them, a trip to the final few seconds of the universe can be taken. Imagine the final bits of matter falling into the universal black hole. The Schwarzschild radius is about 1.44Gpc and the time kickback is about 13.6Ga.

So why would the galaxies continue to expand and not feel the mass signature of this universal black hole? The answer is that the universal black hole is unique, in that it actually ceases to exist very abruptly. The time effect of a universal black hole reduces with time, it does not increase or send gravitational information that may interact or superimpose with alternate time information. When the event horizons of the universal black hole collapse, time flows forward again, and no time kickback exists at this point. So initially, the mass signature is enormous. Then it gradually diminishes as time moves forward from the kickback point. In the future, the universe is already expanding while the current universe is beginning to die. In this way, the two universes are buffered against each other; they should not be able to interact.

So let us analyze what this mass signature seems to be doing. It should appear as a central mass with all of the mass in the universe. Matter (galaxies, proto-galactic nebulae, etc) should be in orbit around this mass. As the mass signature shrinks, these orbiting clumps of matter should maintain their speed and the orbits would expand away from the central mass signature. Using equation 33, the speed of any matter near this universal black hole should close in on c but not quite reach it as $r \rightarrow r_s$. The speed approaches 0.71c.

Let us assume a test mass is twice as far as r_s. Initially, the orbital speed would be .5c. Then let us say the mass decreased to 10% of its initial reading. The speed will

312

still be 0.5c, but only 0.16c is required to stay at that distance, therefore the mass would move outwards into an orbit to compensate for its initial speed, somewhere around 2.88Gpc according to equation 30. As the mass signature of the universal black hole shrinks even further, the mass continues moving outward. This is in effect, the acceleration observed which has been attributed to various ideas including quintessence (dark energy) and the inflationary model of the universe.

In this model, if the outward acceleration is already being observed, then logic dictates that we have already past the point of the maximum universal mass signature. Like dark matter, now dark energy is no longer required to explain our observations.

How would one then measure the galactic curve of such a system? Surely if the model is correct, one could pull a galactic curve model out of it. Actually, there seems to be a way. Take equation 32 and substitute equation 33 solved for mass into it. Then solve for the velocity.

$$t = a \frac{4\sqrt{6\pi}rv^2}{3c^6} \tag{34}$$

$$v = a \frac{3^{1/4}2^{3/4}c^{3/2}\sqrt{t}}{4\pi^{1/4}\sqrt{r}} \tag{35}$$

Here, a is still the constant from equation 29. How does this equation work then? Let us first look at its reduced form with all constants stripped away. This would represent a generic galactic model.

$$v(r) \approx \frac{c^{3/2}\sqrt{t}}{\sqrt{r}} \tag{36}$$

The time kickback observed grows as a function of the radius from the center of the galaxy. Because the model predicts that the further out one is, the more mass signatures will be doubled and the more mass will be consumed before information propagates outwards it is a safe assumption that the time kickback will increase with respect to the radius so this makes sense intuitively if this model holds.

Plotting the $\sqrt{t}$ and the $1/\sqrt{r}$ on top of one another, one can see the potential of pulling out a galactic curve.

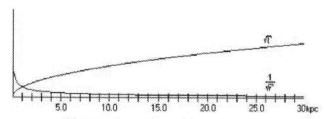

Initially, the $1/\sqrt{r}$ component dominates, and so one would have something small divided by something big, making something small. However, as one approaches the intersection, the result gets larger until finally unity is achieved. Then as you continue going out of the galaxy, one gets something large divided by something small and everything starts to plateau out, gently elevating as the $\sqrt{r}$ term is applied.

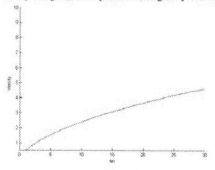

Figure 8: Linear Velocity Galactic Curve.

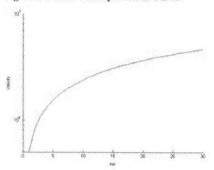

Figure 9: Logarithmic Velocity Galactic Curve.

Looking at the graph of the function $v(r,t)/c^{3/2} = \sqrt{t}/\sqrt{r}$ in both linear and logarithmic format (see figure 9 and figure 10), one can see that the galactic velocity of this model starts out small and increases as one exits the galaxy. Overlapped with an observed galactic curve, the model displays strong similarities. In logarithmic format, this stripped model looks more familiar. A question now arises, what about the hills and

314

valleys in the velocity curves observed in nature, especially the steep one near the galactic core? The answer is that the black hole does not consume matter on a constant basis. It is likely that the galactic black hole consumes matter periodically, and this contributes to the variable gravitational effects seen in the galactic curve. Also, beyond this first order model, there are likely other physical happenings that contribute to the galactic curve shape which are not taken into account. With more computational power, a good fit to the galactic curve will likely be made with this model.

Another way of thinking about this is to envision that two masses fall into a black hole successively, as in figure 11. The first falls in, giving a time kickback of X. The information goes back in time, emerges and is sent out. Then an X amount of time later, a second mass falls in, increasing the time kickback to 2X, the information is now sent out at the same time the information for the first mass is sent out. What is the factor of this overlapping signature? According to equation 33, it is a factor of $\sqrt{t}$ multiplied by some constant and corrected for the speed of light.

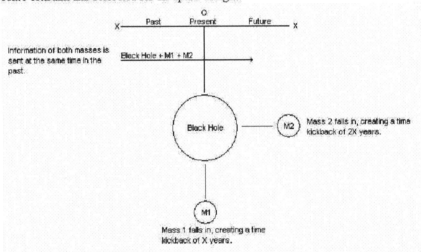

Figure 10: Mass M1 falls into the black hole and the time kickback is X. X time later, mass M2 falls into the black hole and the kickback is now 2X. The result is that the information for M1 and M2 with the initial black hole mass is sent back 2X time.

The problem of causality I dismiss. Causal issues are nicely kept safe within the event horizon. The idea that causal violations cannot occur in nature is not something I subscribe to. While it may be rare, I feel it's a bit much to assume that it can never happen.

Well, that's it for now. Sorry about the length, but I doubt it could get shorter without losing some flavor. But currently this is all I have. I hope you like it, even if it turns out to be nothing, perhaps it makes for some good mind candy and an enjoyable read.

315

Below are some more references I used.

Jacob D. Bekenstein. 1974, Phys. Rev. D, 9, 3292

Shmuel Elitzur, Amit Giveon, David Kutasov & Eliezer Rabinovic. 2002, JHEP, 06, 017

Freedman et al. 2000, ApJ, 553, 47

J.B. Hartle & S.W. Hawking. 1983, Phys. Rev. D, 28, 2960

Hubble, E. 1929, Proc. Nat. Acad. Sci., 15, 168.

J. Khoury, B.A. Ovrut, N. Seiberg, P.J. Steinhardt & N. Turok. 2002, Phys. Rev. D, 65, 086007

Justin Khoury. 2004, in 6th RESCEU Symposium, Frontier in Astroparticle Physics and Cosmology, ed. K. Sato & S. Nagataki (Tokyo, Universal Academic Press, Inc.)

Lawrence M. Krauss. 2003, in Proceedings of the ESO/CERN/ESA Symposium, ESO Astrophysics Symposia, ed. Peter A. Shaver, Luigi DiLella & Alvaro Giménez (Berlin, Springer), 50

J.R. Oppenheimer and G.M. Volkoff. 1939, Phys. Rev., 55, 374

Schwarzschild, K. 1916, Sitzungsberichte der Königlich Preussischen Akademie der Wissenschaften, 1, 189

Nathan Seiberg. 2002, in Francqui Colloquium 5, Strings and Gravity: Tying the Forces Together, Marc Henneaux & Alexander Sevrin (Bruxelles: De Boeck & Larcier)

Andrew J. Tolley & Neil Turok. 2002, Phys. Rev. D, 66, 106005

This email encompasses the thought processes that were inspired by multiple formulae in the first appendix. While this Cosmological model may or may not be true, it should be of interest to readers that several answers may be obtained without consideration of dark energy or dark matter. These two concepts are reminders of the infamous ether concept, and likely since we have not found any of the stuff yet, is simply ether still.

Thus the manuscript is complete. I am of course, naturally curious as to what experiments you come up with after reading this manual. Please email me what you come up with at my email address: lordsomos@yahoo.com with the subject Permutanomicon and I will be sure to read it.

LIST OF REFERENCES

[1] William P. Crummet and Arthur B. Western. "University Physics: Models and Applications" Wm. C. Brown Communications, InC. 1994

[2] Raymond A. Serway and John W. Jewett. "Physics for Scientists and Engineers" 6th Ed. Brooks Cole 2003

[3] Department of Biology, Chemistry, Pharmacy. "http://www.chemie.fu.berlin.de/chemistry/general/si_en.html" Institute of Chemistry, Berlin 2003

[4] Buckingham, E. "On Physically Similar Systems: Illustrations of the Use of Dimensional Equations." Phys. Rev. 4, 345-376, 1914.

[5] Buckingham, E. "Model Experiments and the Form of Empirical Equations." Trans. ASME 37, 263, 1915.

[6] Donald H. Menzel. "Mathematical Physics" Dover 1953

[7] http://www.research.att.com/~njas/sequences

[8] H.K. Kim, *On Regular Polytope Numbers*, Journal: Proc. Amer. Math. Soc. 131 (2003) 65-75

[9] Chung-Chiang Choe and Yuefan Deng. *Decomposing 40 billion integers by four tetrahedral numbers*. Math. Comp. 66 (1997) 893-901.

[10] Sloane, N.J.A. and Plouffe, S. *The Encyclopedia of Integer Sequences*. San Diega, CA: Academic Press, 1995

www.ingramcontent.com/pod-product-compliance
Lightning Source LLC
Chambersburg PA
CBHW080352060326
40689CB00019B/3984